NUMBER THEORY

An Introduction to Mathematics: Part A

NUMBER THEORY

An Introduction to Mathematics: Part A

By

WILLIAM A. COPPEL

 Springer

Library of Congress Control Number: 2005934653

PART A
ISBN-10: 0-387-29851-7   e-ISBN: 0-387-29852-5
ISBN-13: 978-0387-29851-1

PART B
ISBN-10: 0-387-29853-3   e-ISBN: 0-387-29854-1
ISBN-13: 978-0387-29853-5

2-VOLUME SET
ISBN-10: 0-387-30019-8   e-ISBN: 0-387-30529-7
ISBN-13: 978-0387-30019-1

Printed on acid-free paper.

AMS Subject Classifications: 11-xx, 05B20, 33E05

9 8 7 6 5 4 3 2

springer.com

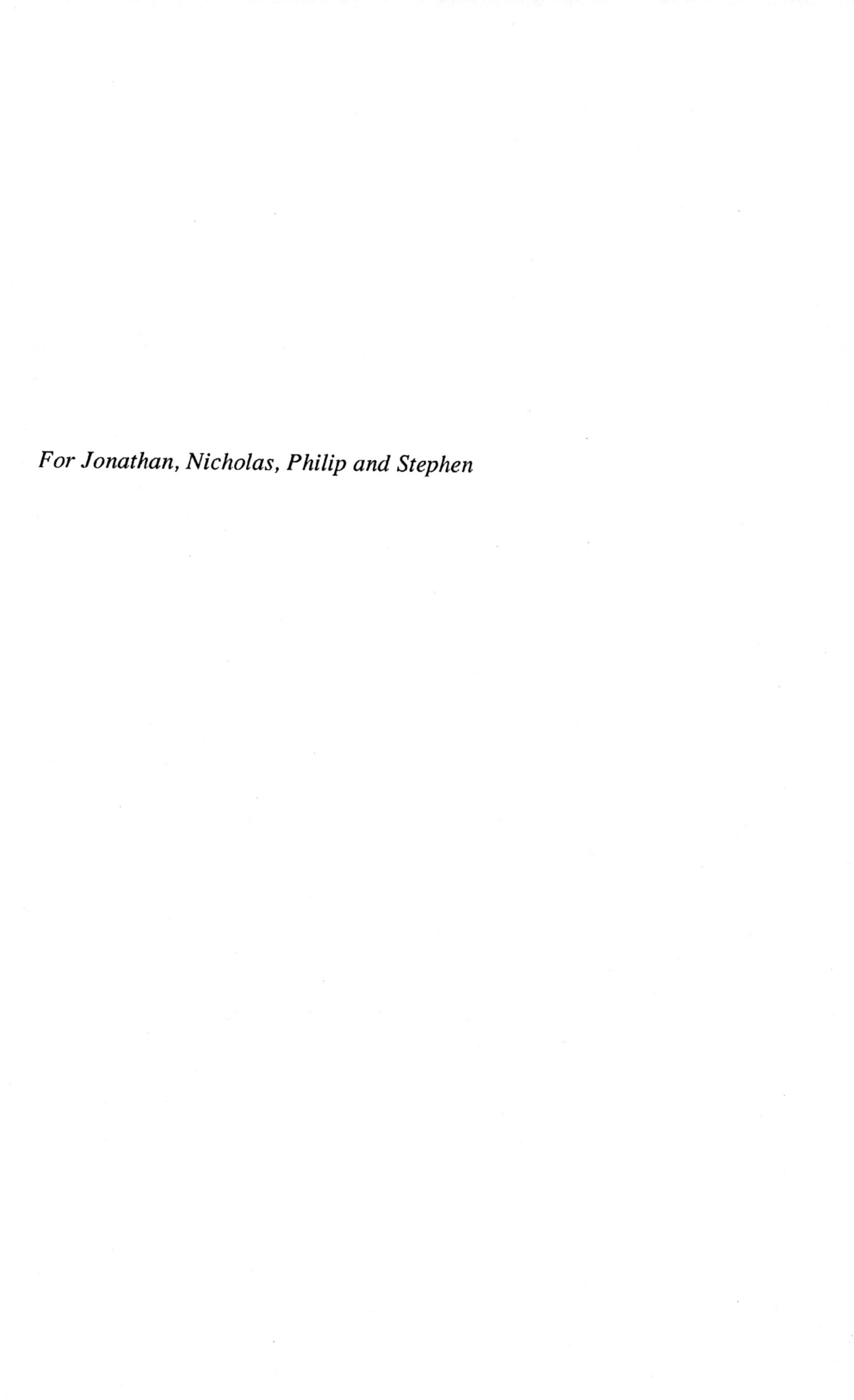

For Jonathan, Nicholas, Philip and Stephen

Contents

Part B

Preface to the revised edition

Undergraduate courses in mathematics are commonly of two types. On the one hand there are courses in subjects, such as linear algebra or real analysis, with which it is considered that every student of mathematics should be acquainted. On the other hand there are courses given by lecturers in their own areas of specialization, which are intended to serve as a preparation for research. There are, I believe, several reasons why students need more than this.

First, although the vast extent of mathematics today makes it impossible for any individual to have a deep knowledge of more than a small part, it is important to have some understanding and appreciation of the work of others. Indeed the sometimes surprising interrelationships and analogies between different branches of mathematics are both the basis for many of its applications and the stimulus for further development. Secondly, different branches of mathematics appeal in different ways and require different talents. It is unlikely that all students at one university will have the same interests and aptitudes as their lecturers. Rather, they will only discover what their own interests and aptitudes are by being exposed to a broader range. Thirdly, many students of mathematics will become, not professional mathematicians, but scientists, engineers or schoolteachers. It is useful for them to have a clear understanding of the nature and extent of mathematics, and it is in the interests of mathematicians that there should be a body of people in the community who have this understanding.

The present book attempts to provide such an understanding of the nature and extent of mathematics. The connecting theme is the theory of numbers, at first sight one of the most abstruse and irrelevant branches of mathematics. Yet by exploring its many connections with other branches, we may obtain a broad picture. The topics chosen are not trivial and demand some effort on the part of the reader. As Euclid already said, there is no royal road. In general I have concentrated attention on those hard-won results which illuminate a wide area. If I am accused of picking the eyes out of some subjects, I have no defence except to say "But what beautiful eyes!"

The book is divided into two parts. Part A, which deals with elementary number theory, should be accessible to a first-year undergraduate. To provide a foundation for subsequent work, Chapter I contains the definitions and basic properties of various mathematical structures.

However, the reader may simply skim through this chapter and refer back to it later as required. Chapter V, on Hadamard's determinant problem, shows that elementary number theory may have unexpected applications.

Part B, which is more advanced, is intended to provide an undergraduate with some idea of the scope of mathematics today. The chapters in this part are largely independent, except that Chapter X depends on Chapter IX and Chapter XIII on Chapter XII.

Although much of the content of the book is common to any introductory work on number theory, I wish to draw attention to the discussion here of quadratic fields and elliptic curves. These are quite special cases of algebraic number fields and algebraic curves, and it may be asked why one should restrict attention to these special cases when the general cases are now well understood and may even be developed in parallel. My answers are as follows. First, to treat the general cases in full rigour requires a commitment of time which many will be unable to afford. Secondly, these special cases are those most commonly encountered and more constructive methods are available for them than for the general cases. There is yet another reason. Sometimes in mathematics a generalization is so simple and far-reaching that the special case is more fully understood as an instance of the generalization. For the topics mentioned, however, the generalization is more complex and is, in my view, more fully understood as a development from the special case.

At the end of each chapter of the book I have added a list of selected references, which will enable readers to travel further in their own chosen directions. Since the literature is voluminous, any such selection must be somewhat arbitrary, but I hope that mine may be found interesting and useful.

The computer revolution has made possible calculations on a scale and with a speed undreamt of a century ago. One consequence has been a considerable increase in 'experimental mathematics' – the search for patterns. This book, on the other hand, is devoted to 'theoretical mathematics' – the explanation of patterns. I do not wish to conceal the fact that the former usually precedes the latter. Nor do I wish to conceal the fact that some of the results here have been proved by the greatest minds of the past only after years of labour, and that their proofs have later been improved and simplified by many other mathematicians. Once obtained, however, a good proof organizes and provides understanding for a mass of computational data. Often it also suggests further developments.

The present book may indeed be viewed as a 'treasury of proofs'. We concentrate attention on this aspect of mathematics, not only because it is a distinctive feature of the subject, but also because we consider its exposition is better suited to a book than to a blackboard or a computer screen. In keeping with this approach, the proofs themselves have been chosen with some care and I hope that a few may be of interest even to those who are no longer students. Proofs which depend on general principles have been given preference over proofs which offer no particular insight.

Mathematics is a part of civilization and an achievement in which human beings may take some pride. It is not the possession of any one national, political or religious group and any attempt to make it so is ultimately destructive. At the present time there are strong pressures to make academic studies more 'relevant'. At the same time, however, staff at some universities are assessed by 'citation counts' and people are paid for giving lectures on chaos, for example, that are demonstrably rubbish.

The theory of numbers provides ample evidence that topics pursued for their own intrinsic interest can later find significant applications. I do not contend that curiosity has been the only driving force. More mundane motives, such as ambition or the necessity of earning a living, have also played a role. It is also true that mathematics pursued for the sake of applications has been of benefit to subjects such as number theory; there is a two-way trade. However, it shows a dangerous ignorance of history and of human nature to promote utility at the expense of spirit.

This book has its origin in a course of lectures which I gave at the Victoria University of Wellington, New Zealand, in 1975. The demands of my own research have hitherto prevented me from completing it, although I have continued to collect material. If it succeeds at all in conveying some idea of the power and beauty of mathematics, the labour of writing it will have been well worthwhile.

As with a previous book, I have to thank Helge Tverberg, who has read most of the manuscript and made many useful suggestions.

In this revised edition of my book, the original edition of which appeared in 2002, I have removed an error in the statement and proof of Proposition II.12 and filled a gap in the proof of Proposition III.12. The statements of the Weil conjectures in Chapter IX and of a result of Heath-Brown in Chapter X have been modified, following comments by J.-P. Serre. I have also corrected a few misprints, made many small expository changes and expanded the index.

Although I have made a few changes to the references, I have not attempted a systematic update. For this I think the Internet has the advantage over a book. The reader is referred to the American Mathematical Society's MathSciNet (www.ams.org/mathscinet) and to The Number Theory Web maintained by Keith Matthews (www.maths.uq.edu.au/~krm/).

Note added (September, 2005) I am grateful to Springer Science for undertaking the commercial publication of my book. I hope you will be also.

I
The expanding universe of numbers

For many people, numbers must seem to be the essence of mathematics. *Number theory*, which is the subject of this book, is primarily concerned with the properties of one particular type of number, the 'whole numbers' or *integers*. However, there are many other types, such as complex numbers and *p*-adic numbers. Somewhat surprisingly, a knowledge of these other types turns out to be necessary for any deeper understanding of the integers.

In this introductory chapter we describe several such types (but defer the study of *p*-adic numbers to Chapter VI). *To embark on number theory proper the reader may proceed to Chapter II now* and refer back to the present chapter, via the Index, only as occasion demands.

When one studies the properties of various types of number, one becomes aware of formal similarities between different types. Instead of repeating the derivations of properties for each individual case, it is more economical – and sometimes actually clearer – to study their common algebraic structure. This algebraic structure may be shared by objects which one would not even consider as numbers.

There is a pedagogic difficulty here. Usually a property is discovered in one context and only later is it realized that it has wider validity. It may be more digestible to prove a result in the context of number theory and then simply point out its wider range of validity. Since this is a book on number theory, and many properties were first discovered in this context, we feel free to adopt this approach. However, to make the statements of such generalizations intelligible, in the latter part of this chapter we describe several basic algebraic structures. We do not attempt to study these structures in depth, but restrict attention to the simplest properties which throw light on the work of later chapters.

0 Sets, relations and mappings

The label '0' given to this section may be interpreted to stand for 'Optional'. We collect here some definitions of a logical nature which have become part of the common language of mathematics. Those who are not already familiar with this language, and who are repelled by its abstraction, should consult this section only when the need arises.

We will not formally define a *set*, but will simply say that it is a collection of objects, which are called its *elements*. We write $a \in A$ if a is an element of the set A and $a \notin A$ if it is not.

A set may be specified by listing its elements. For example, $A = \{a,b,c\}$ is the set whose elements are a,b,c. A set may also be specified by characterizing its elements. For example,

$$A = \{x \in \mathbb{R}: x^2 < 2\}$$

is the set of all real numbers x such that $x^2 < 2$.

If two sets A,B have precisely the same elements, we say that they are *equal* and write $A = B$. (If A and B are not equal, we write $A \neq B$.) For example,

$$\{x \in \mathbb{R}: x^2 = 1\} = \{1, -1\}.$$

Just as it is convenient to admit 0 as a number, so it is convenient to admit the *empty set* $\varnothing$, which has no elements, as a set.

If every element of a set A is also an element of a set B we say that A is a *subset* of B, or that A is *included* in B, or that B *contains* A, and we write $A \subseteq B$. We say that A is a *proper* subset of B, and write $A \subset B$, if $A \subseteq B$ and $A \neq B$.

Thus $\varnothing \subseteq A$ for every set A and $\varnothing \subset A$ if $A \neq \varnothing$. Set inclusion has the following obvious properties:

(i) $A \subseteq A$;
(ii)   *if $A \subseteq B$ and $B \subseteq A$, then $A = B$;*
(iii) *if $A \subseteq B$ and $B \subseteq C$, then $A \subseteq C$.*

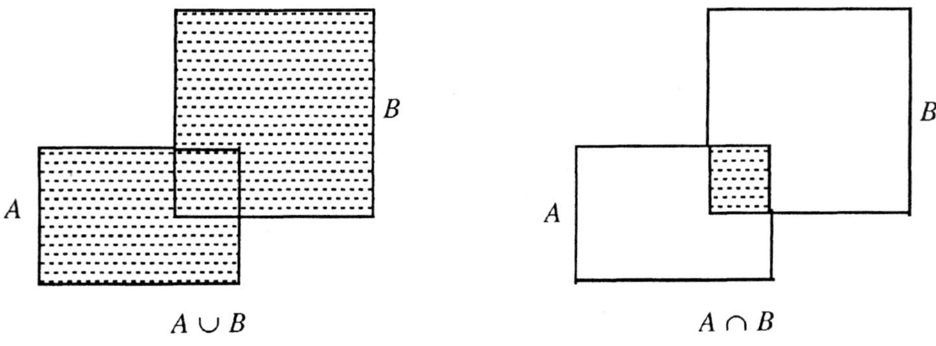

Figure 1: *Union and Intersection*

For any sets A,B, the set whose elements are the elements of A or B (or both) is called the *union* or 'join' of A and B and is denoted by $A \cup B$:

$$A \cup B = \{x : x \in A \text{ or } x \in B\}.$$

The set whose elements are the common elements of A and B is called the *intersection* or 'meet' of A and B and is denoted by $A \cap B$:

$$A \cap B = \{x : x \in A \text{ and } x \in B\}.$$

If $A \cap B = \varnothing$, the sets A and B are said to be *disjoint*.

It is easily seen that union and intersection have the following algebraic properties:

$$A \cup A = A, \quad A \cap A = A,$$
$$A \cup B = B \cup A, \quad A \cap B = B \cap A,$$
$$(A \cup B) \cup C = A \cup (B \cup C), \quad (A \cap B) \cap C = A \cap (B \cap C),$$
$$(A \cup B) \cap C = (A \cap C) \cup (B \cap C), \quad (A \cap B) \cup C = (A \cup C) \cap (B \cup C).$$

Set inclusion could have been defined in terms of either union or intersection, since $A \subseteq B$ is the same as $A \cup B = B$ and also the same as $A \cap B = A$.

For any sets A,B, the set of all elements of B which are not also elements of A is called the *difference* of B from A and is denoted by $B \setminus A$:

$$B \setminus A = \{x : x \in B \text{ and } x \notin A\}.$$

It is easily seen that

$$C \setminus (A \cup B) = (C \setminus A) \cap (C \setminus B), \quad C \setminus (A \cap B) = (C \setminus A) \cup (C \setminus B).$$

An important special case is where all sets under consideration are subsets of a given universal set X. For any $A \subseteq X$, we have

$$\varnothing \cup A = A, \quad \varnothing \cap A = \varnothing,$$
$$X \cup A = X, \quad X \cap A = A.$$

The set $X \setminus A$ is said to be the *complement* of A (in X) and may be denoted by A^c for fixed X. Evidently

$$\varnothing^c = X, \quad X^c = \varnothing,$$
$$A \cup A^c = X, \quad A \cap A^c = \varnothing,$$
$$(A^c)^c = A.$$

By taking $C = X$ in the previous relations for differences, we obtain 'De Morgan's laws':

$$(A \cup B)^c = A^c \cap B^c, \quad (A \cap B)^c = A^c \cup B^c.$$

Since $A \cap B = (A^c \cup B^c)^c$, set intersection can be defined in terms of unions and complements. Alternatively, since $A \cup B = (A^c \cap B^c)^c$, set union can be defined in terms of intersections and complements.

For any sets A,B, the set of all ordered pairs (a,b) with $a \in A$ and $b \in B$ is called the (*Cartesian*) *product* of A by B and is denoted by $A \times B$.

Similarly one can define the product of more than two sets. We mention only one special case. For any positive integer n, we write A^n instead of $A \times ... \times A$ for the set of all (ordered) n-tuples $(a_1,...,a_n)$ with $a_j \in A$ $(1 \leq j \leq n)$. We call a_j the j-th *coordinate* of the n-tuple.

A *binary relation* on a set A is just a subset R of the product set $A \times A$. For any $a,b \in A$, we write aRb if $(a,b) \in R$. A binary relation R on a set A is said to be

reflexive if aRa for every $a \in A$;

symmetric if bRa whenever aRb;

transitive if aRc whenever aRb and bRc.

It is said to be an *equivalence relation* if it is reflexive, symmetric and transitive.

If R is an equivalence relation on a set A and $a \in A$, the *equivalence class* R_a of a is the set of all $x \in A$ such that xRa. Since R is reflexive, $a \in R_a$. Since R is symmetric, $b \in R_a$ implies $a \in R_b$. Since R is transitive, $b \in R_a$ implies $R_b \subseteq R_a$. It follows that, for all $a,b \in A$, either $R_a = R_b$ or $R_a \cap R_b = \varnothing$.

A *partition* $\mathscr{C}$ of a set A is a collection of nonempty subsets of A such that each element of A is an element of exactly one of the subsets in $\mathscr{C}$.

Thus the distinct equivalence classes corresponding to a given equivalence relation on a set A form a partition of A. It is not difficult to see that, conversely, if $\mathscr{C}$ is a partition of A, then an equivalence relation R is defined on A by taking R to be the set of all $(a,b) \in A \times A$ for which a and b are elements of the same subset in the collection $\mathscr{C}$.

Let A and B be nonempty sets. A *mapping* f of A into B is a subset of $A \times B$ with the property that, for each $a \in A$, there is a unique $b \in B$ such that $(a,b) \in f$. We write $f(a) = b$ if $(a,b) \in f$, and say that b is the *image* of a under f or that b is the *value* of f at a. We express that f is a mapping of A into B by writing $f: A \rightarrow B$ and we put

$$f(A) = \{f(a): a \in A\}.$$

The term *function* is often used instead of 'mapping', especially when A and B are sets of real or complex numbers, and 'mapping' itself is often abbreviated to *map*.

If f is a mapping of A into B, and if A' is a nonempty subset of A, then the *restriction* of f to A' is the set of all $(a,b) \in f$ with $a \in A'$.

The *identity map* i_A of a nonempty set A into itself is the set of all ordered pairs (a,a) with $a \in A$.

If f is a mapping of A into B, and g a mapping of B into C, then the *composite mapping* $g \circ f$ of A into C is the set of all ordered pairs (a,c), where $c = g(b)$ and $b = f(a)$. Composition of mappings is associative, i.e. if h is a mapping of C into D, then

$$(h \circ g) \circ f = h \circ (g \circ f).$$

The identity map has the obvious properties $f \circ i_A = f$ and $i_B \circ f = f$.

Let A,B be nonempty sets and $f: A \to B$ a mapping of A into B. The mapping f is said to be 'one-to-one' or *injective* if, for each $b \in B$, there exists at most one $a \in A$ such that $(a,b) \in f$. The mapping f is said to be 'onto' or *surjective* if, for each $b \in B$, there exists at least one $a \in A$ such that $(a,b) \in f$. If f is both injective and surjective, then it is said to be *bijective* or a 'one-to-one correspondence'. The nouns *injection, surjection* and *bijection* are also used instead of the corresponding adjectives.

It is not difficult to see that f is injective if and only if there exists a mapping $g: B \to A$ such that $g \circ f = i_A$, and surjective if and only if there exists a mapping $h: B \to A$ such that $f \circ h = i_B$. Furthermore, if f is bijective, then g and h are unique and equal. Thus, for any bijective map $f: A \to B$, there is a unique *inverse* map $f^{-1}: B \to A$ such that $f^{-1} \circ f = i_A$ and $f \circ f^{-1} = i_B$.

If $f: A \to B$ and $g: B \to C$ are both bijective maps, then $g \circ f: A \to C$ is also bijective and

$$(g \circ f)^{-1} = f^{-1} \circ g^{-1}.$$

1 Natural numbers

The natural numbers are the numbers usually denoted by $1,2,3,4,5,\dots$. However, other notations are also used, e.g. for the chapters of this book. Although one notation may have considerable practical advantages over another, it is the properties of the natural numbers which are basic.

The following system of axioms for the natural numbers was essentially given by Dedekind (1888), although it is usually attributed to Peano (1889):

The natural numbers are the elements of a set $\mathbb{N}$, *with a distinguished element* 1 (*one*) *and map* $S: \mathbb{N} \to \mathbb{N}$, *such that*

(N1) S *is injective, i.e. if* $m,n \in \mathbb{N}$ *and* $m \neq n$, *then* $S(m) \neq S(n)$;
(N2) $1 \notin S(\mathbb{N})$;
(N3) *if* $M \subseteq \mathbb{N}$, $1 \in M$ *and* $S(M) \subseteq M$, *then* $M = \mathbb{N}$.

The element $S(n)$ of $\mathbb{N}$ is called the *successor* of n. The axioms are satisfied by $\{1,2,3,...\}$ if we take $S(n)$ to be the element immediately following the element n.

It follows readily from the axioms that 1 is the only element of $\mathbb{N}$ which is not in $S(\mathbb{N})$. For, if $M = S(\mathbb{N}) \cup \{1\}$, then $M \subseteq \mathbb{N}$, $1 \in M$ and $S(M) \subseteq M$. Hence, by **(N3)**, $M = \mathbb{N}$.

It also follows from the axioms that $S(n) \neq n$ for every $n \in \mathbb{N}$. For let M be the set of all $n \in \mathbb{N}$ such that $S(n) \neq n$. By **(N2)**, $1 \in M$. If $n \in M$ and $n' = S(n)$ then, by **(N1)**, $S(n') \neq n'$. Thus $S(M) \subseteq M$ and hence, by **(N3)**, $M = \mathbb{N}$.

The axioms **(N1)**-**(N3)** actually determine $\mathbb{N}$ up to 'isomorphism'. We will deduce this as a corollary of the following general *recursion theorem*:

PROPOSITION 1 *Given a set A, an element a_1 of A and a map $T: A \to A$, there exists exactly one map* $\varphi: \mathbb{N} \to A$ *such that* $\varphi(1) = a_1$ *and*

$$\varphi(S(n)) = T\varphi(n) \text{ for every } n \in \mathbb{N}.$$

Proof We show first that there is at most one map with the required properties. Let φ_1 and φ_2 be two such maps, and let M be the set of all $n \in \mathbb{N}$ such that

$$\varphi_1(n) = \varphi_2(n).$$

Evidently $1 \in M$. If $n \in M$, then also $S(n) \in M$, since

$$\varphi_1(S(n)) = T\varphi_1(n) = T\varphi_2(n) = \varphi_2(S(n)).$$

Hence, by **(N3)**, $M = \mathbb{N}$. That is, $\varphi_1 = \varphi_2$.

We now show that there exists such a map φ. Let $\mathscr{C}$ be the collection of all subsets C of $\mathbb{N} \times A$ such that $(1,a_1) \in C$ and such that, if $(n,a) \in C$, then also $(S(n),T(a)) \in C$. The collection $\mathscr{C}$ is not empty, since it contains $\mathbb{N} \times A$. Moreover, since every set in $\mathscr{C}$ contains $(1,a_1)$, the intersection D of all sets $C \in \mathscr{C}$ is not empty. It is easily seen that actually $D \in \mathscr{C}$. By its definition, however, no proper subset of D is in $\mathscr{C}$.

Let M be the set of all $n \in \mathbb{N}$ such that $(n,a) \in D$ for exactly one $a \in A$ and, for any $n \in M$, define $\varphi(n)$ to be the unique $a \in A$ such that $(n,a) \in D$. If $M = \mathbb{N}$, then $\varphi(1) = a_1$ and $\varphi(S(n)) = T\varphi(n)$ for all $n \in \mathbb{N}$. Thus we need only show that $M = \mathbb{N}$. As usual, we do this by showing that $1 \in M$ and that $n \in M$ implies $S(n) \in M$.

We have $(1,a_1) \in D$. Assume $(1,a') \in D$ for some $a' \neq a_1$. If $D' = D \setminus \{(1,a')\}$, then $(1,a_1) \in D'$. Moreover, if $(n,a) \in D'$ then $(S(n),T(a)) \in D'$, since $(S(n),T(a)) \in D$ and $(S(n),T(a)) \neq (1,a')$. Hence $D' \in \mathscr{C}$. But this is a contradiction, since D' is a proper subset of D. We conclude that $1 \in M$.

Suppose now that $n \in M$ and let a be the unique element of A such that $(n,a) \in D$. Then $(S(n),T(a)) \in D$, since $D \in \mathscr{C}$. Assume that $(S(n),a'') \in D$ for some $a'' \neq T(a)$ and put $D'' = D \setminus \{(S(n),a'')\}$. Then $(S(n),T(a)) \in D''$ and $(1,a_1) \in D''$. For any $(m,b) \in D''$ we have $(S(m),T(b)) \in D$. If $(S(m),T(b)) = (S(n),a'')$, then $S(m) = S(n)$ and $T(b) = a'' \neq T(a)$, which implies $m = n$ and $b \neq a$. Thus D contains both (n,b) and (n,a), which contradicts $n \in M$. Hence $(S(m),T(b)) \neq (S(n),a'')$, and so $(S(m),T(b)) \in D''$. But then $D'' \in \mathscr{C}$, which is also a contradiction, since D'' is a proper subset of D. We conclude that $S(n) \in M$. $\square$

COROLLARY 2 *If the axioms* **(N1)-(N3)** *are also satisfied by a set* $\mathbb{N}'$ *wth element* $1'$ *and map* S': $\mathbb{N}' \to \mathbb{N}'$, *then there exists a bijective map* φ *of* $\mathbb{N}$ *onto* $\mathbb{N}'$ *such that* $\varphi(1) = 1'$ *and*

$$\varphi(S(n)) = S'\varphi(n) \text{ for every } n \in \mathbb{N}.$$

Proof By taking $A = \mathbb{N}'$, $a_1 = 1'$ and $T = S'$ in Proposition 1, we see that there exists a unique map φ: $\mathbb{N} \to \mathbb{N}'$ such that $\varphi(1) = 1'$ and

$$\varphi(S(n)) = S'\varphi(n) \text{ for every } n \in \mathbb{N}.$$

By interchanging $\mathbb{N}$ and $\mathbb{N}'$, we see also that there exists a unique map ψ: $\mathbb{N}' \to \mathbb{N}$ such that $\psi(1') = 1$ and

$$\psi(S'(n')) = S\psi(n') \text{ for every } n' \in \mathbb{N}'.$$

The composite map $\chi = \psi \circ \varphi$ of $\mathbb{N}$ into $\mathbb{N}$ has the properties $\chi(1) = 1$ and $\chi(S(n)) = S\chi(n)$ for every $n \in \mathbb{N}$. But, by Proposition 1 again, χ is uniquely determined by these properties. Hence $\psi \circ \varphi$ is the identity map on $\mathbb{N}$, and similarly $\varphi \circ \psi$ is the identity map on $\mathbb{N}'$. Consequently φ is a bijection. $\square$

We can also use Proposition 1 to define addition and multiplication of natural numbers. By Proposition 1, for each $m \in \mathbb{N}$ there exists a unique map s_m: $\mathbb{N} \to \mathbb{N}$ such that

$$s_m(1) = S(m),$$
$$s_m(S(n)) = Ss_m(n) \text{ for every } n \in \mathbb{N}.$$

We define the *sum* of m and n to be

$$m + n = s_m(n).$$

It is not difficult to deduce from this definition and the axioms **(N1)**-**(N3)** the usual rules for *addition: for all $a,b,c \in \mathbb{N}$,*

(A1) *if $a + c = b + c$, then $a = b$;*   (cancellation law)

(A2) $a + b = b + a$;   (commutative law)

(A3) $(a + b) + c = a + (b + c)$.   (associative law)

By way of example, we prove the cancellation law. Let M be the set of all $c \in \mathbb{N}$ such that $a + c = b + c$ only if $a = b$. Then $1 \in M$, since $s_a(1) = s_b(1)$ implies $S(a) = S(b)$ and hence $a = b$. Suppose $c \in M$. If $a + S(c) = b + S(c)$, i.e. $s_a(S(c)) = s_b(S(c))$, then $Ss_a(c) = Ss_b(c)$ and hence, by **(N1)**, $s_a(c) = s_b(c)$. Since $c \in M$, this implies $a = b$. Thus also $S(c) \in M$. Hence, by **(N3)**, $M = \mathbb{N}$.

We now show that

$$m + n \neq n \text{ for all } m,n \in \mathbb{N}.$$

For a given $m \in \mathbb{N}$, let M be the set of all $n \in \mathbb{N}$ such that $m + n \neq n$. Then $1 \in M$ since, by **(N2)**, $s_m(1) = S(m) \neq 1$. If $n \in M$, then $s_m(n) \neq n$ and hence, by **(N1)**,

$$s_m(S(n)) = Ss_m(n) \neq S(n).$$

Hence, by **(N3)**, $M = \mathbb{N}$.

By Proposition 1 again, for each $m \in \mathbb{N}$ there exists a unique map $p_m: \mathbb{N} \to \mathbb{N}$ such that

$$p_m(1) = m,$$
$$p_m(S(n)) = s_m(p_m(n)) \text{ for every } n \in \mathbb{N}.$$

We define the *product* of m and n to be

$$m \cdot n = p_m(n).$$

From this definition and the axioms **(N1)**-**(N3)** we may similarly deduce the usual rules for *multiplication: for all $a,b,c \in \mathbb{N}$,*

(M1) *if $a \cdot c = b \cdot c$, then $a = b$;*   (cancellation law)

(M2) $a \cdot b = b \cdot a$;   (commutative law)

(M3)　$(a \cdot b) \cdot c = a \cdot (b \cdot c);$　　　　　　(associative law)

(M4)　$a \cdot 1 = a.$　　　　　　　　(identity element)

Furthermore, addition and multiplication are connected by

(AM1)　$a \cdot (b + c) = (a \cdot b) + (a \cdot c).$　　　(distributive law)

As customary, we will often omit the dot when writing products and we will give multiplication precedence over addition. With these conventions the distributive law becomes simply

$$a(b + c) = ab + ac.$$

We show next how a relation of order may be defined on the set $\mathbb{N}$. For any $m, n \in \mathbb{N}$, we say that *m is less than n*, and write $m < n$, if

$$m + m' = n \quad \text{for some } m' \in \mathbb{N}.$$

Evidently $m < S(m)$ for every $m \in \mathbb{N}$, since $S(m) = m + 1$. Also, if $m < n$, then either $S(m) = n$ or $S(m) < n$. For suppose $m + m' = n$. If $m' = 1$, then $S(m) = n$. If $m' \neq 1$, then $m' = m'' + 1$ for some $m'' \in \mathbb{N}$ and

$$S(m) + m'' = (m + 1) + m'' = m + (1 + m'') = m + m' = n.$$

Again, if $n \neq 1$, then $1 < n$, since the set consisting of 1 and all $n \in \mathbb{N}$ such that $1 < n$ contains 1 and contains $S(n)$ if it contains n.

It will now be shown that the relation '$<$' induces a *total order* on $\mathbb{N}$, which is compatible with both addition and multiplication: *for all $a,b,c \in \mathbb{N}$,*

(O1)　*if $a < b$ and $b < c$, then $a < c$;*　　　(transitive law)

(O2)　*one and only one of the following alternatives holds:*

$$a < b, \ a = b, \ b < a; \quad \text{(law of trichotomy)}$$

(O3)　*$a + c < b + c$ if and only if $a < b$;*

(O4)　*$ac < bc$ if and only if $a < b$.*

The relation **(O1)** follows directly from the associative law for addition. We now prove **(O2)**. If $a < b$ then, for some $a' \in \mathbb{N}$,

$$b = a + a' = a' + a \neq a.$$

Together with **(O1)**, this shows that at most one of the three alternatives in **(O2)** holds.

For a given $a \in \mathbb{N}$, let M be the set of all $b \in \mathbb{N}$ such that at least one of the three alternatives in **(O2)** holds. Then $1 \in M$, since $1 < a$ if $a \neq 1$. Suppose now that $b \in M$. If $a = b$, then $a < S(b)$. If $a < b$, then again $a < S(b)$, by **(O1)**. If $b < a$, then either $S(b) = a$ or $S(b) < a$. Hence also $S(b) \in M$. Consequently, by **(N3)**, $M = \mathbb{N}$. This completes the proof of **(O2)**.

It follows from the associative and commutative laws for addition that, if $a < b$, then $a + c < b + c$. On the other hand, by using also the cancellation law we see that if $a + c < b + c$, then $a < b$.

It follows from the distributive law that, if $a < b$, then $ac < bc$. Finally, suppose $ac < bc$. Then $a \neq b$ and hence, by **(O2)**, either $a < b$ or $b < a$. Since $b < a$ would imply $bc < ac$, by what we have just proved, we must actually have $a < b$.

The law of trichotomy **(O2)** implies that, for given $m,n \in \mathbb{N}$, the equation

$$m + x = n$$

has a solution $x \in \mathbb{N}$ only if $m < n$.

As customary, we write $a \leq b$ to denote either $a < b$ or $a = b$. Also, it is sometimes convenient to write $b > a$ instead of $a < b$, and $b \geq a$ instead of $a \leq b$.

A subset M of $\mathbb{N}$ is said to have a *least element* m' if $m' \in M$ and $m' \leq m$ for every $m \in M$. The least element m' is uniquely determined, if it exists, by **(O2)**. By what we have already proved, 1 is the least element of $\mathbb{N}$.

PROPOSITION 3 *Any nonempty subset M of $\mathbb{N}$ has a least element.*

Proof Assume that some nonempty subset M of $\mathbb{N}$ does not have a least element. Then $1 \notin M$, since 1 is the least element of $\mathbb{N}$. Let L be the set of all $l \in \mathbb{N}$ such that $l < m$ for every $m \in M$. Then L and M are disjoint and $1 \in L$. If $l \in L$, then $S(l) \leq m$ for every $m \in M$. Since M does not have a least element, it follows that $S(l) \notin M$. Thus $S(l) < m$ for every $m \in M$, and so $S(l) \in L$. Hence, by **(N3)**, $L = \mathbb{N}$. Since $L \cap M = \varnothing$, this is a contradiction. $\square$

The method of *proof by induction* is a direct consequence of the axioms defining $\mathbb{N}$. Suppose that with each $n \in \mathbb{N}$ there is associated a proposition P_n. To show that P_n is true for every $n \in \mathbb{N}$, we need only show that P_1 is true and that P_{n+1} is true if P_n is true.

Proposition 3 provides an alternative approach. To show that P_n is true for every $n \in \mathbb{N}$, we need only show that if P_m is false for some m, then P_l is false for some $l < m$. For then the set of all $n \in \mathbb{N}$ for which P_n is false has no least element and consequently is empty.

For any $n \in \mathbb{N}$, we denote by I_n the set of all $m \in \mathbb{N}$ such that $m \leq n$. Thus $I_1 = \{1\}$ and $S(n) \notin I_n$. It is easily seen that

$$I_{S(n)} = I_n \cup \{S(n)\}.$$

Also, for any $p \in I_{S(n)}$, there exists a bijective map f_p of I_n onto $I_{S(n)} \setminus \{p\}$. For, if $p = S(n)$ we can take f_p to be the identity map on I_n, and if $p \in I_n$ we can take f_p to be the map defined by

$$f_p(p) = S(n), \quad f_p(m) = m \text{ if } m \in I_n \setminus \{p\}.$$

PROPOSITION 4 *For any $m,n \in \mathbb{N}$, if a map $f: I_m \to I_n$ is injective and $f(I_m) \neq I_n$, then $m < n$.*

Proof The result certainly holds when $m = 1$, since $I_1 = \{1\}$. Let M be the set of all $m \in \mathbb{N}$ for which the result holds. We need only show that if $m \in M$, then also $S(m) \in M$.

Let $f: I_{S(m)} \to I_n$ be an injective map such that $f(I_{S(m)}) \neq I_n$ and choose $p \in I_n \setminus f(I_{S(m)})$. The restriction g of f to I_m is also injective and $g(I_m) \neq I_n$. Since $m \in M$, it follows that $m < n$. Assume $S(m) = n$. Then there exists a bijective map g_p of $I_{S(m)} \setminus \{p\}$ onto I_m. The composite map $h = g_p \circ f$ maps $I_{S(m)}$ into I_m and is injective. Since $m \in M$, we must have $h(I_m) = I_m$. But, since $h(S(m)) \in I_m$ and h is injective, this is a contradiction. Hence $S(m) < n$ and, since this holds for every f, $S(m) \in M$. $\square$

PROPOSITION 5 *For any $m,n \in \mathbb{N}$, if a map $f: I_m \to I_n$ is not injective and $f(I_m) = I_n$, then $m > n$.*

Proof The result holds vacuously when $m = 1$, since any map $f: I_1 \to I_n$ is injective. Let M be the set of all $m \in \mathbb{N}$ for which the result holds. We need only show that if $m \in M$, then also $S(m) \in M$.

Let $f: I_{S(m)} \to I_n$ be a map such that $f(I_{S(m)}) = I_n$ which is not injective. Then there exist $p,q \in I_{S(m)}$ with $p \neq q$ and $f(p) = f(q)$. We may choose the notation so that $q \in I_m$. If f_p is a bijective map of I_m onto $I_{S(m)} \setminus \{p\}$, then the composite map $h = f \circ f_p$ maps I_m onto I_n. If it is not injective then $m > n$, since $m \in M$, and hence also $S(m) > n$. If h is injective, then it is bijective and has a bijective inverse $h^{-1}: I_n \to I_m$. Since $h^{-1}(I_n)$ is a proper subset of $I_{S(m)}$, it follows from Proposition 4 that $n < S(m)$. Hence $S(m) \in M$. $\square$

Propositions 4 and 5 immediately imply

COROLLARY 6 *For any $n \in \mathbb{N}$, a map $f: I_n \to I_n$ is injective if and only if it is surjective.*
$\square$

COROLLARY 7 *If a map $f: I_m \to I_n$ is bijective, then $m = n$.*

Proof By Proposition 4, $m < S(n)$, i.e. $m \le n$. Replacing f by f^{-1}, we obtain in the same way $n \le m$. Hence $m = n$. $\square$

A set E is said to be *finite* if there exists a bijective map $f\colon E \to I_n$ for some $n \in \mathbb{N}$. Then n is uniquely determined, by Corollary 7. We call it the *cardinality* of E and denote it by $\#(E)$.

It is readily shown that if E is a finite set and F a proper subset of E, then F is also finite and $\#(F) < \#(E)$. Again, if E and F are disjoint finite sets, then their union $E \cup F$ is also finite and $\#(E \cup F) = \#(E) + \#(F)$. Furthermore, for any finite sets E and F, the product set $E \times F$ is also finite and $\#(E \times F) = \#(E) \cdot \#(F)$.

Corollary 6 implies that, for any finite set E, a map $f\colon E \to E$ is injective if and only if it is surjective. This is a precise statement of the so-called *pigeonhole principle*.

A set E is said to be *countably infinite* if there exists a bijective map $f\colon E \to \mathbb{N}$. Any countably infinite set may be bijectively mapped onto a proper subset F, since $\mathbb{N}$ is bijectively mapped onto a proper subset by the successor map S. Thus a map $f\colon E \to E$ of an infinite set E may be injective, but not surjective. It may also be surjective, but not injective; an example is the map $f\colon \mathbb{N} \to \mathbb{N}$ defined by $f(1) = 1$ and, for $n \ne 1$, $f(n) = m$ if $S(m) = n$.

2 Integers and rational numbers

The concept of number will now be extended. The natural numbers $1,2,3,\ldots$ suffice for counting purposes, but for bank balance purposes we require the larger set $\ldots,-2,-1,0,1,2,\ldots$ of integers. (From this point of view, -2 is not so 'unnatural'.) An important reason for extending the concept of number is the greater freedom it gives us. In the realm of natural numbers the equation $a + x = b$ has a solution if and only if $b > a$; in the extended realm of integers it will always have a solution.

Rather than introduce a new set of axioms for the integers, we will define them in terms of natural numbers. Intuitively, an integer is the difference $m - n$ of two natural numbers m,n, with addition and multiplication defined by

$$(m - n) + (p - q) = (m + p) - (n + q),$$
$$(m - n) \cdot (p - q) = (mp + nq) - (mq + np).$$

However, two other natural numbers m',n' may have the same difference as m,n, and anyway what does $m - n$ mean if $m < n$? To make things precise, we proceed in the following way.

Consider the set $\mathbb{N} \times \mathbb{N}$ of all ordered pairs of natural numbers. For any two such ordered pairs, (m,n) and (m',n'), we write

$$(m,n) \sim (m',n') \quad \text{if } m + n' = m' + n.$$

We will show that this is an *equivalence relation*. It follows at once from the definition that $(m,n) \sim (m,n)$ (reflexive law) and that $(m,n) \sim (m',n')$ implies $(m',n') \sim (m,n)$ (symmetric law). It remains to prove the transitive law:

$$(m,n) \sim (m',n') \text{ and } (m',n') \sim (m'',n'') \text{ imply } (m,n) \sim (m'',n'').$$

This follows from the commutative, associative and cancellation laws for addition in $\mathbb{N}$. For we have

$$m + n' = m' + n, \quad m' + n'' = m'' + n',$$

and hence

$$(m + n') + n'' = (m' + n) + n'' = (m' + n'') + n = (m'' + n') + n.$$

Thus

$$(m + n'') + n' = (m'' + n) + n',$$

and so $m + n'' = m'' + n$.

The equivalence class containing $(1,1)$ evidently consists of all pairs (m,n) with $m = n$.

We define an *integer* to be an equivalence class of ordered pairs of natural numbers and, as is now customary, we denote the set of all integers by $\mathbb{Z}$.

Addition of integers is defined componentwise:

$$(m,n) + (p,q) = (m + p, n + q).$$

To justify this definition we must show that it does not depend on the choice of representatives within an equivalence class, i.e. that

$$(m,n) \sim (m',n') \text{ and } (p,q) \sim (p',q') \text{ imply } (m + p, n + q) \sim (m' + p', n' + q').$$

However, if

$$m + n' = m' + n, \quad p + q' = p' + q,$$

then

$$\begin{aligned}
(m + p) + (n' + q') &= (m + n') + (p + q') \\
&= (m' + n) + (p' + q) = (m' + p') + (n + q).
\end{aligned}$$

It follows at once from the corresponding properties of natural numbers that, also in $\mathbb{Z}$, addition satisfies the commutative law **(A2)** and the associative law **(A3)**. Moreover, the equivalence class 0 (*zero*) containing $(1,1)$ is an *identity element* for addition:

(A4) $a + 0 = a$ *for every a.*

Furthermore, the equivalence class containing (n,m) is an *additive inverse* for the equivalence containing (m,n):

(A5) *for each a, there exists* $-a$ *such that* $a + (-a) = 0$.

From these properties we can now obtain

PROPOSITION 8 *For all* $a,b \in \mathbb{Z}$, *the equation* $a + x = b$ *has a unique solution* $x \in \mathbb{Z}$.

Proof It is clear that $x = (-a) + b$ is a solution. Moreover, this solution is unique, since if $a + x = a + x'$ then, by adding $-a$ to both sides, we obtain $x = x'$. $\square$

Proposition 8 shows that the cancellation law **(A1)** is a consequence of **(A2)**-**(A5)**. It also immediately implies

COROLLARY 9 *For each* $a \in \mathbb{Z}$, 0 *is the only element such that* $a + 0 = a$, $-a$ *is uniquely determined by a, and* $a = -(-a)$. $\square$

As usual, we will henceforth write $b - a$ instead of $b + (-a)$.

Multiplication of integers is defined by

$$(m,n) \cdot (p,q) = (mp + nq, mq + np).$$

To justify this definition we must show that $(m,n) \sim (m',n')$ and $(p,q) \sim (p',q')$ imply

$$(mp + nq, mq + np) \sim (m'p' + n'q', m'q' + n'p').$$

From $m + n' = m' + n$, by multiplying by p and q we obtain

$$mp + n'p = m'p + np,$$
$$m'q + nq = mq + n'q,$$

and from $p + q' = p' + q$, by multiplying by m' and n' we obtain

$$m'p + m'q' = m'p' + m'q,$$
$$n'p' + n'q = n'p + n'q'.$$

Adding these four equations and cancelling the terms common to both sides, we get

$$(mp + nq) + (m'q' + n'p') = (m'p' + n'q') + (mq + np),$$

as required.

It is easily verified that, also in $\mathbb{Z}$, multiplication satisfies the commutative law **(M2)** and the associative law **(M3)**. Moreover, the distributive law **(AM1)** holds and, if 1 is the equivalence class containing $(1+1,1)$, then **(M4)** also holds. (In practice it does not cause confusion to denote identity elements of $\mathbb{N}$ and $\mathbb{Z}$ by the same symbol.)

PROPOSITION 10 *For every* $a \in \mathbb{Z}$, $a \cdot 0 = 0$.

Proof We have

$$a \cdot 0 = a \cdot (0 + 0) = a \cdot 0 + a \cdot 0.$$

Adding $-(a \cdot 0)$ to both sides, we obtain the result. $\square$

Proposition 10 could also have been derived directly from the definitions, but we prefer to view it as a consequence of the properties which have been labelled.

COROLLARY 11 *For all* $a,b \in \mathbb{Z}$,

$$a(-b) = -(ab), \quad (-a)(-b) = ab.$$

Proof The first relation follows from

$$ab + a(-b) = a \cdot 0 = 0,$$

and the second relation follows from the first, since $c = -(-c)$. $\square$

By the definitions of 0 and 1 we also have

(AM2) $1 \neq 0$.

(In fact $1 = 0$ would imply $a = 0$ for every a, since $a \cdot 1 = a$ and $a \cdot 0 = 0$.)

We will say that an integer a is *positive* if it is represented by an ordered pair (m,n) with $n < m$. This definition does not depend on the choice of representative. For if $n < m$ and $m + n' = m' + n$, then $m + n' < m' + m$ and hence $n' < m'$.

We will denote by P the set of all positive integers. The law of trichotomy **(O2)** for natural numbers immediately implies

(P1) *for every a, one and only one of the following alternatives holds:*

$$a \in P, \quad a = 0, \quad -a \in P.$$

We say that an integer is *negative* if it has the form $-a$, where $a \in P$, and we denote by $-P$ the set of all negative integers. Since $a = -(-a)$, **(P1)** says that $\mathbb{Z}$ is the disjoint union of the sets P, $\{0\}$ and $-P$.

From the property **(O3)** of natural numbers we immediately obtain

(P2) *if $a \in P$ and $b \in P$, then $a + b \in P$.*

Furthermore, we have

(P3) *if $a \in P$ and $b \in P$, then $a \cdot b \in P$.*

To prove this we need only show that if m,n,p,q are natural numbers such that $n < m$ and $q < p$, then

$$mq + np \; < \; mp + nq.$$

Since $q < p$, there exists a natural number q' such that $q + q' = p$. But then $nq' < mq'$, since $n < m$, and hence

$$mq + np = (m + n)q + nq' < (m + n)q + mq' = mp + nq.$$

We may write **(P2)** and **(P3)** symbolically in the form

$$P + P \subseteq P, \quad P \cdot P \subseteq P.$$

We now show that there are no *divisors of zero* in $\mathbb{Z}$:

PROPOSITION 12 *If $a \neq 0$ and $b \neq 0$, then $ab \neq 0$.*

Proof By **(P1)**, either a or $-a$ is positive, and either b or $-b$ is positive. If $a \in P$ and $b \in P$ then $ab \in P$, by **(P3)**, and hence $ab \neq 0$, by **(P1)**. If $a \in P$ and $-b \in P$, then $a(-b) \in P$. Hence $ab = -(a(-b)) \in -P$ and $ab \neq 0$. Similarly if $-a \in P$ and $b \in P$. Finally, if $-a \in P$ and $-b \in P$, then $ab = (-a)(-b) \in P$ and again $ab \neq 0$. $\square$

The proof of Proposition 12 also shows that any nonzero square is positive:

PROPOSITION 13 *If $a \neq 0$, then $a^2 := aa \in P$.* $\square$

It follows that $1 \in P$, since $1 \neq 0$ and $1^2 = 1$.

The set P of positive integers induces an order relation in $\mathbb{Z}$. Write

$$a < b \ \text{ if } \ b - a \in P,$$

so that $a \in P$ if and only if $0 < a$. From this definition and the properties of P it follows that the order properties **(O1)-(O3)** hold also in $\mathbb{Z}$, and that **(O4)** holds in the modified form:

(O4)' *if $0 < c$, then $ac < bc$ if and only if $a < b$.*

We now show that we can represent any $a \in \mathbb{Z}$ in the form $a = b - c$, where $b,c \in P$. In fact, if $a = 0$, we can take $b = 1$ and $c = 1$; if $a \in P$, we can take $b = a + 1$ and $c = 1$; and if $-a \in P$, we can take $b = 1$ and $c = 1 - a$.

An element a of $\mathbb{Z}$ is said to be a *lower bound* for a subset X of $\mathbb{Z}$ if $a \leq x$ for every $x \in X$. Proposition 3 immediately implies that if a subset of $\mathbb{Z}$ has a lower bound, then it has a least element.

For any $n \in \mathbb{N}$, let n' be the integer represented by $(n + 1,1)$. Then $n' \in P$. We are going to study the map $n \to n'$ of $\mathbb{N}$ into P. The map is injective, since $n' = m'$ implies $n = m$. It is also surjective, since if $a \in P$ is represented by (m,n), where $n < m$, then it is also represented by $(p + 1,1)$, where $p \in \mathbb{N}$ satisfies $n + p = m$. It is easily verified that the map preserves sums and products:

$$(m + n)' = m' + n', \quad (mn)' = m'n'.$$

Since $1' = 1$, it follows that $S(n)' = n' + 1$. Furthermore, we have

$$m' < n' \ \text{ if and only if } \ m < n.$$

Thus the map $n \to n'$ establishes an 'isomorphism' of $\mathbb{N}$ with P. In other words, P is a copy of $\mathbb{N}$ situated within $\mathbb{Z}$. By identifying n with n', we may regard $\mathbb{N}$ itself as a subset of $\mathbb{Z}$ (and stop talking about P). Then 'natural number' is the same as 'positive integer' and any integer is the difference of two natural numbers.

Number theory, in its most basic form, is the study of the properties of the set $\mathbb{Z}$ of integers. It will be considered in some detail in later chapters of this book, but to relieve the abstraction of the preceding discussion we consider here the *division algorithm*:

PROPOSITION 14 *For any integers a,b with $a > 0$, there exist unique integers q,r such that*

$$b = qa + r, \ 0 \leq r < a.$$

Proof We consider first uniqueness. Suppose

$$qa + r = q'a + r', \ \ 0 \leq r,r' < a.$$

If $r < r'$, then from

$$(q - q')a = r' - r,$$

we obtain first $q > q'$ and then $r' - r \geq a$, which is a contradiction. If $r' < r$, we obtain a contradiction similarly. Hence $r = r'$, which implies $q = q'$.

We consider next existence. Let S be the set of all integers $y \geq 0$ which can be represented in the form $y = b - xa$ for some $x \in \mathbb{Z}$. The set S is not empty, since it contains $b - 0$ if $b \geq 0$ and $b - ba$ if $b < 0$. Hence S contains a least element r. Then $b = qa + r$, where $q,r \in \mathbb{Z}$ and $r \geq 0$. Since $r - a = b - (q + 1)a$ and r is the least element in S, we must also have $r < a$. $\square$

The concept of number will now be further extended to include 'fractions' or 'rational numbers'. For measuring lengths the integers do not suffice, since the length of a given segment may not be an exact multiple of the chosen unit of length. Similarly for measuring weights, if we find that three identical coins balance five of the chosen unit weights, then we ascribe to each coin the weight 5/3. In the realm of integers the equation $ax = b$ frequently has no solution; in the extended realm of rational numbers it will always have a solution if $a \neq 0$.

Intuitively, a rational number is the ratio or 'quotient' a/b of two integers a,b, where $b \neq 0$, with addition and multiplication defined by

$$a/b + c/d = (ad + cb)/bd,$$
$$a/b \cdot c/d = ac/bd.$$

However, two other integers a',b' may have the same ratio as a,b, and anyway what does a/b mean? To make things precise, we proceed in much the same way as before.

Put $\mathbb{Z}^\times = \mathbb{Z} \setminus \{0\}$ and consider the set $\mathbb{Z} \times \mathbb{Z}^\times$ of all ordered pairs (a,b) with $a \in \mathbb{Z}$ and $b \in \mathbb{Z}^\times$. For any two such ordered pairs, (a,b) and (a',b'), we write

$$(a,b) \sim (a',b') \quad \text{if } ab' = a'b.$$

To show that this is an equivalence relation it is again enough to verify that $(a,b) \sim (a',b')$ and $(a',b') \sim (a'',b'')$ imply $(a,b) \sim (a'',b'')$. The same calculation as before, with addition replaced by multiplication, shows that $(ab'')b' = (a''b)b'$. Since $b' \neq 0$, it follows that $ab'' = a''b$.

The equivalence class containing $(0,1)$ evidently consists of all pairs $(0,b)$ with $b \neq 0$, and the equivalence class containing $(1,1)$ consists of all pairs (b,b) with $b \neq 0$.

We define a *rational number* to be an equivalence class of elements of $\mathbb{Z} \times \mathbb{Z}^\times$ and, as is now customary, we denote the set of all rational numbers by $\mathbb{Q}$.

Addition of rational numbers is defined by

$$(a,b) + (c,d) = (ad + cb, bd),$$

where $bd \neq 0$ since $b \neq 0$ and $d \neq 0$. To justify the definition we must show that

$$(a,b) \sim (a',b') \text{ and } (c,d) \sim (c',d') \text{ imply } (ad + cb, bd) \sim (a'd' + c'b', b'd').$$

But if $ab' = a'b$ and $cd' = c'd$, then

$$(ad + cb)(b'd') = (ab')(dd') + (cd')(bb')$$
$$= (a'b)(dd') + (c'd)(bb') = (a'd' + c'b')(bd).$$

It is easily verified that, also in $\mathbb{Q}$, addition satisfies the commutative law **(A2)** and the associative law **(A3)**. Moreover **(A4)** and **(A5)** also hold, the equivalence class 0 containing $(0,1)$ being an identity element for addition and the equivalence class containing $(-b,c)$ being the additive inverse of the equivalence class containing (b,c).

Multiplication of rational numbers is defined componentwise:

$$(a,b) \cdot (c,d) = (ac, bd).$$

To justify the definition we must show that

$$(a,b) \sim (a',b') \text{ and } (c,d) \sim (c',d') \text{ imply } (ac, bd) \sim (a'c', b'd').$$

But if $ab' = a'b$ and $cd' = c'd$, then

$$(ac)(b'd') = (ab')(cd') = (a'b)(c'd) = (a'c')(bd).$$

It is easily verified that, also in $\mathbb{Q}$, multiplication satisfies the commutative law **(M2)** and the associative law **(M3)**. Moreover **(M4)** also holds, the equivalence class 1 containing $(1,1)$ being an identity element for multiplication. Furthermore, addition and multiplication are connected by the distributive law **(AM1)**, and **(AM2)** also holds since $(0,1)$ is not equivalent to $(1,1)$.

Unlike the situation for $\mathbb{Z}$, however, every nonzero element of $\mathbb{Q}$ has a *multiplicative inverse*:

(M5) *for each $a \neq 0$, there exists a^{-1} such that $aa^{-1} = 1$.*

In fact, if a is represented by (b,c), then a^{-1} is represented by (c,b).

It follows that, for all $a,b \in \mathbb{Q}$ with $a \neq 0$, the equation $ax = b$ has a unique solution $x \in \mathbb{Q}$, namely $x = a^{-1}b$. Hence, if $a \neq 0$, then 1 is the only solution of $ax = a$, a^{-1} is uniquely determined by a, and $a = (a^{-1})^{-1}$.

We will say that a rational number a is *positive* if it is represented by an ordered pair (b,c) of integers for which $bc > 0$. This definition does not depend on the choice of representative. For suppose $0 < bc$ and $bc' = b'c$. Then $bc' \neq 0$, since $b \neq 0$ and $c' \neq 0$, and hence $0 < (bc')^2$. Since $(bc')^2 = (bc)(b'c')$ and $0 < bc$, it follows that $0 < b'c'$.

Our previous use of P having been abandoned in favour of $\mathbb{N}$, we will now denote by P the set of all positive rational numbers and by $-P$ the set of all rational numbers $-a$, where $a \in P$. From the corresponding result for $\mathbb{Z}$, it follows that **(P1)** continues to hold in $\mathbb{Q}$. We will show that **(P2)** and **(P3)** also hold.

To see that the sum of two positive rational numbers is again positive, we observe that if a,b,c,d are integers such that $0 < ab$ and $0 < cd$, then also

$$0 < (ab)d^2 + (cd)b^2 = (ad + cb)(bd).$$

To see that the product of two positive rational numbers is again positive, we observe that if a,b,c,d are integers such that $0 < ab$ and $0 < cd$, then also

$$0 < (ab)(cd) = (ac)(bd).$$

Since **(P1)**-**(P3)** all hold, it follows as before that Propositions 12 and 13 also hold in $\mathbb{Q}$. Hence $1 \in P$ and **(O4)$'$** now implies that $a^{-1} \in P$ if $a \in P$. If $a,b \in P$ and $a < b$, then $b^{-1} < a^{-1}$, since $bb^{-1} = 1 = aa^{-1} < ba^{-1}$.

The set P of positive elements now induces an order relation on $\mathbb{Q}$. We write $a < b$ if $b - a \in P$, so that $a \in P$ if and only if $0 < a$. Then the order relations **(O1)**-**(O3)** and **(O4)$'$** continue to hold in $\mathbb{Q}$.

Unlike the situation for $\mathbb{Z}$, however, the ordering of $\mathbb{Q}$ is *dense*, i.e. if $a,b \in \mathbb{Q}$ and $a < b$, then there exists $c \in \mathbb{Q}$ such that $a < c < b$. For example, we can take c to be the solution of $(1 + 1)c = a + b$.

Let $\mathbb{Z}'$ denote the set of all rational numbers a' which can be represented by $(a,1)$ for some $a \in \mathbb{Z}$. For every $c \in \mathbb{Q}$, there exist $a',b' \in \mathbb{Z}'$ with $b' \neq 0$ such that $c = a'b'^{-1}$. In fact, if c is represented by (a,b), we can take a' to be represented by $(a,1)$ and b' by $(b,1)$. Instead of $c = a'b'^{-1}$, we also write $c = a'/b'$.

For any $a \in \mathbb{Z}$, let a' be the rational number represented by $(a,1)$. The map $a \to a'$ of $\mathbb{Z}$ into $\mathbb{Z}'$ is clearly bijective. Moreover, it preserves sums and products:

$$(a + b)' = a' + b', \quad (ab)' = a'b'.$$

Furthermore,

$$a' < b' \text{ if and only if } a < b.$$

Thus the map $a \rightarrow a'$ establishes an 'isomorphism' of $\mathbb{Z}$ with $\mathbb{Z}'$, and $\mathbb{Z}'$ is a copy of $\mathbb{Z}$ situated within $\mathbb{Q}$. By identifying a with a', we may regard $\mathbb{Z}$ itself as a subset of $\mathbb{Q}$. Then any rational number is the ratio of two integers.

By way of illustration, we show that if a and b are positive rational numbers, then there exists a positive integer l such that $la > b$. For if $a = m/n$ and $b = p/q$, where m,n,p,q are positive integers, then

$$(np + 1)a > pm \geq p \geq b.$$

3 Real numbers

It was discovered by the ancient Greeks that even rational numbers do not suffice for the measurement of lengths. If x is the length of the hypotenuse of a right-angled triangle whose other two sides have unit length then, by Pythagoras' theorem, $x^2 = 2$. But it was proved, probably by a disciple of Pythagoras, that there is no rational number x such that $x^2 = 2$. (A more general result is proved in Book X, Proposition 9 of Euclid's *Elements*.) We give here a somewhat different proof from the classical one.

Assume that such a rational number x exists. Since x may be replaced by $-x$, we may suppose that $x = m/n$, where $m,n \in \mathbb{N}$. Then $m^2 = 2n^2$. Among all pairs m,n of positive integers with this property, there exists one for which n is least. If we put

$$p = 2n - m, \quad q = m - n,$$

then p and q are positive integers, since clearly $n < m < 2n$. But

$$p^2 = 4n^2 - 4mn + m^2 = 2(m^2 - 2mn + n^2) = 2q^2.$$

Since $q < n$, this contradicts the minimality of n.

If we think of the rational numbers as measuring distances of points on a line from a given origin O on the line (with distances on one side of O positive and distances on the other side negative), this means that, even though a dense set of points is obtained in this way, not all points of the line are accounted for. In order to fill in the gaps the concept of number will now be extended from 'rational number' to 'real number'.

It is possible to define real numbers as infinite decimal expansions, the rational numbers being those whose decimal expansions are eventually periodic. However, the choice of base 10 is arbitrary and carrying through this approach is awkward.

There are two other commonly used approaches, one based on *order* and the other on *distance*. The first was proposed by Dedekind (1872), the second by Méray (1869) and Cantor (1872). We will follow Dedekind's approach, since it is conceptually simpler. However, the second method is also important and in a sense more general. In Chapter VI we will use it to extend the rational numbers to the *p-adic numbers*.

It is convenient to carry out Dedekind's construction in two stages. We will first define 'cuts' (which are just the positive real numbers), and then pass from cuts to arbitrary real numbers in the same way that we passed from the natural numbers to the integers.

Intuitively, a cut is the set of all rational numbers which represent points of the line between the origin O and some other point. More formally, we define a *cut* to be a nonempty proper subset A of the set P of all positive rational numbers such that

(i)   *if $a \in A$, $b \in P$ and $b < a$, then $b \in A$;*
(ii) *if $a \in A$, then there exists $a' \in A$ such that $a < a'$.*

For example, the set I of all positive rational numbers $a < 1$ is a cut. Similarly, the set T of all positive rational numbers a such that $a^2 < 2$ is a cut. We will denote the set of all cuts by $\mathscr{P}$.

For any $A, B \in \mathscr{P}$ we write $A < B$ if A is a proper subset of B. We will show that this induces a *total order* on $\mathscr{P}$.

It is clear that if $A < B$ and $B < C$, then $A < C$. It remains to show that, for any $A, B \in \mathscr{P}$, one and only one of the following alternatives holds:

$$A < B, \quad A = B, \quad B < A.$$

It is obvious from the definition by set inclusion that at most one holds. Now suppose that neither $A < B$ nor $A = B$. Then there exists $a \in A \setminus B$. It follows from (i), applied to B, that every $b \in B$ satisfies $b < a$ and then from (i), applied to A, that $b \in A$. Thus $B < A$.

Let $\mathscr{S}$ be any nonempty collection of cuts. A cut B is said to be an *upper bound* for $\mathscr{S}$ if $A \leq B$ for every $A \in \mathscr{S}$, and a *lower bound* for $\mathscr{S}$ if $B \leq A$ for every $A \in \mathscr{S}$. An upper bound for $\mathscr{S}$ is said to be a *least upper bound* or *supremum* for $\mathscr{S}$ if it is a lower bound for the collection of all upper bounds. Similarly, a lower bound for $\mathscr{S}$ is said to be a *greatest lower bound* or *infimum* for $\mathscr{S}$ if it is an upper bound for the collection of all lower bounds. Clearly, $\mathscr{S}$ has at most one supremum and at most one infimum.

The set $\mathscr{P}$ has the following basic property:

(P4) *if a nonempty subset $\mathscr{S}$ has an upper bound, then it has a least upper bound.*

Proof Let C be the union of all sets $A \in \mathscr{S}$. By hypothesis there exists a cut B such that $A \subseteq B$ for every $A \in \mathscr{S}$. Since $C \subseteq B$ for any such B, and $A \subseteq C$ for every $A \in \mathscr{S}$, we need only show that C is a cut.

Evidently C is a nonempty proper subset of P, since $B \neq P$. Suppose $c \in C$. Then $c \in A$ for some $A \in \mathscr{S}$. If $d \in P$ and $d < c$, then $d \in A$, since A is a cut. Furthermore $c < a'$ for some $a' \in A$. Since $A \subseteq C$, this proves that C is a cut. $\square$

In the set P of positive rational numbers, the subset T of all $x \in P$ such that $x^2 < 2$ has an upper bound, but no least upper bound. Thus **(P4)** shows that there is a difference between the total order on P and that on $\mathscr{P}$.

We now define addition of cuts. For any $A,B \in \mathscr{P}$, let $A + B$ denote the set of all rational numbers $a + b$, with $a \in A$ and $b \in B$. We will show that also $A + B \in \mathscr{P}$. Evidently $A + B$ is a nonempty subset of P. It is also a proper subset. For choose $c \in P \backslash A$ and $d \in P \backslash B$. Then, by (i), $a < c$ for all $a \in A$ and $b < d$ for all $b \in B$. Since $a + b < c + d$ for all $a \in A$ and $b \in B$, it follows that $c + d \notin A + B$.

Suppose now that $a \in A$, $b \in B$ and that $c \in P$ satisfies $c < a + b$. If $c > b$, then $c = b + d$ for some $d \in P$, and $d < a$. Hence, by (i), $d \in A$ and $c = d + b \in A + B$. Similarly, $c \in A + B$ if $c > a$. Finally, if $c \leq a$ and $c \leq b$, choose $e \in P$ so that $e < c$. Then $e \in A$ and $c = e + f$ for some $f \in P$. Then $f \in B$, since $f < c$, and $c = e + f \in A + B$.

Thus $A + B$ has the property (i). It is trivial that $A + B$ also has the property (ii), since if $a \in A$ and $b \in B$, there exists $a' \in A$ such that $a < a'$ and then $a + b < a' + b$. This completes the proof that $A + B$ is a cut.

It follows at once from the corresponding properties of rational numbers that addition of cuts satisfies the commutative law **(A2)** and the associative law **(A3)**.

We consider next the connection between addition and order.

LEMMA 15 *For any cut A and any $c \in P$, there exists $a \in A$ such that $a + c \notin A$.*

Proof If $c \notin A$, then $a + c \notin A$ for every $a \in A$, since $c < a + c$. Thus we may suppose $c \in A$. Choose $b \in P \backslash A$. For some positive integer n we have $b < nc$ and hence $nc \notin A$. If n is the least positive integer such that $nc \notin A$, then $n > 1$ and $(n - 1)c \in A$. Consequently we can take $a = (n - 1)c$. $\square$

PROPOSITION 16 *For any cuts A,B, there exists a cut C such that $A + C = B$ if and only if $A < B$.*

Proof We prove the necessity of the condition by showing that $A < A + C$ for any cuts A,C. If $a \in A$ and $c \in C$, then $a < a + c$. Since $A + C$ is a cut, it follows that $a \in A + C$. Consequently $A \leq A + C$, and Lemma 15 implies that $A \neq A + C$.

Suppose now that A and B are cuts such that $A < B$, and let C be the set of all $c \in P$ such that $c + d \in B$ for some $d \in P \setminus A$. We are going to show that C is a cut and that $A + C = B$.

The set C is not empty. For choose $b \in B \setminus A$ and then $b' \in B$ with $b < b'$. Then $b' = b + c'$ for some $c' \in P$, which implies $c' \in C$. On the other hand, $C \leq B$, since $c + d \in B$ and $d \in P$ imply $c \in B$. Thus C is a proper subset of P.

Suppose $c \in C, p \in P$ and $p < c$. We have $c + d \in B$ for some $d \in P \setminus A$ and $c = p + e$ for some $e \in P$. Since $d + e \in P \setminus A$ and $p + (d + e) = c + d \in B$, it follows that $p \in C$.

Suppose now that $c \in C$, so that $c + d \in B$ for some $d \in P \setminus A$. Choose $b \in B$ so that $c + d < b$. Then $b = c + d + e$ for some $e \in P$. If we put $c' = c + e$, then $c < c'$. Moreover $c' \in C$, since $c' + d = b$. This completes the proof that C is a cut.

Suppose $a \in A$ and $c \in C$. Then $c + d \in B$ for some $d \in P \setminus A$. Hence $a < d$. It follows that $a + c < c + d$, and so $a + c \in B$. Thus $A + C \leq B$.

It remains to show that $B \leq A + C$. Pick any $b \in B$. If $b \in A$, then also $b \in A + C$, since $A < A + C$. Thus we now assume $b \notin A$. Choose $b' \in B$ with $b < b'$. Then $b' = b + d$ for some $d \in P$. By Lemma 15, there exists $a \in A$ such that $a + d \notin A$. Moreover $a < b$, since $b \notin A$, and hence $b = a + c$ for some $c \in P$. Since $c + (a + d) = b + d = b'$, it follows that $c \in C$. Thus $b \in A + C$ and $B \leq A + C$. $\square$

We can now show that addition of cuts satisfies the order relation **(O3)**. Suppose first that $A < B$. Then, by Proposition 16, there exists a cut D such that $A + D = B$. Hence, for any cut C,

$$A + C \;<\; (A + C) + D \;=\; B + C.$$

Suppose next that $A + C < B + C$. Then $A \neq B$. Since $B < A$ would imply $B + C < A + C$, by what we have just proved, it follows from the law of trichotomy that $A < B$.

From **(O3)** and the law of trichotomy, it follows that addition of cuts satisfies the cancellation law **(A1)**.

We next define multiplication of cuts. For any $A,B \in \mathscr{P}$, let AB denote the set of all rational numbers ab, with $a \in A$ and $b \in B$. In the same way as for $A + B$, it may be shown that $AB \in \mathscr{P}$. We note only that if $a \in A$, $b \in B$ and $c < ab$, then $b^{-1}c < a$. Hence $b^{-1}c \in A$ and $c = (b^{-1}c)b \in AB$.

It follows from the corresponding properties of rational numbers that multiplication of cuts satisfies the commutative law **(M2)** and the associative law **(M3)**. Moreover **(M4)** holds, the identity element for multiplication being the cut I consisting of all positive rational numbers less than 1.

We now show that the distributive law (**AM1**) also holds. The distributive law for rational numbers shows at once that

$$A(B + C) \leq AB + AC.$$

It remains to show that $a_1 b + a_2 c \in A(B + C)$ if $a_1, a_2 \in A$, $b \in B$ and $c \in C$. But

$$a_1 b + a_2 c \leq a_2(b + c) \quad \text{if } a_1 \leq a_2,$$

and

$$a_1 b + a_2 c \leq a_1(b + c) \quad \text{if } a_2 \leq a_1.$$

In either event it follows that $a_1 b + a_2 c \in A(B + C)$.

We can now show that multiplication of cuts satisfies the order relation (**O4**). If $A < B$, then there exists a cut D such that $A + D = B$ and hence $AC < AC + DC = BC$. Conversely, suppose $AC < BC$. Then $A \neq B$. Since $B < A$ would imply $BC < AC$, it follows that $A < B$.

From (**O4**) and the law of trichotomy (**O2**) it follows that multiplication of cuts satisfies the cancellation law (**M1**).

We next prove the existence of multiplicative inverses. The proof will use the following multiplicative analogue of Lemma 15:

LEMMA 17 *For any cut A and any $c \in P$ with $c > 1$, there exists $a \in A$ such that $ac \notin A$.*

Proof Choose any $b \in A$. We may suppose $bc \in A$, since otherwise we can take $a = b$. Since $b < bc$, we have $bc = b + d$ for some $d \in P$. By Lemma 15 we can choose $a \in A$ so that $a + d \notin A$. Since $b + d \in A$, it follows that $b + d < a + d$, and so $b < a$. Hence $ab^{-1} > 1$ and

$$a + d < a + (ab^{-1})d = ab^{-1}(b + d) = ac.$$

Since $a + d \notin A$, it follows that $ac \notin A$. $\quad\square$

PROPOSITION 18 *For any $A \in \mathscr{P}$, there exists $A^{-1} \in \mathscr{P}$ such that $AA^{-1} = I$.*

Proof Let A^{-1} be the set of all $b \in P$ such that $b < c^{-1}$ for some $c \in P \setminus A$. It is easily verified that A^{-1} is a cut. We note only that $a^{-1} \notin A^{-1}$ if $a \in A$ and that, if $b < c^{-1}$, then also $b < d^{-1}$ for some $d > c$.

We now show that $AA^{-1} = I$. If $a \in A$ and $b \in A^{-1}$ then $ab < 1$, since $a \geq b^{-1}$ would imply $a > c$ for some $c \in P \setminus A$. Thus $AA^{-1} \leq I$. On the other hand, if $0 < d < 1$ then, by Lemma 17, there exists $a \in A$ such that $ad^{-1} \notin A$. Choose $a' \in A$ so that $a < a'$, and put

$b = (a')^{-1}d$. Then $b < a^{-1}d$. Since $a^{-1}d = (ad^{-1})^{-1}$, it follows that $b \in A^{-1}$ and consequently $d = a'b \in AA^{-1}$. Thus $I \leq AA^{-1}$. $\square$

For any positive rational number a, the set A_a consisting of all positive rational numbers c such that $c < a$ is a cut. The map $a \to A_a$ of P into $\mathcal{P}$ is injective and preserves sums and products:

$$A_{a+b} = A_a + A_b, \quad A_{ab} = A_a A_b.$$

Moreover, $A_a < A_b$ if and only if $a < b$.

By identifying a with A_a we may regard P as a subset of $\mathcal{P}$. It is a proper subset, since **(P4)** does not hold in P.

This completes the first stage of Dedekind's construction. In the second stage we pass from cuts to real numbers. Intuitively, a real number is the difference of two cuts. We will deal with the second stage rather briefly since, as has been said, it is completely analogous to the passage from the natural numbers to the integers.

On the set $\mathcal{P} \times \mathcal{P}$ of all ordered pairs of cuts an equivalence relation is defined by

$$(A,B) \sim (A',B') \quad \text{if } A + B' = A' + B.$$

We define a *real number* to be an equivalence class of ordered pairs of cuts and, as is now customary, we denote the set of all real numbers by $\mathbb{R}$.

Addition and multiplication are unambiguously defined by

$$(A,B) + (C,D) = (A + C, B + D),$$
$$(A,B) \cdot (C,D) = (AC + BD, AD + BC).$$

They obey the laws **(A2)-(A5)**, **(M2)-(M5)** and **(AM1)-(AM2)**.

A real number represented by (A,B) is said to be *positive* if $B < A$. If we denote by $\mathcal{P}'$ the set of all positive real numbers, then **(P1)-(P3)** continue to hold with $\mathcal{P}'$ in place of P. An order relation, satisfying **(O1)-(O3)**, is induced on $\mathbb{R}$ by writing $a < b$ if $b - a \in \mathcal{P}'$. Moreover, any $a \in \mathbb{R}$ may be written in the form $a = b - c$, where $b,c \in \mathcal{P}'$. It is easily seen that $\mathcal{P}$ is isomorphic with $\mathcal{P}'$. By identifying $\mathcal{P}$ with $\mathcal{P}'$, we may regard both $\mathcal{P}$ and $\mathbb{Q}$ as subsets of $\mathbb{R}$. An element of $\mathbb{R} \setminus \mathbb{Q}$ is said to be an *irrational* real number.

Upper and lower bounds, and suprema and infima, may be defined for subsets of $\mathbb{R}$ in the same way as for subsets of $\mathcal{P}$. Moreover, the least upper bound property **(P4)** continues to hold in $\mathbb{R}$. By applying **(P4)** to the subset $-\mathcal{S} = \{-a : a \in \mathcal{S}\}$ we see that if a nonempty subset $\mathcal{S}$ of $\mathbb{R}$ has a lower bound, then it has a greatest lower bound.

The least upper bound property implies the so-called *Archimedean property*:

PROPOSITION 19 *For any positive real numbers a,b, there exists a positive integer n such that na > b.*

Proof Assume, on the contrary, that $na \leq b$ for every $n \in \mathbb{N}$. Then b is an upper bound for the set $\{na: n \in \mathbb{N}\}$. Let c be a least upper bound for this set. From $na \leq c$ for every $n \in \mathbb{N}$ we obtain $(n + 1)a \leq c$ for every $n \in \mathbb{N}$. But this implies $na \leq c - a$ for every $n \in \mathbb{N}$. Since $c - a < c$ and c is a least upper bound, we have a contradiction. $\square$

PROPOSITION 20 *For any real numbers a,b with a < b, there exists a rational number c such that a < c < b.*

Proof Suppose first that $a \geq 0$. By Proposition 19 there exists a positive integer n such that $n(b - a) > 1$. Then $b > a + n^{-1}$. There exists also a positive integer m such that $mn^{-1} > a$. If m is the least such positive integer, then $(m - 1)n^{-1} \leq a$ and hence $mn^{-1} \leq a + n^{-1} < b$. Thus we can take $c = mn^{-1}$.

If $a < 0$ and $b > 0$ we can take $c = 0$. If $a < 0$ and $b \leq 0$, then $- b < d < - a$ for some rational d and we can take $c = - d$. $\square$

PROPOSITION 21 *For any positive real number a, there exists a unique positive real number b such that $b^2 = a$.*

Proof Let S be the set of all positive real numbers x such that $x^2 \leq a$. The set S is not empty, since it contains a if $a \leq 1$ and 1 if $a > 1$. If $y > 0$ and $y^2 > a$, then y is an upper bound for S. In particular, $1 + a$ is an upper bound for S. Let b be the least upper bound for S. Then $b^2 = a$, since $b^2 < a$ would imply $(b + 1/n)^2 < a$ for sufficiently large $n > 0$ and $b^2 > a$ would imply $(b - 1/n)^2 > a$ for sufficiently large $n > 0$. Finally, if $c^2 = a$ and $c > 0$, then $c = b$, since

$$(c - b)(c + b) = c^2 - b^2 = 0. \quad \square$$

The unique positive real number b in the statement of Proposition 21 is said to be a *square root* of a and is denoted by $\sqrt{a}$ or $a^{1/2}$. In the same way it may be shown that, for any positive real number a and any positive integer n, there exists a unique positive real number b such that $b^n = a$, where $b^n = b \cdots b$ (n times). We say that b is an *n-th root* of a and write $b = \sqrt[n]{a}$ or $a^{1/n}$.

A set is said to be a *field* if two binary operations, addition and multiplication, are defined on it with the properties **(A2)-(A5)**, **(M2)-(M5)** and **(AM1)-(AM2)**. A field is said to be *ordered* if it contains a subset P of 'positive' elements with the properties **(P1)-(P3)**. An ordered field is said to be *complete* if, with the order induced by P, it has the property **(P4)**.

Propositions 19-21 hold in any complete ordered field, since only the above properties were used in their proofs. By construction, the set $\mathbb{R}$ of all real numbers is a complete ordered field. In fact, any complete ordered field F is isomorphic to $\mathbb{R}$, i.e. there exists a bijective map $\varphi: F \to \mathbb{R}$ such that, for all $a, b \in F$,

$$\varphi(a + b) = \varphi(a) + \varphi(b), \quad \varphi(ab) = \varphi(a)\varphi(b),$$

and $\varphi(a) > 0$ if and only if $a \in P$. We sketch the proof.

Let e be the identity element for multiplication in F and, for any positive integer n, let $ne = e + \ldots + e$ (n summands). Since F is ordered, ne is positive and so has a multiplicative inverse. For any rational number m/n, where $m, n \in \mathbb{Z}$ and $n > 0$, write $(m/n)e = m(ne)^{-1}$ if $m > 0$, $= -(-m)(ne)^{-1}$ if $m < 0$, and $= 0$ if $m = 0$. The elements $(m/n)e$ form a subfield of F isomorphic to $\mathbb{Q}$ and we define $\varphi((m/n)e) = m/n$. For any $a \in F$, we define $\varphi(a)$ to be the least upper bound of all rational numbers m/n such that $(m/n)e \leq a$. One verifies first that the map $\varphi: F \to \mathbb{R}$ is bijective and that $\varphi(a) < \varphi(b)$ if and only if $a < b$. One then deduces that φ preserves sums and products.

Actually, any bijective map $\varphi: F \to \mathbb{R}$ which preserves sums and products is also order-preserving. For, by Proposition 21, $b > a$ if and only if $b - a = c^2$ for some $c \neq 0$, and then

$$\varphi(b) - \varphi(a) \ = \ \varphi(b - a) \ = \ \varphi(c^2) \ = \ \varphi(c)^2 > 0.$$

Those whose primary interest lies in real analysis may *define* $\mathbb{R}$ to be a complete ordered field and omit the tour through $\mathbb{N}, \mathbb{Z}, \mathbb{Q}$ and $\mathscr{P}$. That is, one takes as axioms the 14 properties above which define a complete ordered field and simply assumes that they are consistent.

The notion of convergence can be defined in any totally ordered set. A sequence $\{a_n\}$ is said to *converge*, with *limit* l, if for any l', l'' such that $l' < l < l''$, there exists a positive integer $N = N(l', l'')$ such that

$$l' < a_n < l'' \quad \text{for every } n \geq N.$$

The limit l of the convergent sequence $\{a_n\}$ is clearly uniquely determined; we write

$$\lim_{n \to \infty} a_n = l,$$

or $a_n \to l$ as $n \to \infty$.

It is easily seen that any convergent sequence is *bounded*, i.e. it has an upper bound and a lower bound. A trivial example of a convergent sequence is the *constant* sequence $\{a_n\}$, where $a_n = a$ for every n; its limit is again a.

In the set $\mathbb{R}$ of real numbers, or in any totally ordered set in which each bounded sequence has a least upper bound and a greatest lower bound, the definition of convergence can be reformulated. For, let $\{a_n\}$ be a bounded sequence. Then, for any positive integer m, the subsequence $\{a_n\}_{n \geq m}$ has a greatest lower bound b_m and a least upper bound c_m:

$$b_m = \inf_{n \geq m} a_n, \quad c_m = \sup_{n \geq m} a_n.$$

The sequences $\{b_m\}_{m \geq 1}$ and $\{c_m\}_{m \geq 1}$ are also bounded and, for any positive integer m,

$$b_m \leq b_{m+1} \leq c_{m+1} \leq c_m.$$

If we define the *lower limit* and *upper limit* of the sequence $\{a_n\}$ by

$$\underline{\lim}_{n \to \infty} a_n := \sup_{m \geq 1} b_m, \quad \overline{\lim}_{n \to \infty} a_n := \inf_{m \geq 1} c_m,$$

then $\underline{\lim}_{n \to \infty} a_n \leq \overline{\lim}_{n \to \infty} a_n$, and it is readily shown that $\lim_{n \to \infty} a_n = l$ if and only if

$$\underline{\lim}_{n \to \infty} a_n = l = \overline{\lim}_{n \to \infty} a_n.$$

A sequence $\{a_n\}$ is said to be *nondecreasing* if $a_n \leq a_{n-1}$ for every n and *nonincreasing* if $a_{n+1} \leq a_n$ for every n. It is said to be *monotonic* if it is either nondecreasing or nonincreasing.

PROPOSITION 22 *Any bounded monotonic sequence of real numbers is convergent.*

Proof Let $\{a_n\}$ be a bounded monotonic sequence and suppose, for definiteness, that it is nondecreasing: $a_1 \leq a_2 \leq a_3 \leq \dots$. In this case, in the notation used above we have $b_m = a_m$ and $c_m = c_1$ for every m. Hence

$$\underline{\lim}_{n \to \infty} a_n = \sup_{m \geq 1} a_m = c_1 = \overline{\lim}_{n \to \infty} a_n. \quad \square$$

Proposition 22 may be applied to the centuries-old algorithm for calculating square roots, which is commonly used today in pocket calculators. Take any real number $a > 1$ and put

$$x_1 = (1 + a)/2.$$

Then $x_1 > 1$ and $x_1^2 > a$, since $(a - 1)^2 > 0$. Define the sequence $\{x_n\}$ recursively by

$$x_{n+1} = (x_n + a/x_n)/2 \quad (n \geq 1).$$

It is easily verified that if $x_n > 1$ and $x_n^2 > a$, then $x_{n+1} > 1$, $x_{n+1}^2 > a$ and $x_{n+1} < x_n$. Since the inequalities hold for $n = 1$, it follows that they hold for all n. Thus the sequence $\{x_n\}$ is nonincreasing and bounded, and therefore convergent. If $x_n \to b$, then $a/x_n \to a/b$ and $x_{n+1} \to b$. Hence $b = (b + a/b)/2$, which simplifies to $b^2 = a$.

We consider now sequences of real numbers which are not necessarily monotonic.

LEMMA 23 *Any sequence $\{a_n\}$ of real numbers has a monotonic subsequence.*

Proof Let M be the set of all positive integers m such that $a_m \geq a_n$ for every $n > m$. If M contains infinitely many positive integers $m_1 < m_2 < \ldots$, then $\{a_{m_k}\}$ is a nonincreasing subsequence of $\{a_n\}$. If M is empty or finite, there is a positive integer n_1 such that no positive integer $n \geq n_1$ is in M. Then $a_{n_2} > a_{n_1}$ for some $n_2 > n_1$, $a_{n_3} > a_{n_2}$ for some $n_3 > n_2$, and so on. Thus $\{a_{n_k}\}$ is a nondecreasing subsequence of $\{a_n\}$. $\square$

It is clear from the proof that Lemma 23 also holds for sequences of elements of any totally ordered set. In the case of $\mathbb{R}$, however, it follows at once from Lemma 23 and Proposition 22 that

PROPOSITION 24 *Any bounded sequence of real numbers has a convergent subsequence.*
$\square$

Proposition 24 is often called the Bolzano–Weierstrass theorem. It was stated by Bolzano (c. 1830) in work which remained unpublished until a century later. It became generally known through the lectures of Weierstrass (c. 1874).

A sequence $\{a_n\}$ of real numbers is said to be a *fundamental sequence*, or 'Cauchy sequence', if for each $\varepsilon > 0$ there exists a positive integer $N = N(\varepsilon)$ such that

$$-\varepsilon < a_p - a_q < \varepsilon \quad \text{for all } p,q \geq N.$$

Any fundamental sequence $\{a_n\}$ is bounded, since any finite set is bounded and

$$a_N - \varepsilon < a_p < a_N + \varepsilon \quad \text{for } p \geq N.$$

Also, any convergent sequence is a fundamental sequence. For suppose $a_n \to l$ as $n \to \infty$. Then, for any $\varepsilon > 0$, there exists a positive integer N such that

$$l - \varepsilon/2 < a_n < l + \varepsilon/2 \quad \text{for every } n \geq N.$$

It follows that

$$-\varepsilon < a_p - a_q < \varepsilon \quad \text{for } p \geq q \geq N.$$

The definitions of convergent sequence and fundamental sequence, and the preceding result that 'convergent' implies 'fundamental', hold also for sequences of rational numbers, and even for sequences with elements from any ordered field. However, for sequences of real numbers there is a converse result:

PROPOSITION 25 *Any fundamental sequence of real numbers is convergent.*

Proof If $\{a_n\}$ is a fundamental sequence of real numbers, then $\{a_n\}$ is bounded and, for any $\varepsilon > 0$, there exists a positive integer $m = m(\varepsilon)$ such that

$$-\varepsilon/2 < a_p - a_q < \varepsilon/2 \quad \text{for all } p,q \geq m.$$

But, by Proposition 24, the sequence $\{a_n\}$ has a convergent subsequence $\{a_{n_k}\}$. If l is the limit of this subsequence, then there exists a positive integer $N \geq m$ such that

$$l - \varepsilon/2 < a_{n_k} < l + \varepsilon/2 \quad \text{for } n_k \geq N.$$

It follows that

$$l - \varepsilon < a_n < l + \varepsilon \quad \text{for } n \geq N.$$

Thus the sequence $\{a_n\}$ converges with limit l. $\square$

Proposition 25 was known to Bolzano (1817) and was clearly stated in the influential *Cours d'analyse* of Cauchy (1821). However, a rigorous proof was impossible until the real numbers themselves had been precisely defined.

The Méray–Cantor method of constructing the real numbers from the rationals is based on Proposition 25. We define two fundamental sequences $\{a_n\}$ and $\{a'_n\}$ of rational numbers to be equivalent if $a_n - a'_n \to 0$ as $n \to \infty$. This is indeed an equivalence relation, and we define a real number to be an equivalence class of fundamental sequences. The set of all real numbers acquires the structure of a field if addition and multiplication are defined by

$$\{a_n\} + \{b_n\} = \{a_n + b_n\}, \quad \{a_n\} \cdot \{b_n\} = \{a_n b_n\}.$$

It acquires the structure of a complete ordered field if the fundamental sequence $\{a_n\}$ is said to be positive when it has a positive lower bound. The field $\mathbb{Q}$ of rational numbers may be regarded as a subfield of the field thus constructed by identifying the rational number a with the equivalence class containing the constant sequence $\{a_n\}$, where $a_n = a$ for every n.

It is not difficult to show that an ordered field is complete if every bounded monotonic sequence is convergent, or if every bounded sequence has a convergent subsequence. In this sense, Propositions 22 and 24 state equivalent forms for the least upper bound property. This is not true, however, for Proposition 25. An ordered field need not have the least upper bound property, even though every fundamental sequence is convergent. It is true, however, that an ordered field has the least upper bound property if and only if it has the Archimedean property (Proposition 19) *and* every fundamental sequence is convergent.

In a course of real analysis one would now define continuity and prove those properties of continuous functions which, in the 18th century, were assumed as 'geometrically obvious'. For example, for given $a,b \in \mathbb{R}$ with $a < b$, let $I = [a,b]$ be the *interval* consisting of all $x \in \mathbb{R}$ such that $a \le x \le b$. If $f: I \to \mathbb{R}$ is continuous, then it attains its supremum, i.e. there exists $c \in I$ such that $f(x) \le f(c)$ for every $x \in I$. Also, if $f(a)f(b) < 0$, then $f(d) = 0$ for some $d \in I$ (the intermediate-value theorem). Real analysis is not our primary concern, however, and we do not feel obliged to establish even those properties which we may later use.

4 Metric spaces

The notion of convergence is meaningful not only for points on a line, but also for points in space, where there is no natural relation of order. We now reformulate our previous definition, so as to make it more generally applicable.

The *absolute value* $|a|$ of a real number a is defined by

$$|a| = a \quad \text{if } a \ge 0,$$
$$|a| = -a \quad \text{if } a < 0.$$

It is easily seen that absolute values have the following properties:

$|0| = 0, \ |a| > 0$ if $a \neq 0$;
$|a| = |-a|$;
$|a + b| \le |a| + |b|$.

The first two properties follow at once from the definition. To prove the third, we observe first that $a + b \le |a| + |b|$, since $a \le |a|$ and $b \le |b|$. Replacing a by $-a$ and b by $-b$, we obtain also $-(a + b) \le |a| + |b|$. But $|a + b|$ is either $a + b$ or $-(a + b)$.

The *distance* between two real numbers a and b is defined to be the real number

$$d(a,b) = |a - b|.$$

From the preceding properties of absolute values we obtain their counterparts for distances:

(D1) $d(a,a) = 0, \ d(a,b) > 0$ *if* $a \neq b$;

(D2) $d(a,b) = d(b,a)$;

(D3) $d(a,b) \le d(a,c) + d(c,b)$.

The third property is known as the *triangle inequality*, since it may be interpreted as saying that, in any triangle, the length of one side does not exceed the sum of the lengths of the other two.

Fréchet (1906) recognized these three properties as the essential characteristics of any measure of distance and introduced the following general concept. A set E is a *metric space* if with each ordered pair (a,b) of elements of E there is associated a real number $d(a,b)$, so that the properties **(D1)-(D3)** hold for all $a,b,c \in E$.

We note first some simple consequences of these properties. For all $a,b,a',b' \in E$ we have

$$|d(a,b) - d(a',b')| \le d(a,a') + d(b,b') \qquad (*)$$

since, by **(D2)** and **(D3)**,

$$d(a,b) \le d(a,a') + d(a',b') + d(b,b'),$$
$$d(a',b') \le d(a,a') + d(a,b) + d(b,b').$$

Taking $b = b'$ in (*), we obtain from **(D1)**,

$$|d(a,b) - d(a',b)| \le d(a,a'). \qquad (**)$$

In any metric space there is a natural *topology*. A subset G of a metric space E is *open* if for each $x \in G$ there is a positive real number $\delta = \delta(x)$ such that G also contains the whole open ball $\beta_\delta(x) = \{y \in E: d(x,y) < \delta\}$. A set $F \subseteq E$ is *closed* if its complement $E \setminus F$ is open.

For any set $A \subseteq E$, its *closure* $\overline{A}$ is the intersection of all closed sets containing it, and its *interior* int A is the union of all open sets contained in it.

A subset F of E is *connected* if it is not contained in the union of two open subsets of E whose intersections with F are disjoint and nonempty. A subset F of E is (sequentially) *compact* if every sequence of elements of F has a subsequence converging to an element of F (and *locally compact* if this holds for every bounded sequence of elements of F).

A map $f: X \to Y$ from one metric space X to another metric space Y is *continuous* if, for each open subset G of Y, the set of all $x \in X$ such that $f(x) \in G$ is an open subset of X. The two properties stated at the end of §3 admit far-reaching generalizations for continuous maps between subsets of metric spaces, namely that the image under a continuous map of a compact set is again compact, and the image of a connected set is again connected.

There are many examples of metric spaces:

(i)   Let $E = \mathbb{R}^n$ be the set of all n-tuples $a = (\alpha_1,...,\alpha_n)$ of real numbers and define

$$d(b,c) = |b - c|,$$

where $b - c = (\beta_1 - \gamma_1,...,\beta_n - \gamma_n)$ if $b = (\beta_1,...,\beta_n)$ and $c = (\gamma_1,...,\gamma_n)$, and

$$|a| = \max_{1 \leq j \leq n} |\alpha_j|.$$

Alternatively, one can replace the *norm* $|a|$ by either

$$|a|_1 = \sum_{j=1}^{n} |\alpha_j|$$

or

$$|a|_2 = (\sum_{j=1}^{n} |\alpha_j|^2)^{1/2}.$$

In the latter case, $d(b,c)$ is the *Euclidean distance* between b and c. The triangle inequality in this case follows from the *Cauchy–Schwarz inequality*: for any real numbers β_j, γ_j $(j = 1,...,n)$

$$(\sum_{j=1}^{n} \beta_j \gamma_j)^2 \leq (\sum_{j=1}^{n} \beta_j^2)(\sum_{j=1}^{n} \gamma_j^2).$$

(ii)   Let $E = \mathbb{F}_2^n$ be the set of all n-tuples $a = (\alpha_1,...,\alpha_n)$, where $\alpha_j = 0$ or 1 for each j, and define the *Hamming distance* $d(b,c)$ between $b = (\beta_1,...,\beta_n)$ and $c = (\gamma_1,...,\gamma_n)$ to be the number of j such that $\beta_j \neq \gamma_j$. This metric space plays a basic role in the theory of *error-correcting codes*.

(iii)   Let $E = \mathscr{C}(I)$ be the set of all continuous functions $f: I \to \mathbb{R}$, where

$$I = [a,b] = \{x \in \mathbb{R}: a \leq x \leq b\}$$

is an interval of $\mathbb{R}$, and define $d(g,h) = |g - h|$, where

$$|f| = \sup_{a \leq x \leq b} |f(x)|.$$

(A well-known property of continuous functions ensures that f is bounded on I.) Alternatively, one can replace the *norm* $|f|$ by either

$$|f|_1 = \int_a^b |f(x)|\, dx$$

or

$$|f|_2 = (\int_a^b |f(x)|^2 dx)^{1/2}.$$

(iv)   Let $E = \mathscr{C}(\mathbb{R})$ be the set of all continuous functions $f: \mathbb{R} \to \mathbb{R}$ and define

$$d(g,h) = \sum_{N \geq 1} d_N(g,h)/2^N[1 + d_N(g,h)],$$

where $d_N(g,h) = \sup_{|x| \leq N} |g(x) - h(x)|$. The triangle inequality **(D3)** follows from the inequality

$$|\alpha + \beta|/[1 + |\alpha + \beta|] \leq |\alpha|/[1 + |\alpha|] + |\beta|/[1 + |\beta|]$$

for arbitrary real numbers α, β.

The metric here has the property that $d(f_n, f) \to 0$ if and only if $f_n(x) \to f(x)$ uniformly on every bounded subinterval of $\mathbb{R}$. It may be noted that, even though E is a vector space, the metric is not derived from a norm since, if $\lambda \in \mathbb{R}$, one may have $d(\lambda g, \lambda h) \neq |\lambda|\, d(g, h)$.

(v) Let E be the set of all measurable functions $f: I \to \mathbb{R}$, where $I = [a, b]$ is an interval of $\mathbb{R}$, and define

$$d(g, h) \;=\; \int_a^b |g(x) - h(x)|\,(1 + |g(x) - h(x)|)^{-1} dx.$$

In order to obtain **(D1)**, we identify functions which take the same value at all points of I, except for a set of measure zero.

Convergence with respect to this metric coincides with *convergence in measure*, which plays a role in the theory of probability.

(vi) Let $E = \mathbb{F}_2^\infty$ be the set of all infinite sequences $a = (\alpha_1, \alpha_2, \dots\,)$, where $\alpha_j = 0$ or 1 for every j, and define $d(a, a) = 0$, $d(a, b) = 2^{-k}$ if $a \neq b$, where $b = (\beta_1, \beta_2, \dots\,)$ and k is the least positive integer such that $\alpha_k \neq \beta_k$.

Here the triangle inequality holds in the stronger form

$$d(a, b) \;\leq\; \max[d(a, c),\, d(c, b)].$$

This metric space plays a basic role in the theory of *dynamical systems*.

(vii) A connected *graph* can be given the structure of a metric space by defining the distance between two vertices to be the number of edges on the shortest path joining them.

Let E be an arbitrary metric space and $\{a_n\}$ a sequence of elements of E. The sequence $\{a_n\}$ is said to *converge*, with *limit $a \in E$*, if

$$d(a_n, a) \;\to\; 0 \quad \text{as } n \to \infty,$$

i.e. if for each real $\varepsilon > 0$ there is a corresponding positive integer $N = N(\varepsilon)$ such that $d(a_n, a) < \varepsilon$ for every $n \geq N$.

The limit a is uniquely determined, since if also $d(a_n, a') \to 0$, then

$$d(a, a') \;\leq\; d(a_n, a) + d(a_n, a'),$$

and the right side can be made arbitrarily small by taking n sufficiently large. We write

$$\lim_{n \to \infty} a_n = a,$$

or $a_n \to a$ as $n \to \infty$. If the sequence $\{a_n\}$ has limit a, then so also does any (infinite) subsequence.

If $a_n \to a$ and $b_n \to b$, then $d(a_n,b_n) \to d(a,b)$, as one sees by taking $a' = a_n$ and $b' = b_n$ in (*).

The sequence $\{a_n\}$ is said to be a *fundamental sequence*, or 'Cauchy sequence', if for each real $\varepsilon > 0$ there is a corresponding positive integer $N = N(\varepsilon)$ such that $d(a_m,a_n) < \varepsilon$ for all $m,n \geq N$.

If $\{a_n\}$ and $\{b_n\}$ are fundamental sequences then, by (*), the sequence $\{d(a_n,b_n)\}$ of real numbers is a fundamental sequence, and therefore convergent.

A set $S \subseteq E$ is said to be *bounded* if the set of all real numbers $d(a,b)$ with $a,b \in S$ is a bounded subset of $\mathbb{R}$.

Any fundamental sequence $\{a_n\}$ is bounded, since if

$$d(a_m,a_n) < 1 \quad \text{for all } m,n \geq N,$$

then

$$d(a_m,a_n) < 1 + \delta \quad \text{for all } m,n \in \mathbb{N},$$

where $\delta = \max_{1 \leq j < k \leq N} d(a_j,a_k)$.

Furthermore, any convergent sequence $\{a_n\}$ is a fundamental sequence, as one sees by taking $a = \lim_{n \to \infty} a_n$ in the inequality

$$d(a_m,a_n) \leq d(a_m,a) + d(a_n,a).$$

A metric space is said to be *complete* if, conversely, every fundamental sequence is convergent.

By generalizing the Méray–Cantor method of extending the rational numbers to the real numbers, Hausdorff (1913) showed that any metric space can be embedded in a complete metric space. To state his result precisely, we introduce some definitions.

A subset F of a metric space E is said to be *dense* in E if, for each $a \in E$ and each real $\varepsilon > 0$, there exists some $b \in F$ such that $d(a,b) < \varepsilon$.

A map σ from one metric space E to another metric space E' is necessarily injective if it is distance-preserving, i.e. if

$$d'(\sigma(a),\sigma(b)) = d(a,b) \quad \text{for all } a,b \in E.$$

If the map σ is also surjective, then it is said to be an *isometry* and the metric spaces E and E' are said to be *isometric*.

A metric space $\overline{E}$ is said to be a *completion* of a metric space E if $\overline{E}$ is complete and E is isometric to a dense subset of $\overline{E}$. It is easily seen that any two completions of a given metric space are isometric.

Hausdorff's result says that *any metric space E has a completion $\overline{E}$*. We sketch the proof. Define two fundamental sequences $\{a_n\}$ and $\{a'_n\}$ in E to be equivalent if

$$\lim_{n\to\infty} d(a_n,a'_n) = 0.$$

It is easily shown that this is indeed an equivalence relation. Moreover, if the fundamental sequences $\{a_n\},\{b_n\}$ are equivalent to the fundamental sequences $\{a'_n\},\{b'_n\}$ respectively, then

$$\lim_{n\to\infty} d(a_n,b_n) = \lim_{n\to\infty} d(a'_n,b'_n).$$

We can give the set $\overline{E}$ of all equivalence classes of fundamental sequences the structure of a metric space by defining

$$\overline{d}(\{a_n\},\{b_n\}) = \lim_{n\to\infty} d(a_n,b_n).$$

For each $a \in E$, let $\overline{a}$ be the equivalence class in $\overline{E}$ which contains the fundamental sequence $\{a_n\}$ such that $a_n = a$ for every n. Since

$$\overline{d}(\overline{a},\overline{b}) = d(a,b) \quad \text{for all } a,b \in E,$$

E is isometric to the set $E' = \{\overline{a} : a \in E\}$. It is not difficult to show that E' is dense in $\overline{E}$ and that $\overline{E}$ is complete.

Which of the previous examples of metric spaces are complete? In example (i), the completeness of $\mathbb{R}^n$ with respect to the first definition of distance follows directly from the completeness of $\mathbb{R}$. It is also complete with respect to the two alternative definitions of distance, since a sequence which converges with respect to one of the three metrics also converges with respect to the other two. Indeed it is easily shown that, for every $a \in \mathbb{R}^n$,

$$|a| \le |a|_2 \le |a|_1$$

and

$$|a|_1 \le n^{1/2} |a|_2, \quad |a|_2 \le n^{1/2} |a|.$$

In example (ii), the completeness of $\mathbb{F}_2^n$ is trivial, since any fundamental sequence is ultimately constant.

In example (iii), the completeness of $\mathscr{C}(I)$ with respect to the first definition of distance follows from the completeness of $\mathbb{R}$ and the fact that the limit of a uniformly convergent sequence of continuous functions is again a continuous function.

However, $\mathscr{C}(I)$ is not complete with respect to either of the two alternative definitions of distance. It is possible also for a sequence to converge with respect to the two alternative definitions of distance, but not with respect to the first definition. Similarly, a sequence may converge in the first alternative metric, but not even be a fundamental sequence in the second.

The completions of the metric space $\mathscr{C}(I)$ with respect to the two alternative metrics may actually be identified with spaces of functions. The completion for the first alternative metric is the set $L(I)$ of all *Lebesgue measurable* functions $f\colon I \to \mathbb{R}$ such that

$$\int_a^b |f(x)|\, dx < \infty,$$

functions which take the same value at all points of I, except for a set of measure zero, being identified. The completion $L^2(I)$ for the second alternative metric is obtained by replacing $\int_a^b |f(x)|\, dx$ by $\int_a^b |f(x)|^2\, dx$ in this statement.

It may be shown that the metric spaces of examples (iv)-(vi) are all complete. In example (vi), the strong triangle inequality implies that $\{a_n\}$ is a fundamental sequence if (and only if) $d(a_{n+1},a_n) \to 0$ as $n \to \infty$.

Let E be an arbitrary metric space and $f\colon E \to E$ a map of E into itself. A point $\bar{x} \in E$ is said to be a *fixed point* of f if $f(\bar{x}) = \bar{x}$. A useful property of complete metric spaces is the following *contraction principle*, which was first established in the present generality by Banach (1922), but was previously known in more concrete situations.

PROPOSITION 26 *Let E be a complete metric space and let $f\colon E \to E$ be a map of E into itself. If there exists a real number θ, with $0 < \theta < 1$, such that*

$$d(f(x'),f(x'')) \le \theta\, d(x',x'') \quad \text{for all } x',x'' \in E,$$

then the map f has a unique fixed point $\bar{x} \in E$.

Proof It is clear that there is at most one fixed point, since $0 \le d(x',x'') \le \theta\, d(x',x'')$ implies $x' = x''$. To prove that a fixed point exists we use the *method of successive approximations*.

Choose any $x_0 \in E$ and define the sequence $\{x_n\}$ recursively by

$$x_n = f(x_{n-1}) \quad (n \ge 1).$$

For any $k \ge 1$ we have

$$d(x_{k+1},x_k) = d(f(x_k),f(x_{k-1})) \le \theta\, d(x_k,x_{k-1}).$$

Applying this k times, we obtain

$$d(x_{k+1},x_k) \le \theta^k\, d(x_1,x_0).$$

Consequently, if $n > m \ge 0$,

$$d(x_n, x_m) \leq d(x_n, x_{n-1}) + d(x_{n-1}, x_{n-2}) + \ldots + d(x_{m+1}, x_m)$$
$$\leq (\theta^{n-1} + \theta^{n-2} + \ldots + \theta^m) \, d(x_1, x_0)$$
$$\leq \theta^m \, (1 - \theta)^{-1} \, d(x_1, x_0),$$

since $0 < \theta < 1$. It follows that $\{x_n\}$ is a fundamental sequence and so a convergent sequence, since E is complete. If $\bar{x} = \lim_{n \to \infty} x_n$, then

$$d(f(\bar{x}), \bar{x}) \leq d(f(\bar{x}), x_{n+1}) + d(x_{n+1}, \bar{x}) \leq \theta \, d(\bar{x}, x_n) + d(\bar{x}, x_{n+1}).$$

Since the right side can be made less than any given positive real number by taking n large enough, we must have $f(\bar{x}) = \bar{x}$. The proof shows also that, for any $m \geq 0$,

$$d(\bar{x}, x_m) \leq \theta^m \, (1 - \theta)^{-1} \, d(x_1, x_0). \quad \square$$

The contraction principle is surprisingly powerful, considering the simplicity of its proof. We give two significant applications: an inverse function theorem and an existence theorem for ordinary differential equations. In both cases we will use the notion of differentiability for functions of several real variables. The unambitious reader may simply take $n = 1$ in the following discussion (so that 'invertible' means 'nonzero'). Functions of several variables are important, however, and it is remarkable that the proper definition of differentiability in this case was first given by Stolz (1887).

A map $\varphi: U \to \mathbb{R}^m$, where $U \subseteq \mathbb{R}^n$ is a *neighbourhood* of $x_0 \in \mathbb{R}^n$ (i.e., U contains some open ball $\{x \in \mathbb{R}^n: |x - x_0| < \rho\}$), is said to be *differentiable* at x_0 if there exists a linear map $A: \mathbb{R}^n \to \mathbb{R}^m$ such that

$$|\varphi(x) - \varphi(x_0) - A(x - x_0)|/|x - x_0| \to 0 \quad \text{as } |x - x_0| \to 0.$$

(The inequalities between the various norms show that it is immaterial which norm is used.) The linear map A, which is then uniquely determined, is called the *derivative* of φ at x_0 and will be denoted by $\varphi'(x_0)$.

This definition is a natural generalization of the usual definition when $m = n = 1$, since it says that the difference $\varphi(x_0 + h) - \varphi(x_0)$ admits the linear approximation Ah for $|h| \to 0$.

Evidently, if φ_1 and φ_2 are differentiable at x_0, then so also is $\varphi = \varphi_1 + \varphi_2$ and

$$\varphi'(x_0) = \varphi_1'(x_0) + \varphi_2'(x_0).$$

It also follows directly from the definition that derivatives satisfy the *chain rule*: If $\varphi: U \to \mathbb{R}^m$, where U is a neighbourhood of $x_0 \in \mathbb{R}^n$, is differentiable at x_0, and if $\psi: V \to \mathbb{R}^l$, where V is a neighbourhood of $y_0 = \varphi(x_0) \in \mathbb{R}^m$, is differentiable at y_0, then the composite map $\chi = \psi \circ \varphi$: $U \to \mathbb{R}^l$ is differentiable at x_0 and

$$\chi'(x_0) \;=\; \psi'(y_0)\,\varphi'(x_0),$$

the right side being the composite linear map.

We will also use the notion of norm of a linear map. If $A: \mathbb{R}^n \to \mathbb{R}^m$ is a linear map, its *norm* $|A|$ is defined by

$$|A| \;=\; \sup_{|x|\le 1} |Ax|.$$

Evidently

$$|A_1 + A_2| \;\le\; |A_1| + |A_2|.$$

Furthermore, if $B: \mathbb{R}^m \to \mathbb{R}^l$ is another linear map, then

$$|BA| \;\le\; |B||A|.$$

Hence, if $m = n$ and $|A| < 1$, then the linear map $I - A$ is invertible, its inverse being given by the geometric series

$$(I - A)^{-1} \;=\; I + A + A^2 + \dots \;.$$

It follows that, for any linear maps $A, B: \mathbb{R}^n \to \mathbb{R}^n$ with A invertible, if $|B - A| < |A^{-1}|^{-1}$, then B is also invertible and $|B^{-1} - A^{-1}| \to 0$ as $|B - A| \to 0$.

If $\varphi: U \to \mathbb{R}^m$ is differentiable at $x_0 \in \mathbb{R}^n$, then it is also continuous at x_0, since

$$|\varphi(x) - \varphi(x_0)| \;\le\; |\varphi(x) - \varphi(x_0) - \varphi'(x_0)(x - x_0)| + |\varphi'(x_0)||x - x_0|.$$

We say that φ is *continuously differentiable* in U if it is differentiable at each point of U and if the derivative $\varphi'(x)$ is a continuous function of x in U. The *inverse function theorem* says:

PROPOSITION 27 *Let U_0 be a neighbourhood of $x_0 \in \mathbb{R}^n$ and let $\varphi: U_0 \to \mathbb{R}^n$ be a continuously differentiable map for which $\varphi'(x_0)$ is invertible.*

Then, for some $\delta > 0$, the ball $U = \{x \in \mathbb{R}^n: |x - x_0| < \delta\}$ is contained in U_0 and

(i)   *the restriction of φ to U is injective;*

(ii)   *$V := \varphi(U)$ is open, i.e. if $\eta \in V$, then V contains all $y \in \mathbb{R}^n$ near η;*

(iii) *the inverse map $\psi: V \to U$ is also continuously differentiable and, if $y = \varphi(x)$, then $\psi'(y)$ is the inverse of $\varphi'(x)$.*

Proof To simplify notation, assume $x_0 = \varphi(x_0) = 0$ and write $A = \varphi'(0)$. For any $y \in \mathbb{R}^n$, put

$$f_y(x) \;=\; x + A^{-1}[y - \varphi(x)].$$

Evidently x is a fixed point of f_y if and only if $\varphi(x) = y$. The map f_y is also continuously differentiable and

$$f_y'(x) = I - A^{-1}\varphi'(x) = A^{-1}[A - \varphi'(x)].$$

Since $\varphi'(x)$ is continuous, we can choose $\delta > 0$ so that the ball $U = \{x \in \mathbb{R}^n : |x| < \delta\}$ is contained in U_0 and

$$|f_y'(x)| \leq 1/2 \quad \text{for } x \in U.$$

If $x_1, x_2 \in U$, then

$$\begin{aligned}
|f_y(x_2) - f_y(x_1)| &= \left| \int_0^1 f'((1-t)x_1 + tx_2)(x_2 - x_1)\, dt \right| \\
&\leq |x_2 - x_1|/2.
\end{aligned}$$

It follows that f_y has at most one fixed point in U. Since this holds for arbitrary $y \in \mathbb{R}^n$, the restriction of φ to U is injective.

Suppose next that $\eta = \varphi(\xi)$ for some $\xi \in U$. We wish to show that, if y is near η, the map f_y has a fixed point near ξ.

Choose $r = r(\xi) > 0$ so that the closed ball $B_r = \{x \in \mathbb{R}^n : |x - \xi| \leq r\}$ is contained in U, and fix $y \in \mathbb{R}^n$ so that $|y - \eta| < r/2|A^{-1}|$. Then

$$|f_y(\xi) - \xi| = |A^{-1}(y - \eta)| \leq |A^{-1}||y - \eta| < r/2.$$

Hence if $|x - \xi| \leq r$, then

$$\begin{aligned}
|f_y(x) - \xi| &\leq |f_y(x) - f_y(\xi)| + |f_y(\xi) - \xi| \\
&\leq |x - \xi|/2 + r/2 \leq r.
\end{aligned}$$

Thus $f_y(B_r) \subseteq B_r$. Also, if $x_1, x_2 \in B_r$, then

$$|f_y(x_2) - f_y(x_1)| \leq |x_2 - x_1|/2.$$

But B_r is a complete metric space, with the same metric as $\mathbb{R}^n$, since it is a closed subset (if $x_n \in B_r$ and $x_n \to x$ in $\mathbb{R}^n$, then also $x \in B_r$). Consequently, by the contraction principle (Proposition 26), f_y has a fixed point $x \in B_r$. Then $\varphi(x) = y$, which proves (ii).

Suppose now that $y, \eta \in V$. Then $y = \varphi(x)$, $\eta = \varphi(\xi)$ for unique $x, \xi \in U$. Since

$$|f_y(x) - f_y(\xi)| \leq |x - \xi|/2$$

and

$$f_y(x) - f_y(\xi) = x - \xi - A^{-1}(y - \eta),$$

we have

$$|A^{-1}(y - \eta)| \geq |x - \xi|/2.$$

Thus

$$|x - \xi| \le 2|A^{-1}||y - \eta|.$$

If $F = \varphi'(\xi)$ and $G = F^{-1}$, then

$$\psi(y) - \psi(\eta) - G(y - \eta) = x - \xi - G(y - \eta)$$
$$= -G[\varphi(x) - \varphi(\xi) - F(x - \xi)].$$

Hence

$$|\psi(y) - \psi(\eta) - G(y - \eta)|/|y - \eta| \le 2|A^{-1}||G||\varphi(x) - \varphi(\xi) - F(x - \xi)|/|x - \xi|.$$

If $|y - \eta| \to 0$, then $|x - \xi| \to 0$ and the right side tends to 0. Consequently ψ is differentiable at η and $\psi'(\eta) = G = F^{-1}$.

Thus ψ is differentiable in U and, *a fortiori*, continuous. In fact ψ is continuously differentiable, since F is a continuous function of ξ (by hypothesis), since $\xi = \psi(\eta)$ is a continuous function of η, and since F^{-1} is a continuous function of F. $\quad\Box$

To bring out the meaning of Proposition 27 we add some remarks:

(i)   The invertibility of $\varphi'(x_0)$ is necessary for the existence of a differentiable inverse map, but not for the existence of a continuous inverse map. For example, the continuously differentiable map $\varphi: \mathbb{R} \to \mathbb{R}$ defined by $\varphi(x) = x^3$ is bijective and has the continuous inverse $\psi(y) = y^{1/3}$, although $\varphi'(0) = 0$.

(ii)   The hypothesis that φ is *continuously* differentiable cannot be totally dispensed with. For example, the map $\varphi: \mathbb{R} \to \mathbb{R}$ defined by

$$\varphi(x) = x + x^2 \sin(1/x) \ \text{ if } x \ne 0, \ \ \varphi(0) = 0,$$

is everywhere differentiable and $\varphi'(0) \ne 0$, but φ is not injective in any neighbourhood of 0.

(iii)   The inverse map may not be defined throughout U_0. For example, the map $\varphi: \mathbb{R}^2 \to \mathbb{R}^2$ defined by

$$\varphi_1(x_1,x_2) = x_1{}^2 - x_2{}^2, \ \ \varphi_2(x_1,x_2) = 2x_1x_2,$$

is everywhere continuously differentiable and has an invertible derivative at every point except the origin. Thus the hypotheses of Proposition 27 are satisfied in any connected open set $U_0 \subseteq \mathbb{R}^2$ which does not contain the origin, and yet $\varphi(1,1) = \varphi(-1,-1)$.

It was first shown by Cauchy (c. 1844) that, under quite general conditions, an ordinary differential equation has local solutions. The method of successive approximations (i.e., the contraction principle) was used for this purpose by Picard (1890):

PROPOSITION 28 *Let $t_0 \in \mathbb{R}$, $\xi_0 \in \mathbb{R}^n$ and let U be a neighbourhood of (t_0,ξ_0) in $\mathbb{R} \times \mathbb{R}^n$. If $\varphi: U \to \mathbb{R}^n$ is a continuous map with a derivative φ' with respect to x that is continuous in U, then the differential equation*

$$dx/dt = \varphi(t,x) \tag{1}$$

has a unique solution $x(t)$ which satisfies the initial condition

$$x(t_0) = \xi_0 \tag{2}$$

and is defined in some interval $|t - t_0| \le \delta$, where $\delta > 0$.

Proof If $x(t)$ is a solution of the differential equation (1) which satisfies the initial condition (2), then by integration we get

$$x(t_0) = \xi_0 + \int_{t_0}^{t} \varphi[\tau,x(\tau)] \, d\tau.$$

Conversely, if a *continuous* function $x(t)$ satisfies this relation then, since φ is continuous, $x(t)$ is actually differentiable and is a solution of (1) that satisfies (2). Hence we need only show that the map $\mathscr{F}$ defined by

$$(\mathscr{F}x)(t) = \xi_0 + \int_{t_0}^{t} \varphi[\tau,x(\tau)] \, d\tau$$

has a unique fixed point in the space of continuous functions.

There exist positive constants M,L such that

$$|\varphi(t,\xi)| \le M, \quad |\varphi'(t,\xi)| \le L$$

for all (t,ξ) in a neighbourhood of (t_0,ξ_0), which we may take to be U. If $(t,\xi_1) \in U$ and $(t,\xi_2) \in U$, then

$$|\varphi(t,\xi_2) - \varphi(t,\xi_1)| = \left| \int_0^1 \varphi'(t,(1 - u)\xi_1 + u\xi_2)(\xi_2 - \xi_1) \, du \right|$$

$$\le L|\xi_2 - \xi_1|.$$

Choose $\delta > 0$ so that the box $|t - t_0| \le \delta$, $|\xi - \xi_0| \le M\delta$ is contained in U and also $L\delta < 1$. Take $I = [t_0 - \delta, t_0 + \delta]$ and let $\mathscr{C}(I)$ be the complete metric space of all continuous functions $x: I \to \mathbb{R}^n$ with the distance function

$$d(x_1,x_2) = \sup_{t \in I} |x_1(t) - x_2(t)|.$$

The constant function $x_0(t) = \xi_0$ is certainly in $\mathscr{C}(I)$. Let E be the subset of all $x \in \mathscr{C}(I)$ such that $x(t_0) = \xi_0$ and $d(x,x_0) \le M\delta$. Evidently if $x_n \in E$ and $x_n \to x$ in $\mathscr{C}(I)$, then $x \in E$. Hence E is also a complete metric space with the same metric. Moreover $\mathscr{F}(E) \subseteq E$, since if $x \in E$ then $(\mathscr{F}x)(t_0) = \xi_0$ and, for all $t \in I$,

$$\left|(\mathscr{F}x)(t) - \xi_0\right| = \left|\int_{t_0}^{t} \varphi[\tau,x(\tau)]\, d\tau\right| \le M\delta.$$

Furthermore, if $x_1,x_2 \in E$, then $d(\mathscr{F}x_1,\mathscr{F}x_2) \le L\delta\, d(x_1,x_2)$, since for all $t \in I$,

$$\left|(\mathscr{F}x_1)(t) - (\mathscr{F}x_2)(t)\right| = \left|\int_{t_0}^{t} \{\varphi[\tau,x_1(\tau)] - \varphi[\tau,x_2(\tau)]\}\, d\tau\right| \le L\delta\, d(x_1,x_2).$$

Since $L\delta < 1$, the result now follows from Proposition 26. $\square$

Proposition 28 only guarantees the local existence of solutions, but this is in the nature of things. For example, if $n = 1$, the unique solution of the differential equation

$$dx/dt = x^2$$

such that $x(t_0) = \xi_0 > 0$ is given by

$$x(t) = \{1 - (t - t_0)\xi_0\}^{-1}\, \xi_0.$$

Thus the solution is defined only for $t < t_0 + \xi_0^{-1}$, even though the differential equation itself has exemplary behaviour everywhere.

To illustrate Proposition 28, take $n = 1$ and let $E(t)$ be the solution of the (linear) differential equation

$$dx/dt = x \tag{3}$$

which satisfies the initial condition $E(0) = 1$. Then $E(t)$ is defined for $|t| < R$, for some $R > 0$. If $|\tau| < R/2$ and $x_1(t) = E(t + \tau)$, then $x_1(t)$ is the solution of the differential equation (3) which satisfies the initial condition $x_1(0) = E(\tau)$. But $x_2(t) = E(\tau)E(t)$ satisfies the same differential equation and the same initial condition. Hence we must have $x_1(t) = x_2(t)$ for $|t| < R/2$, i.e.

$$E(t + \tau) = E(t)E(\tau). \tag{4}$$

In particular,

$$E(t)E(-t) = 1, \quad E(2t) = E(t)^2.$$

The last relation may be used to extend the definition of $E(t)$, so that it is continuously differentiable and a solution of (3) also for $|t| < 2R$. It follows that the solution $E(t)$ is defined for all $t \in \mathbb{R}$ and satisfies the *addition theorem* (4) for all $t,\tau \in \mathbb{R}$.

It is instructive to carry through the method of successive approximations explicitly in this case. If we take $x_0(t)$ to be the constant 1, then

$$x_1(t) = 1 + \int_0^t x_0(\tau)\, d\tau = 1 + t,$$

$$x_2(t) = 1 + \int_0^t x_1(\tau)\, d\tau = 1 + t + t^2/2,$$

$$\cdots\cdots$$

By induction we obtain, for every $n \geq 1$,

$$x_n(t) = 1 + t + t^2/2! + \ldots + t^n/n!.$$

Since $x_n(t) \to E(t)$ as $n \to \infty$, we obtain for the solution $E(t)$ the infinite series representation

$$E(t) = 1 + t + t^2/2! + t^3/3! + \ldots ,$$

valid actually for every $t \in \mathbb{R}$. In particular,

$$e: = E(1) = 1 + 1 + 1/2! + 1/3! + \ldots = 2.7182818\ldots .$$

Of course $E(t) = e^t$ is the *exponential function*. We will now adopt the usual notation, but we remark that the definition of e^t as a solution of a differential equation provides a meaning for irrational t, as well as a simple proof of both the addition theorem and the exponential series.

The power series for e^t shows that

$$e^t > 1 + t > 1 \quad \text{for every } t > 0.$$

Since $e^{-t} = (e^t)^{-1}$, it follows that $0 < e^t < 1$ for every $t < 0$. Thus $e^t > 0$ for all $t \in \mathbb{R}$. Hence, by (3), e^t is a strictly increasing function. But $e^t \to +\infty$ as $t \to +\infty$ and $e^t \to 0$ as $t \to -\infty$. Consequently, since it is certainly continuous, the exponential function maps the real line $\mathbb{R}$ bijectively onto the positive half-line $\mathbb{R}_+ = \{x \in \mathbb{R}: x > 0\}$. For any $x > 0$, the unique $t \in \mathbb{R}$ such that $e^t = x$ is denoted by $\ln x$ (the *natural logarithm* of x) or simply $\log x$.

5 Complex numbers

By extending the rational numbers to the real numbers, we ensured that every positive number had a square root. By further extending the real numbers to the complex numbers, we will now ensure that all numbers have square roots.

The first use of complex numbers, by Cardano (1545), may have had its origin in the solution of cubic, rather than quadratic, equations. The cubic polynomial

$$f(x) = x^3 - 3px - 2q$$

has three real roots if $d := q^2 - p^3 < 0$ since then, for large $X > 0$,

$$f(-X) < 0, \ f(-p^{1/2}) > 0, \ f(p^{1/2}) < 0, \ f(X) > 0.$$

Cardano's formula for the three roots,

$$f(x) = \sqrt[3]{(q + \sqrt{d})} + \sqrt[3]{(q - \sqrt{d})},$$

gives real values, even though d is negative, because the two summands are conjugate complex numbers. This was explicitly stated by Bombelli (1572). It is a curious fact, first proved by Hölder (1891), that if a cubic equation has three distinct real roots, then it is impossible to represent these roots solely by real radicals.

Intuitively, complex numbers are expressions of the form $a + ib$, where a and b are real numbers and $i^2 = -1$. But what is i? Hamilton (1835) defined complex numbers as ordered pairs of real numbers, with appropriate rules for addition and multiplication. Although this approach is similar to that already used in this chapter, and actually was its first appearance, we now choose a different method.

We define a *complex number* to be a 2×2 matrix of the form

$$A = \begin{pmatrix} a & b \\ -b & a \end{pmatrix},$$

where a and b are real numbers. The set of all complex numbers is customarily denoted by $\mathbb{C}$. We may define addition and multiplication in $\mathbb{C}$ to be matrix addition and multiplication, since $\mathbb{C}$ is closed under these operations: if

$$B = \begin{pmatrix} c & d \\ -d & c \end{pmatrix},$$

then

$$A + B = \begin{pmatrix} a+c & b+d \\ -(b+d) & a+c \end{pmatrix}, \ AB = \begin{pmatrix} ac-bd & ad+bc \\ -(ad+bc) & ac-bd \end{pmatrix}.$$

Furthermore $\mathbb{C}$ contains

$$0 = \begin{pmatrix} 0 & 0 \\ 0 & 0 \end{pmatrix}, \ 1 = \begin{pmatrix} 1 & 0 \\ 0 & 1 \end{pmatrix},$$

and $A \in \mathbb{C}$ implies $-A \in \mathbb{C}$.

It follows from the properties of matrix addition and multiplication that addition and multiplication of complex numbers have the properties (A2)-(A5), (M2)-(M4) and (AM1)-(AM2), with *0* and *1* as identity elements for addition and multiplication respectively. The property (M5) also holds, since if a and b are not both zero, and if

$$a' = a/(a^2 + b^2), \quad b' = -b/(a^2 + b^2),$$

then

$$A^{-1} = \begin{pmatrix} a' & b' \\ -b' & a' \end{pmatrix}$$

is a multiplicative inverse of A. Thus $\mathbb{C}$ satisfies the axioms for a *field*.

The set $\mathbb{C}$ also contains the matrix

$$i = \begin{pmatrix} 0 & 1 \\ -1 & 0 \end{pmatrix},$$

for which $i^2 = -1$, and any $A \in \mathbb{C}$ can be represented in the form

$$A = a1 + bi,$$

where $a,b \in \mathbb{R}$. The multiples $a1$, where $a \in \mathbb{R}$, form a subfield of $\mathbb{C}$ isomorphic to the real field $\mathbb{R}$. By identifying the real number a with the complex number $a1$, we may regard $\mathbb{R}$ itself as contained in $\mathbb{C}$.

Thus we will now stop using matrices and use only the fact that $\mathbb{C}$ is a field containing $\mathbb{R}$ such that every $z \in \mathbb{C}$ can be represented in the form

$$z = x + iy,$$

where $x,y \in \mathbb{R}$ and $i \in \mathbb{C}$ satisfies $i^2 = -1$. The representation is necessarily unique, since $i \notin \mathbb{R}$. We call x and y the *real* and *imaginary parts* of z and denote them by $\mathfrak{R}z$ and $\mathfrak{I}z$ respectively. Complex numbers of the form iy, where $y \in \mathbb{R}$, are said to be *pure imaginary*.

It is worth noting that $\mathbb{C}$ cannot be given the structure of an *ordered* field, since in an ordered field any nonzero square is positive, whereas $i^2 + 1^2 = (-1) + 1 = 0$.

It is often suggestive to regard complex numbers as points of a plane, the complex number $z = x + iy$ being the point with coordinates (x,y) in some chosen system of rectangular coordinates.

The *complex conjugate* of the complex number $z = x + iy$, where $x,y \in \mathbb{R}$, is the complex number $\bar{z} = x - iy$. In the geometrical representation of complex numbers, $\bar{z}$ is the reflection of z in the x-axis. From the definition we at once obtain

$$\mathfrak{R}z = (z + \bar{z})/2, \quad \mathfrak{I}z = (z - \bar{z})/2i.$$

It is easily seen also that

$$\overline{z+w} = \bar{z} + \bar{w}, \quad \overline{zw} = \bar{z}\,\bar{w}, \quad \bar{\bar{z}} = z.$$

Moreover, $\bar{z} = z$ if and only if $z \in \mathbb{R}$. Thus the map $z \to \bar{z}$ is an 'involutory automorphism' of the field $\mathbb{C}$, with the subfield $\mathbb{R}$ as its set of fixed points. It follows that $\overline{-z} = -\bar{z}$.

If $z = x + iy$, where $x, y \in \mathbb{R}$, then

$$z\bar{z} = (x + iy)(x - iy) = x^2 + y^2.$$

Hence $z\bar{z}$ is a positive real number for any nonzero $z \in \mathbb{C}$. The *absolute value* $|z|$ of the complex number z is defined by

$$|0| = 0, \quad |z| = \sqrt{(z\bar{z})} \ \text{ if } z \neq 0,$$

(with the positive value for the square root). This agrees with the definition in §4 if $z = x$ is a real number.

It follows at once from the definition that $|\bar{z}| = |z|$ for every $z \in \mathbb{C}$, and $z^{-1} = \bar{z}/|z|^2$ if $z \neq 0$.

Absolute values have the following properties: *for all $z, w \in \mathbb{C}$,*

(i) $|0| = 0, \ |z| > 0 \ if \ z \neq 0$;

(ii) $|zw| = |z||w|$;

(iii) $|z + w| \leq |z| + |w|$.

The first property follows at once from the definition. To prove (ii), observe that both sides are non-negative and that

$$|zw|^2 = zw\,\overline{zw} = zw\,\bar{z}\,\bar{w} = z\bar{z}\,w\bar{w} = |z|^2\,|w|^2.$$

To prove (iii), we first evaluate $|z + w|^2$:

$$|z + w|^2 = (z + w)(\bar{z} + \bar{w}) = z\bar{z} + (z\bar{w} + w\bar{z}) + w\bar{w} = |z|^2 + 2\,\mathfrak{R}(z\bar{w}) + |w|^2.$$

Since $\mathfrak{R}(z\bar{w}) \leq |z\bar{w}| = |z||w|$, this yields

$$|z + w|^2 \leq |z|^2 + 2\,|z||w| + |w|^2 = (|z| + |w|)^2,$$

and (iii) follows by taking square roots.

Several other properties are consequences of these three, although they may also be verified directly. By taking $z = w = 1$ in (ii) and using (i), we obtain $|1| = 1$. By taking

$z = w = -1$ in (ii) and using (i), we now obtain $|-1| = 1$. Taking $w = -1$ and $w = z^{-1}$ in (ii), we further obtain

$$|-z| = |z|, \quad |z^{-1}| = |z|^{-1} \ \text{ if } z \ne 0.$$

Again, by replacing z by $z - w$ in (iii), we obtain

$$||z| - |w|| \le |z - w|.$$

This shows that $|z|$ is a continuous function of z. In fact $\mathbb{C}$ is a metric space, with the metric $d(z,w) = |z - w|$. By considering real and imaginary parts separately, one verifies that this metric space is complete, i.e. every fundamental sequence is convergent, and that the Bolzano–Weierstrass property continues to hold, i.e. any bounded sequence of complex numbers has a convergent subsequence.

It will now be shown that any complex number has a square root. If $w = u + iv$ and $z = x + iy$, then $z^2 = w$ is equivalent to

$$x^2 - y^2 = u, \quad 2xy = v.$$

Since

$$(x^2 + y^2)^2 = (x^2 - y^2)^2 + (2xy)^2,$$

these equations imply

$$x^2 + y^2 = \surd(u^2 + v^2).$$

Hence

$$x^2 = \{u + \surd(u^2 + v^2)\}/2.$$

Since the right side is positive if $v \ne 0$, x is then uniquely determined apart from sign and $y = v/2x$ is uniquely determined by x. If $v = 0$, then $x = \pm \surd u$ and $y = 0$ when $u > 0$; $x = 0$ and $y = \pm \surd(-u)$ when $u < 0$, and $x = y = 0$ when $u = 0$.

It follows that any quadratic polynomial

$$q(z) = az^2 + bz + c,$$

where $a,b,c \in \mathbb{C}$ and $a \ne 0$, has two complex roots, given by the well-known formula

$$z = \{-b \pm \surd(b^2 - 4ac)\}/2a.$$

However, much more is true. The so-called *fundamental theorem of algebra* asserts that any polynomial

$$f(z) = a_0 z^n + a_1 z^{n-1} + \ldots + a_n,$$

where $n \geq 1$, $a_0 \neq 0$ and $a_0, a_1, \ldots, a_n \in \mathbb{C}$, has a complex root. Thus by adjoining to the real field $\mathbb{R}$ a root of the polynomial $z^2 + 1$ we ensure that every non-constant polynomial has a root. Today the fundamental theorem of algebra is considered to belong to analysis, rather than to algebra. It is useful to retain the name, however, as a reminder that our own pronouncements may seem equally quaint in the future.

Our proof of the theorem will use the fact that any polynomial is differentiable, since sums and products of differentiable functions are again differentiable, and hence also continuous. We first prove

PROPOSITION 29 *Let $G \subseteq \mathbb{C}$ be an open set and E a proper subset (possibly empty) of G such that each point of G has a neighbourhood containing at most one point of E. If $f : G \to \mathbb{C}$ is a continuous map which at every point of $G \setminus E$ is differentiable and has a nonzero derivative, then $f(G)$ is an open subset of $\mathbb{C}$.*

Proof   Evidently $G \setminus E$ is an open set. We show first that $f(G \setminus E)$ is also an open set. Let $\zeta \in G \setminus E$. Then f is differentiable at ζ and $\rho = |f'(\zeta)| > 0$. We can choose $\delta > 0$ so that the closed disc $B = \{z \in \mathbb{C} : |z - \zeta| \leq \delta\}$ contains no point of E, is contained in G and

$$|f(z) - f(\zeta)| \geq \rho |z - \zeta|/2 \quad \text{for every } z \in B.$$

In particular, if $S = \{z \in \mathbb{C} : |z - \zeta| = \delta\}$ is the boundary of B, then

$$|f(z) - f(\zeta)| \geq \rho\delta/2 \quad \text{for every } z \in S.$$

Choose $w \in \mathbb{C}$ so that $|w - f(\zeta)| < \rho\delta/4$ and consider the minimum in the compact set B of the continuous real-valued function $\phi(z) = |f(z) - w|$. On the boundary S we have

$$\phi(z) \geq |f(z) - f(\zeta)| - |f(\zeta) - w| \geq \rho\delta/2 - \rho\delta/4 = \rho\delta/4.$$

Since $\phi(\zeta) < \rho\delta/4$, it follows that ϕ attains its minimum value in B at an interior point z_0. Since $z_0 \notin E$, we can take

$$z = z_0 - h[f'(z_0)]^{-1}\{f(z_0) - w\},$$

where $h > 0$ is so small that $|z - \zeta| < \delta$. Then

$$f(z) - w = f(z_0) - w + f'(z_0)(z - z_0) + o(h) = (1 - h)\{f(z_0) - w\} + o(h).$$

If $f(z_0) \neq w$ then, for sufficiently small $h > 0$,

$$|f(z) - w| \leq (1 - h/2)|f(z_0) - w| < |f(z_0) - w|,$$

which contradicts the definition of z_0. We conclude that $f(z_0) = w$. Thus $f(G \setminus E)$ contains not only $f(\zeta)$, but also an open disc $\{w \in \mathbb{C}: |w - f(\zeta)| < \rho\delta/4\}$ surrounding it. Since this holds for every $\zeta \in G \setminus E$, it follows that $f(G \setminus E)$ is an open set.

Now let $\zeta \in E$ and assume that $f(G)$ does not contain any open neighbourhood of $\omega: = f(\zeta)$. Then $f(z) \neq \omega$ for every $z \in G \setminus E$. Choose $\delta > 0$ so small that the closed disc $B = \{z \in \mathbb{C}: |z - \zeta| \leq \delta\}$ is contained in G and contains no point of E except ζ. If $S = \{z \in \mathbb{C}: |z - \zeta| = \delta\}$ is the boundary of B, there exists an open disc U with centre ω that contains no point of $f(S)$. It follows that if $A = \{z \in \mathbb{C}: 0 < |z - \zeta| < \delta\}$ is the annulus $B \setminus (S \cup \{\zeta\})$, then $U \setminus \{\omega\}$ is the union of the disjoint nonempty open sets $U \cap \{\mathbb{C} \setminus f(B)\}$ and $U \cap f(A)$. Since $U \setminus \{\omega\}$ is a connected set (in fact it is *path-connected*), this is a contradiction. $\square$

From Proposition 29 we readily obtain

THEOREM 30 *If*

$$f(z) = z^n + a_1 z^{n-1} + \dots + a_n$$

is a polynomial of degree $n \geq 1$ with complex coefficients $a_1,\dots,a_n$, then $f(\zeta) = 0$ for some $\zeta \in \mathbb{C}$.

Proof We will prove the apparently stronger statement that $f(\mathbb{C}) = \mathbb{C}$. The set E of all $z \in \mathbb{C}$ such that $f'(z) = 0$ is finite. (In fact E contains at most $n - 1$ points, by Proposition II.15.) Hence $\mathbb{C} \setminus E$ is open and, by Proposition 29, $D: = f(\mathbb{C})$ is an open set.

We will show that D is also a closed set. Let $\{\zeta_k\}$ be a sequence of points such that $f(\zeta_k) \to \omega$. The sequence $\{\zeta_k\}$ is necessarily bounded, since

$$f(z)/z^n = 1 + a_1/z + \dots + a_n/z^n \to 1 \quad \text{as } |z| \to \infty,$$

and hence has a convergent subsequence. If ζ is the limit of this subsequence, then $f(\zeta) = \omega$.

If $D \neq \mathbb{C}$, then $\mathbb{C}$ is the union of the disjoint nonempty open sets D and $\mathbb{C} \setminus D$. Since $\mathbb{C}$ is connected, this is a contradiction. Hence $D = \mathbb{C}$. $\square$

The first 'proof' of the fundamental theorem of algebra was given by d'Alembert (1746). Assuming the convergence of what are now called Puiseux expansions, he showed that if a polynomial assumes a value $w \neq 0$, then it also assumes a value w' such that $|w'| < |w|$. A much simpler way of reaching this conclusion, which required only the existence of k-th roots of complex numbers, was given by Argand (1814). Cauchy (1820) gave a similar proof and, with latter-day rigour, it is still reproduced in textbooks. The proof we have given rests on the same general principle, but uses neither the existence of k-th roots nor the continuity of the derivative. These may be called *differential calculus proofs*.

The basis for an *algebraic proof* was given by Euler (1749). His proof was completed by Lagrange (1772) and then simplified by Laplace (1795). The algebraic proof starts from the facts that $\mathbb{R}$ is an ordered field, that any positive element of $\mathbb{R}$ has a square root in $\mathbb{R}$ and that any polynomial of odd degree with coefficients from $\mathbb{R}$ has a root in $\mathbb{R}$. It then shows that any polynomial of degree $n \geq 1$ with coefficients from $\mathbb{C} = \mathbb{R}(i)$, where $i^2 = -1$, has a root in $\mathbb{C}$ by using induction on the highest power of 2 which divides n.

Gauss (1799) objected to this proof, because it assumed that there were 'roots' and then proved that these roots were complex numbers. The difficulty disappears if one uses the result, due to Kronecker (1887), that a polynomial with coefficients from an arbitrary field K decomposes into linear factors in a field L which is a finite extension of K. This general result, which is not difficult to prove, is actually all that is required for many of the previous applications of the fundamental theorem of algebra.

It is often said that the first rigorous proof of the fundamental theorem of algebra was given by Gauss (1799). Like d'Alembert, however, Gauss assumed properties of algebraic curves which were unknown at the time. The gaps in this proof of Gauss were filled by Ostrowski (1920).

There are also *topological proofs* of the fundamental theorem of algebra, e.g. using the notion of topological degree. This type of proof is intuitively appealing, but not so easy to make rigorous. Finally, there are *complex analysis proofs*, which depend ultimately on Cauchy's theorem on complex line integrals. (The latter proofs are more closely related to either the differential calculus proofs or the topological proofs than they seem to be at first sight.)

The *exponential function* e^z may be defined, for any complex value of z, as the sum of the everywhere convergent power series

$$\sum_{n\geq 0} z^n/n! = 1 + z + z^2/2! + z^3/3! + \dots .$$

It is easily verified that $w(z) = e^z$ is a solution of the differential equation $dw/dz = w$ satisfying the initial condition $w(0) = 1$.

For any $\zeta \in \mathbb{C}$, put $\varphi(z) = e^{\zeta-z} e^z$. Differentiating by the product rule, we obtain

$$\varphi'(z) = -e^{\zeta-z} e^z + e^{\zeta-z} e^z = 0.$$

Since this holds for all $z \in \mathbb{C}$, $\varphi(z)$ is a constant. Thus $\varphi(z) = \varphi(0) = e^\zeta$. Replacing ζ by $\zeta + z$, we obtain the *addition theorem* for the exponential function:

$$e^{\zeta}\, e^{z} \;=\; e^{\zeta+z} \quad \text{for all } z,\zeta \in \mathbb{C}.$$

In particular, $e^{-z}\, e^{z} = 1$ and hence $e^{z} \neq 0$ for every $z \in \mathbb{C}$.

The power series for e^{z} shows that, for any real y, e^{-iy} is the complex conjugate of e^{iy} and hence

$$|e^{iy}|^{2} \;=\; e^{iy}\, e^{-iy} \;=\; 1.$$

It follows that, for all real x,y,

$$|e^{x+iy}| \;=\; |e^{x}|\,|e^{iy}| \;=\; e^{x}.$$

The *trigonometric functions* $\cos z$ and $\sin z$ may be defined, for any complex value of z, by the formulas of Euler (1740):

$$\cos z = (e^{iz} + e^{-iz})/2, \quad \sin z = (e^{iz} - e^{-iz})/2i.$$

It follows at once that

$$e^{iz} \;=\; \cos z + i \sin z,$$
$$\cos 0 = 1, \quad \sin 0 = 0,$$
$$\cos (-z) = \cos z, \quad \sin (-z) = -\sin z,$$

and the relation $e^{iz}\, e^{-iz} = 1$ implies that

$$\cos^{2} z + \sin^{2} z \;=\; 1.$$

From the power series for e^{z} we obtain, for every $z \in \mathbb{C}$,

$$\cos z \;=\; \textstyle\sum_{n\geq 0} (-1)^{n} z^{2n}/(2n)! \;=\; 1 - z^{2}/2! + z^{4}/4! + \dots ,$$

$$\sin z \;=\; \textstyle\sum_{n\geq 0} (-1)^{n} z^{2n+1}/(2n+1)! \;=\; z - z^{3}/3! + z^{5}/5! + \dots .$$

From the differential equation we obtain, for every $z \in \mathbb{C}$,

$$d\,(\cos z)/dz = -\sin z, \quad d\,(\sin z)/dz = \cos z.$$

From the addition theorem we obtain, for all $z,\zeta \in \mathbb{C}$,

$$\cos (z + \zeta) \;=\; \cos z \cos \zeta - \sin z \sin \zeta,$$

$$\sin (z + \zeta) \;=\; \sin z \cos \zeta + \cos z \sin \zeta.$$

We now consider periodicity properties. By the addition theorem for the exponential function, $e^{z+h} = e^{z}$ if and only if $e^{h} = 1$. Thus the exponential function has period h if and only if $e^h = 1$. Since $e^h = 1$ implies $h = ix$ for some real x, and since $\cos x$ and $\sin x$ are real for real x, the periods correspond to those real values of x for which

$$\cos x = 1, \quad \sin x = 0.$$

In fact, the second relation follows from the first, since $\cos^2 x + \sin^2 x = 1$.

By bracketing the power series for $\cos x$ in the form

$$\cos x \;=\; (1 - x^2/2! + x^4/4!) - (1 - x^2/7{\cdot}8)x^6/6! - (1 - x^2/11{\cdot}12)x^{10}/10! - \ldots$$

and taking $x = 2$, we see that $\cos 2 < 0$. Since $\cos 0 = 1$ and $\cos x$ is a continuous function of x, there is a least positive value ξ of x such that $\cos \xi = 0$. Then $\sin^2 \xi = 1$. In fact $\sin \xi = 1$, since $\sin 0 = 0$ and $\sin' x = \cos x > 0$ for $0 \leq x < \xi$. Thus

$$0 < \sin x < 1 \quad \text{for } 0 < x < \xi$$

and

$$e^{i\xi} \;=\; \cos \xi + i \sin \xi \;=\; i.$$

As usual, we now write $\pi = 2\xi$. From $e^{\pi i/2} = i$, we obtain

$$e^{2\pi i} = i^4 = (-1)^2 = 1.$$

Thus the exponential function has period $2\pi i$. It follows that it also has period $2n\pi i$, for every $n \in \mathbb{Z}$. We will show that there are no other periods.

Suppose $e^{ix'} = 1$ for some $x' \in \mathbb{R}$ and choose $n \in \mathbb{Z}$ so that $n \leq x'/2\pi < n + 1$. If $x = x' - 2n\pi$, then $e^{ix} = 1$ and $0 \leq x < 2\pi$. If $x \neq 0$, then $0 < x/4 < \pi/2$ and hence $0 < \sin x/4 < 1$. Thus $e^{ix/4} \neq \pm 1, \pm i$. But this is a contradiction, since

$$(e^{ix/4})^4 \;=\; e^{ix} \;=\; 1.$$

We show next that the map $x \to e^{ix}$ maps the interval $0 \leq x < 2\pi$ bijectively onto the *unit circle*, i.e. the set of all complex numbers w such that $|w| = 1$. We already know that $|e^{ix}| = 1$ if $x \in \mathbb{R}$. If $e^{ix} = e^{ix'}$, where $0 \leq x \leq x' < 2\pi$, then $e^{i(x'-x)} = 1$. Since $0 \leq x' - x < 2\pi$, this implies $x' = x$.

It remains to show that if $u, v \in \mathbb{R}$ and $u^2 + v^2 = 1$, then

$$u = \cos x, \quad v = \sin x$$

for some x such that $0 \leq x < 2\pi$. If $u, v > 0$, then also $u, v < 1$. Hence $u = \cos x$ for some x such that $0 < x < \pi/2$. It follows that $v = \sin x$, since $\sin^2 x = 1 - u^2 = v^2$ and $\sin x > 0$. The other possible sign combinations for u,v may be reduced to the case $u,v > 0$ by means of the relations

$$\sin (x + \pi/2) = \cos x, \quad \cos (x + \pi/2) = - \sin x.$$

If z is any nonzero complex number, then $r = |z| > 0$ and $|z/r| = 1$. It follows that any nonzero complex number z can be uniquely expressed in the form

$$z = re^{i\theta},$$

where r,θ are real numbers such that $r > 0$ and $0 \le \theta < 2\pi$. We call r,θ the *polar coordinates* of z and θ the *argument* of z. If $z = x + iy$, where $x,y \in \mathbb{R}$, then $r = \sqrt{(x^2 + y^2)}$ and

$$x = r \cos \theta, \quad y = r \sin \theta.$$

Hence, in the geometrical representation of complex numbers by points of a plane, r is the distance of z from O and θ measures the angle between the positive x-axis and the ray $\overrightarrow{Oz}$.

We now show that the exponential function assumes every nonzero complex value w. Since $|w| > 0$, we have $|w| = e^x$ for some $x \in \mathbb{R}$. If $w' = w/|w|$, then $|w'| = 1$ and so $w' = e^{iy}$ for some $y \in \mathbb{R}$. Consequently,

$$w = |w| \, w' = e^x \, e^{iy} = e^{x+iy}.$$

It follows that, for any positive integer n, a nonzero complex number w has n distinct n-th roots. In fact, if $w = e^z$, then w has the distinct n-th roots

$$\zeta_k = \zeta\omega^k \quad (k = 0,1,...,n - 1),$$

where $\zeta = e^{z/n}$ and $\omega = e^{2\pi i/n}$. In the geometrical representation of complex numbers by points of a plane, the n-th roots of w are the vertices of an n-sided regular polygon.

It remains to show that π has its usual geometric significance. Since the continuously differentiable function $z(t) = e^{it}$ describes the unit circle as t increases from 0 to 2π, the length of the unit circle is

$$L = \int_0^{2\pi} |z'(t)| \, dt.$$

But $|z'(t)| = 1$, since $z'(t) = ie^{it}$, and hence $L = 2\pi$.

In a course of complex analysis one would now define complex line integrals, prove Cauchy's theorem and deduce its numerous consequences. The miracle is that, if $D = \{z \in \mathbb{C}: |z| < \rho\}$ is a disc with centre the origin, then any differentiable function $f: D \to \mathbb{C}$ can be represented by a *power series*,

$$f(z) = c_0 + c_1 z + c_2 z^2 + ... \, ,$$

which is convergent for $|z| < \rho$. It follows that, if f vanishes at a sequence of distinct points converging to 0, then it vanishes everywhere. This is the basis for *analytic continuation*.

A complex-valued function f is said to be *holomorphic* at $a \in \mathbb{C}$ if, in some neighbourhood of a, it can be represented as the sum of a convergent power series (its 'Taylor' series):

$$f(z) = c_0 + c_1(z - a) + c_2(z - a)^2 + \dots .$$

It is said to be *meromorphic* at $a \in \mathbb{C}$ if, for some integer n, it can be represented near a as the sum of a convergent series (its 'Laurent' series):

$$f(z) = c_0(z - a)^{-n} + c_1(z - a)^{-n+1} + c_2(z - a)^{-n+2} + \dots .$$

If $c_0 \neq 0$, then $(z - a)f'(z)/f(z) \to -n$ as $z \to a$. If also $n > 0$ we say that a is a *pole* of f of *order n*. If $n = 1$, the pole is *simple* with *residue* c_0.

Let G be a nonempty connected open subset of $\mathbb{C}$. From what has been said, if $f: G \to \mathbb{C}$ is differentiable throughout G, then it is also holomorphic throughout G. If f_1 and f_2 are holomorphic throughout G and f_2 is not identically zero, then the quotient $f = f_1/f_2$ is meromorphic throughout G. Conversely, it may be shown that if f is meromorphic throughout G, then $f = f_1/f_2$ for some functions f_1, f_2 which are holomorphic throughout G.

The behaviour of many functions is best understood by studying them in the complex domain, as the exponential and trigonometric functions already illustrate. Complex numbers, when they first appeared, were called 'impossible' numbers. They are now indispensable.

6 Quaternions and octonions

Quaternions were invented by Hamilton (1843) in order to be able to 'multiply' points of 3-dimensional space, in the same way that complex numbers enable one to multiply points of a plane. The definition of quaternions adopted here will be analogous to our definition of complex numbers.

We define a *quaternion* to be a 2×2 matix of the form

$$A = \begin{pmatrix} a & b \\ -\bar{b} & \bar{a} \end{pmatrix},$$

where a and b are complex numbers and the bar denotes complex conjugation. The set of all quaternions will be denoted by $\mathbb{H}$. We may define addition and multiplication in $\mathbb{H}$ to be matrix addition and multiplication, since $\mathbb{H}$ is closed under these operations. Furthermore $\mathbb{H}$ contains

$$0 = \begin{pmatrix} 0 & 0 \\ 0 & 0 \end{pmatrix}, \quad 1 = \begin{pmatrix} 1 & 0 \\ 0 & 1 \end{pmatrix},$$

and $A \in \mathbb{H}$ implies $-A \in \mathbb{H}$.

It follows from the properties of matrix addition and multiplication that addition and multiplication of quaternions have the properties **(A2)-(A5)** and **(M3)-(M4)**, with 0 and 1 as identity elements for addition and multiplication respectively. However, **(M2)** no longer holds, since multiplication is not always commutative. For example,

$$\begin{pmatrix} 0 & 1 \\ -1 & 0 \end{pmatrix} \begin{pmatrix} 0 & i \\ i & 0 \end{pmatrix} \neq \begin{pmatrix} 0 & i \\ i & 0 \end{pmatrix} \begin{pmatrix} 0 & 1 \\ -1 & 0 \end{pmatrix}.$$

On the other hand, there are now two distributive laws: *for all $A,B,C \in \mathbb{H}$,*

$$A(B + C) = AB + AC, \quad (B + C)A = BA + CA.$$

It is easily seen that $A \in \mathbb{H}$ is in the *centre* of $\mathbb{H}$, i.e. $AB = BA$ for every $B \in \mathbb{H}$, if and only if $A = \lambda 1$ for some real number λ. Since the map $\lambda \to \lambda 1$ preserves sums and products, we can regard $\mathbb{R}$ as contained in $\mathbb{H}$ by identifying the real number λ with the quaternion $\lambda 1$.

We define the *conjugate* of the quaternion

$$A = \begin{pmatrix} a & b \\ -\overline{b} & \overline{a} \end{pmatrix},$$

to be the quaternion

$$\overline{A} = \begin{pmatrix} \overline{a} & -b \\ \overline{b} & a \end{pmatrix}.$$

It is easily verified that

$$\overline{A+B} = \overline{A} + \overline{B}, \quad \overline{AB} = \overline{B}\,\overline{A}, \quad \overline{\overline{A}} = A.$$

Furthermore,

$$\overline{A}A = A\overline{A} = n(A), \quad A + \overline{A} = t(A),$$

where the *norm* $n(A)$ and *trace* $t(A)$ are both real:

$$n(A) = a\overline{a} + b\overline{b}, \quad t(A) = a + \overline{a}.$$

Moreover, $n(A) > 0$ if $A \neq 0$. It follows that any quaternion $A \neq 0$ has a multiplicative inverse: if $A^{-1} = n(A)^{-1}\overline{A}$, then

$$A^{-1}A = AA^{-1} = 1.$$

Norms and traces have the following properties: *for all $A, B \in \mathbb{H}$,*

$$t(\overline{A}) = t(A), \quad n(\overline{A}) = n(A),$$
$$t(A + B) = t(A) + t(B),$$
$$n(AB) = n(A)n(B).$$

Only the last property is not immediately obvious, and it can be proved in one line:

$$n(AB) = \overline{AB}AB = \overline{B}\,\overline{A}\,AB = n(A)\overline{B}B = n(A)n(B).$$

Furthermore, for any $A \in \mathbb{H}$ we have

$$A^2 - t(A)A + n(A) = 0,$$

since the left side can be written in the form $A^2 - (A + \overline{A})A + \overline{A}A$. (The relation is actually just a special case of the 'Cayley–Hamilton theorem' of linear algebra.) It follows that the quadratic polynomial $x^2 + 1$ has not two, but infinitely many quaternionic roots.

If we put

$$I = \begin{pmatrix} 0 & 1 \\ -1 & 0 \end{pmatrix}, \quad J = \begin{pmatrix} 0 & i \\ i & 0 \end{pmatrix}, \quad K = \begin{pmatrix} i & 0 \\ 0 & -i \end{pmatrix},$$

then

$$I^2 = J^2 = K^2 = -1,$$
$$IJ = K = -JI, \quad JK = I = -KJ, \quad KI = J = -IK.$$

Moreover, any quaternion A can be uniquely represented in the form

$$A = \alpha_0 + \alpha_1 I + \alpha_2 J + \alpha_3 K,$$

where $\alpha_0, \ldots, \alpha_3 \in \mathbb{R}$. In fact this is equivalent to the previous representation with

$$a = \alpha_0 + i\alpha_3, \quad b = \alpha_1 + i\alpha_2.$$

The corresponding representation of the conjugate quaternion is

$$\overline{A} = \alpha_0 - \alpha_1 I - \alpha_2 J - \alpha_3 K.$$

Hence $\overline{A} = A$ if and only if $\alpha_1 = \alpha_2 = \alpha_3 = 0$ and $\overline{A} = -A$ if and only if $\alpha_0 = 0$.

A quaternion A is said to be *pure* if $\overline{A} = -A$. Thus any quaternion can be uniquely represented as the sum of a real number and a pure quaternion.

It follows from the multiplication table for the units I,J,K that $A = \alpha_0 + \alpha_1 I + \alpha_2 J + \alpha_3 K$ has norm

$$n(A) = \alpha_0^2 + \alpha_1^2 + \alpha_2^2 + \alpha_3^2.$$

Consequently the relation $n(A)n(B) = n(AB)$ may be written in the form

$$(\alpha_0^2 + \alpha_1^2 + \alpha_2^2 + \alpha_3^2)(\beta_0^2 + \beta_1^2 + \beta_2^2 + \beta_3^2) = \gamma_0^2 + \gamma_1^2 + \gamma_2^2 + \gamma_3^2,$$

where

$$\begin{aligned}
\gamma_0 &= \alpha_0\beta_0 - \alpha_1\beta_1 - \alpha_2\beta_2 - \alpha_3\beta_3, \\
\gamma_1 &= \alpha_0\beta_1 + \alpha_1\beta_0 + \alpha_2\beta_3 - \alpha_3\beta_2, \\
\gamma_2 &= \alpha_0\beta_2 - \alpha_1\beta_3 + \alpha_2\beta_0 + \alpha_3\beta_1, \\
\gamma_3 &= \alpha_0\beta_3 + \alpha_1\beta_2 - \alpha_2\beta_1 + \alpha_3\beta_0.
\end{aligned}$$

This '4-squares identity' was already known to Euler (1770).

An important application of quaternions is to the parametrization of rotations in 3-dimensional space. In describing this application it will be convenient to denote quaternions now by lower case letters. In particular, we will write i,j,k in place of I,J,K.

Let u be a quaternion with norm $n(u) = 1$, and consider the mapping $T\colon \mathbb{H} \to \mathbb{H}$ defined by

$$Tx = uxu^{-1}.$$

Evidently

$$\begin{aligned}
T(x + y) &= Tx + Ty, \\
T(xy) &= (Tx)(Ty), \\
T(\lambda x) &= \lambda Tx \quad \text{if } \lambda \in \mathbb{R}.
\end{aligned}$$

Moreover, since $u^{-1} = \bar{u}$,

$$T\bar{x} = \overline{Tx}.$$

It follows that

$$n(Tx) = n(x),$$

since

$$n(Tx) = Tx\overline{Tx} = Tx\,T\bar{x} = T(x\bar{x}) = n(x)\,T1 = n(x).$$

Furthermore, T maps pure quaternions into pure quaternions, since $\bar{x} = -x$ implies

$$\overline{Tx} = T\bar{x} = -Tx.$$

If we write

$$x = \xi_1 i + \xi_2 j + \xi_3 k,$$

then

$$Tx = y = \eta_1 i + \eta_2 j + \eta_3 k,$$

where $\eta_\mu = \sum_{v=1}^{3} \beta_{\mu v} \xi_v$ for some $\beta_{\mu v} \in \mathbb{R}$. Since

$$\eta_1^2 + \eta_2^2 + \eta_3^2 = \xi_1^2 + \xi_2^2 + \xi_3^2,$$

the matrix $V = (\beta_{\mu v})$ is *orthogonal*: $V^{-1} = V^t$.

Thus with every quaternion u with norm 1 there is associated a 3×3 orthogonal matrix $V = (\beta_{\mu v})$. Explicitly, if

$$u = \alpha_0 + \alpha_1 i + \alpha_2 j + \alpha_3 k,$$

where

$$\alpha_0^2 + \alpha_1^2 + \alpha_2^2 + \alpha_3^2 = 1,$$

then

$$\beta_{11} = \alpha_0^2 + \alpha_1^2 - \alpha_2^2 - \alpha_3^2, \quad \beta_{12} = 2(\alpha_1\alpha_2 - \alpha_0\alpha_3), \quad \beta_{13} = 2(\alpha_1\alpha_3 + \alpha_0\alpha_2),$$

$$\beta_{21} = 2(\alpha_1\alpha_2 + \alpha_0\alpha_3), \quad \beta_{22} = \alpha_0^2 - \alpha_1^2 + \alpha_2^2 - \alpha_3^2, \quad \beta_{23} = 2(\alpha_2\alpha_3 - \alpha_0\alpha_1),$$

$$\beta_{31} = 2(\alpha_1\alpha_3 - \alpha_0\alpha_2), \quad \beta_{32} = 2(\alpha_2\alpha_3 + \alpha_0\alpha_1), \quad \beta_{33} = \alpha_0^2 - \alpha_1^2 - \alpha_2^2 + \alpha_3^2.$$

This parametrization of orthogonal transformations was first discovered by Euler (1770).

We now consider the dependence of V on u, and consequently write $V(u)$ in place of V. Since

$$u_1 u_2 x (u_1 u_2)^{-1} = u_1 (u_2 x u_2^{-1}) u_1^{-1},$$

we have

$$V(u_1 u_2) = V(u_1)V(u_2).$$

Thus the map $u \to V(u)$ is a 'homomorphism' of the multiplicative group of all quaternions of norm 1 into the group of all 3×3 real orthogonal matrices. In particular, $V(\bar{u}) = V(u)^{-1}$.

We show next that two quaternions u_1, u_2 of norm 1 yield the same orthogonal matrix if and only if $u_2 = \pm u_1$. Put $u = u_2^{-1} u_1$. Then $u_1 x u_1^{-1} = u_2 x u_2^{-1}$ if and only if $ux = xu$. This holds for every pure quaternion x if and only if u is real, i.e. if and only if $u = \pm 1$, since $n(u) = 1$.

The question arises whether all 3×3 orthogonal matrices may be represented in the above way. It follows readily from the preceding formulas for $\beta_{\mu v}$ that the orthogonal matrix $-I$ cannot be so represented. Consequently, if an orthogonal matrix V is represented, then $-V$ is not. On the other hand, suppose u is a pure quaternion, so that $\alpha_0 = 0$. Then $ux + xu = ux + \bar{x}\,\bar{u}$ is real, and given by

$$ux + xu = -2(\alpha_1\xi_1 + \alpha_2\xi_2 + \alpha_3\xi_3) = 2<\bar{u},x>,$$

with the notation of §10 for inner products in $\mathbb{R}^3$. It follows that

$$y \; = \; ux\bar{u} \; = \; 2\langle\bar{u},x\rangle\bar{u} \, - x.$$

But the mapping $x \to x - 2\langle\bar{u},x\rangle\bar{u}$ is a *reflection* in the plane orthogonal to the unit vector u. Hence, for every reflection $R, -R$ is represented. It may be shown that every orthogonal transformation of $\mathbb{R}^3$ is a product of reflections. (Indeed, this is a special case of a more general result which will be proved in Proposition 17 of Chapter VII.) It follows that an orthogonal matrix V is represented if and only if V is a product of an even number of reflections (or, equivalently, if and only if V has determinant 1, as defined in Chapter V, §1).

Finally, according to our initial definition of quaternions, the quaternions of norm 1 are precisely the 2×2 *unitary* matrices with determinant 1. Thus our results may be summarized by saying that there is a homomorphism of the *special unitary group* $SU_2(\mathbb{C})$ onto the *special orthogonal group* $SO_3(\mathbb{R})$, with kernel $\{\pm I\}$. (Here 'special' signifies 'determinant 1'.)

Since the quaternions of norm 1 may be identified with the points of the unit sphere S^3 in $\mathbb{R}^4$ it follows that, as a topological space, $SO_3(\mathbb{R})$ is homeomorphic to S^3 with antipodal points identified, i.e. to the projective space $P^3(\mathbb{R})$. Similarly (cf. Chapter X, §8), the topological group $SU_2(\mathbb{C})$ is the *simply-connected covering space* of the topological group $SO_3(\mathbb{R})$.

Again, by considering the map $T: \mathbb{H} \to \mathbb{H}$ defined by $Tx = vxu^{-1}$, where u,v are quaternions with norm 1, it may be seen that that there is a homomorphism of the direct product $SU_2(\mathbb{C}) \times SU_2(\mathbb{C})$ onto the special orthogonal group $SO_4(\mathbb{R})$ of 4×4 real orthogonal matrices with determinant 1, the kernel being $\{\pm (I,I)\}$.

Almost immediately after Hamilton's invention of quaternions Graves (1844), in a letter to Hamilton, and Cayley (1845) invented 'octonions', also known as 'octaves' or 'Cayley numbers'. We define an *octonion* to be an ordered pair (a_1,a_2) of quaternions, with addition and multiplication defined by

$$(a_1,a_2) + (b_1,b_2) \; = \; (a_1 + b_1, a_2 + b_2),$$
$$(a_1,a_2) \cdot (b_1,b_2) \; = \; (a_1b_1 - \bar{b_2}\,a_2, b_2a_1 + a_2\bar{b_1}).$$

Then the set $\mathbb{O}$ of all octonions is a commutative group under addition, i.e. the laws **(A2)-(A5)** hold, with $0 = (0,0)$ as identity element, and multiplication is both left and right distributive with respect to addition. The octonion $1 = (1,0)$ is a two-sided identity element for multiplication, and the octonion $\varepsilon = (0,1)$ has the property $\varepsilon^2 = -1$.

It is easily seen that $\alpha \in \mathbb{O}$ is in the *centre* of $\mathbb{O}$, i.e. $\alpha\beta = \beta\alpha$ for every $\beta \in \mathbb{O}$, if and only if $\alpha = (c,0)$ for some $c \in \mathbb{R}$.

Since the map $a \to (a,0)$ preserves sums and products, we may regard $\mathbb{H}$ as contained in $\mathbb{O}$ by identifying the quaternion a with the octonion $(a,0)$. This shows that multiplication of octonions is in general not commutative. It is also in general not even associative; for example,

$$(ij)\varepsilon = k\varepsilon = (0,k), \quad i(j\varepsilon) = i(0,j) = (0,-k).$$

It is for this reason that we defined octonions as ordered pairs, rather than as matrices. It should be mentioned, however, that we could have used precisely the same construction to define complex numbers as ordered pairs of real numbers, and quaternions as ordered pairs of complex numbers, but the verification of the associative law for multiplication would then have been more laborious.

Although multiplication is non-associative, $\mathbb{O}$ does inherit some other properties from $\mathbb{H}$. If we define the *conjugate* of the octonion $\alpha = (a_1,a_2)$ to be the octonion $\overline{\alpha} = (\overline{a_1},-a_2)$, then it is easily verified that

$$\overline{\alpha+\beta} = \overline{\alpha} + \overline{\beta}, \quad \overline{\alpha\beta} = \overline{\beta}\,\overline{\alpha}, \quad \overline{\overline{\alpha}} = \alpha.$$

Furthermore,

$$\alpha\overline{\alpha} = \overline{\alpha}\alpha = n(\alpha),$$

where the *norm* $n(\alpha) = a_1\overline{a_1} + a_2\overline{a_2}$ is real. Moreover $n(\alpha) > 0$ if $\alpha \neq 0$, and $n(\overline{\alpha}) = n(\alpha)$.

It will now be shown that if $\alpha,\beta \in \mathbb{O}$ and $\alpha \neq 0$, then the equation

$$\xi\alpha = \beta$$

has a unique solution $\xi \in \mathbb{O}$. Writing $\alpha = (a_1,a_2)$, $\beta = (b_1,b_2)$ and $\xi = (x_1,x_2)$, we have to solve the simultaneous quaternionic equations

$$x_1a_1 - \overline{a_2}x_2 = b_1,$$
$$a_2x_1 + x_2\overline{a_1} = b_2.$$

If we multiply the second equation on the right by a_1 and replace x_1a_1 by its value from the first equation, we get

$$n(\alpha)x_2 = b_2a_1 - a_2b_1.$$

Similarly, if we multiply the first equation on the right by $\overline{a_1}$ and replace $x_2\overline{a_1}$ by its value from the second equation, we get

$$n(\alpha)x_1 = b_1\overline{a_1} + \overline{a_2}b_2.$$

It follows that the equation $\xi\alpha = \beta$ has the unique solution

$$\xi = n(\alpha)^{-1}\beta\overline{\alpha}.$$

Since the equation $\alpha\eta = \beta$ is equivalent to $\overline{\eta}\,\overline{\alpha} = \overline{\beta}$, it has the unique solution $\eta = n(\alpha)^{-1}\,\overline{\alpha}\beta$. Thus $\mathbb{O}$ is a *division algebra*. It should be noted that, since $\mathbb{O}$ is non-associative, it is not enough to verify that every nonzero element has a multiplicative inverse.

It follows from the preceding discussion that, *for all $\alpha,\beta \in \mathbb{O}$,*

$$(\beta\,\overline{\alpha})\alpha = n(\alpha)\beta = \alpha(\overline{\alpha}\beta).$$

Consequently the norm is multiplicative: *for all $\alpha,\beta \in \mathbb{O}$,*

$$n(\alpha\beta) = n(\alpha)n(\beta).$$

For, putting $\gamma = \alpha\beta$, we have

$$n(\gamma)\,\overline{\alpha} = (\overline{\alpha}\gamma)\,\overline{\gamma} = (\overline{\alpha}(\alpha\beta))\,\overline{\gamma} = n(\alpha)\beta\,\overline{\gamma} = n(\alpha)\beta(\overline{\beta}\,\overline{\alpha}) = n(\alpha)n(\beta)\,\overline{\alpha}.$$

This establishes the result when $\alpha \neq 0$, and when $\alpha = 0$ it is obvious.

Every $\alpha \in \mathbb{O}$ has a unique representation $\alpha = a_1 + a_2\varepsilon$, where $a_1,a_2 \in \mathbb{H}$, and hence a unique representation

$$\alpha = c_0 + c_1 i + c_2 j + c_3 k + c_4\varepsilon + c_5 i\varepsilon + c_6 j\varepsilon + c_7 k\varepsilon,$$

where $c_0,\dots,c_7 \in \mathbb{R}$. Since $\overline{\alpha} = \overline{a_1} - a_2\varepsilon$ and $n(\alpha) = a_1\overline{a_1} + a_2\overline{a_2}$, it follows that

$$\overline{\alpha} = c_0 - c_1 i - c_2 j - c_3 k - c_4\varepsilon - c_5 i\varepsilon - c_6 j\varepsilon - c_7 k\varepsilon$$

and

$$n(\alpha) = c_0^2 + \dots + c_7^2.$$

Consequently the relation $n(\alpha)n(\beta) = n(\alpha\beta)$ may be written in the form

$$(c_0^2 + \dots + c_7^2)(d_0^2 + \dots + d_7^2) = e_0^2 + \dots + e_7^2,$$

where $e_i = \sum_{j=0}^{7}\sum_{k=0}^{7}\rho_{ijk}c_j d_k$ for some real constants ρ_{ijk} which do not depend on the c's and d's. An '8-squares identity' of this type was first found by Degen (1818).

7 Groups

A nonempty set G is said is said to be a *group* if a binary operation φ, i.e. a mapping $\varphi\colon G{\times}G \to G$, is defined with the properties

(i) $\varphi(\varphi(a,b),c) = \varphi(a,\varphi(b,c))$ for all $a,b,c \in G$;   (associative law)

(ii)   there exists $e \in G$ such that $\varphi(e,a) = a$ for every $a \in G$;   (identity element)

(iii) for each $a \in G$, there exists $a^{-1} \in G$ such that $\varphi(a^{-1},a) = e$. (inverse elements)

If, in addition,

(iv) $\varphi(a,b) = \varphi(b,a)$ for all $a,b \in G$,   (commutative law)

then the group G is said to be *commutative* or *abelian*.

For example, the set $\mathbb{Z}$ of all integers is a commutative group under addition, i.e. with $\varphi(a,b) = a + b$, with 0 as identity element and $-a$ as the inverse of a. Similarly, the set $\mathbb{Q}^{\times}$ of all nonzero rational numbers is a commutative group under multiplication, i.e. with $\varphi(a,b) = ab$, with 1 as identity element and a^{-1} as the inverse of a.

We now give an example of a noncommutative group. The set $\mathcal{S}_A$ of all bijective maps $f: A \to A$ of a nonempty set A to itself is a group under composition, i.e. with $\varphi(a,b) = a \circ b$, with the identity map i_A as identity element and the inverse map f^{-1} as the inverse of f. If A contains at least 3 elements, then $\mathcal{S}_A$ is a noncommutative group. For suppose a,b,c are distinct elements of A, let $f: A \to A$ be the bijective map defined by

$$f(a) = b, \ f(b) = a, \ f(x) = x \text{ if } x \neq a,b,$$

and let $g: A \to A$ be the bijective map defined by

$$g(a) = c, \ g(c) = a, \ g(x) = x \text{ if } x \neq a,c.$$

Then $f \circ g \neq g \circ f$, since $(f \circ g)(a) = c$ and $(g \circ f)(a) = b$.

For arbitrary groups, instead of $\varphi(a,b)$ we usually write $a{\cdot}b$ or simply ab. For commutative groups, instead of $\varphi(a,b)$ we often write $a + b$.

Since, by the associative law,

$$(ab)c = a(bc),$$

we will usually dispense with brackets.

We now derive some simple properties possessed by all groups. By (iii) we have $a^{-1}a = e$. In fact also $aa^{-1} = e$. This may be seen by multiplying on the left, by the inverse of a^{-1}, the relation

$$a^{-1}aa^{-1} = ea^{-1} = a^{-1}.$$

By (ii) we have $ea = a$. It now follows that also $ae = a$, since

$$ae = aa^{-1}a = ea.$$

For all elements a,b of the group G, the equation $ax = b$ has the solution $x = a^{-1}b$ and the equation $ya = b$ has the solution $y = ba^{-1}$. Moreover, these solutions are unique. For from $ax = ax'$ we obtain $x = x'$ by multiplying on the left by a^{-1}, and from $ya = y'a$ we obtain $y = y'$ by multiplying on the right by a^{-1}.

In particular, the identity element e is unique, since it is the solution of $ea = a$, and the inverse a^{-1} of a is unique, since it is the solution of $a^{-1}a = e$. It follows that the inverse of a^{-1} is a and the inverse of ab is $b^{-1}a^{-1}$.

As the preceding argument suggests, in the definition of a group we could have replaced left identity and left inverse by right identity and right inverse, i.e. we could have required $ae = a$ and $aa^{-1} = e$, instead of $ea = a$ and $a^{-1}a = e$. (However, left identity and right inverse, or right identity and left inverse, would not give the same result.)

If H,K are nonempty subsets of a group G, we denote by HK the subset of G consisting of all elements hk, where $h \in H$ and $k \in K$. If L is also a nonempty subset of G, then evidently

$$(HK)L \;=\; H(KL).$$

A subset H of a group G is said to be a *subgroup* of G if it is a group under the same group operation as G itself. A nonempty subset H is a subgroup of G if and only if it is 'closed' under multiplication and inversion, i.e. $a,b \in H$ implies $ab \in H$ and $a^{-1} \in H$. Indeed the necessity of the conditions is obvious. They are also sufficient, since they imply $e \in H$ and the associative law in H is inherited from G.

We now show that a nonempty *finite* subset H of a group G is a subgroup of G if it is closed under multiplication only. For, if $a \in H$, then $ha \in H$ for all $h \in H$. Since H is finite and the mapping $h \to ha$ of H into itself is injective, it is also surjective by the pigeonhole principle (Corollary I.6). Hence $ha = a$ for some $h \in H$, which shows that H contains the identity element of G. It now further follows that $ha = e$ for some $h \in H$, which shows that H is also closed under inversion.

A group is said to be *finite* if it contains only finitely many elements and to be of *order n* if it contains exactly n elements.

In order to give an important example of a subgroup we digress briefly. Let n be a positive integer and let A be the set $\{1,2,\dots,n\}$ with the elements in their natural order. Since we regard A as ordered, a bijective map $\alpha\colon A \to A$ will be called a *permutation*. The set of all permutations of A is a group under composition, the *symmetric group* $\mathcal{S}_n$.

Suppose now that $n > 1$. An inversion of order induced by the permutation α is a pair (i,j) with $i < j$ for which $\alpha(i) > \alpha(j)$. The permutation α is said to be *even* or *odd* according as the total number of inversions of order is even or odd. For example, the permutation $\{1,2,3,4,5\}$ $\rightarrow \{3,5,4,1,2\}$ is odd, since there are $2 + 3 + 2 = 7$ inversions of order.

The *sign* of the permutation α is defined by

$$\mathrm{sgn}(\alpha) \;=\; 1 \text{ or} -1 \text{ according as } \alpha \text{ is even or odd.}$$

Evidently we can write

$$\mathrm{sgn}(\alpha) \;=\; \Pi_{1 \le i < j \le n} \{\alpha(j) - \alpha(i)\}/(j - i),$$

from which it follows that

$$\mathrm{sgn}(\alpha\beta) \;=\; \mathrm{sgn}(\alpha)\,\mathrm{sgn}(\beta).$$

Since the sign of the identity permutation is 1, this implies

$$\mathrm{sgn}(\alpha^{-1}) \;=\; \mathrm{sgn}(\alpha).$$

Thus $\mathrm{sgn}(\rho^{-1}\alpha\rho) = \mathrm{sgn}(\alpha)$ for any permutation ρ of A, and so $\mathrm{sgn}(\alpha)$ is actually independent of the ordering of A.

Since the product of two even permutations is again an even permutation, the even permutations form a subgroup of $\mathcal{S}_n$, the *alternating group* $\mathcal{A}_n$. The order of $\mathcal{A}_n$ is $n!/2$. For let τ be the permutation $\{1,2,3,...,n\} \rightarrow \{2,1,3,...,n\}$. Since there is only one inversion of order, τ is odd. Since $\tau\tau$ is the identity permutation, a permutation is odd if and only if it has the form $\alpha\tau$, where α is even. Hence the number of odd permutations is equal to the number of even permutations.

It may be mentioned that the sign of a permutation can also be determined without actually counting the total number of inversions. In fact any $\alpha \in \mathcal{S}_n$ may be written as a product of ν disjoint cycles, and α is even or odd according as $n - \nu$ is even or odd.

We now return to the main story. Let H be a subgroup of an arbitrary group G and let a,b be elements of G. We write $a \sim_r b$ if $ba^{-1} \in H$. We will show that this is an equivalence relation.

The relation is certainly reflexive, since $e \in H$. It is also symmetric, since if $c = ba^{-1} \in H$, then $c^{-1} = ab^{-1} \in H$. Furthermore it is transitive, since if $ba^{-1} \in H$ and $cb^{-1} \in H$, then also $ca^{-1} = (cb^{-1})(ba^{-1}) \in H$.

The equivalence class which contains a is the set Ha of all elements ha, where $h \in H$. We call any such equivalence class a *right coset* of the subgroup H, and any element of a given coset is said to be a *representative* of that coset.

It follows from the remarks in §0 about arbitrary equivalence relations that, for any two cosets Ha and Ha', either $Ha = Ha'$ or $Ha \cap Ha' = \varnothing$. Moreover, the distinct right cosets form a partition of G.

If H is a subgroup of a finite group G, then H is also finite and the number of distinct right cosets is finite. Moreover, each right coset Ha contains the same number of elements as H, since the mapping $h \to ha$ of H to Ha is bijective. It follows that the order of the subgroup H divides the order of the whole group G, a result usually known as *Lagrange's theorem*. The quotient of the orders, i.e. the number of distinct cosets, is called the *index* of H in G.

Suppose again that H is a subgroup of an arbitrary group G and that $a,b \in G$. By writing $a \sim_l b$ if $a^{-1}b \in H$, we obtain another equivalence relation. The equivalence class which contains a is now the set aH of all elements ah, where $h \in H$. We call any such equivalence class a *left coset* of the subgroup H. Again, two left cosets either coincide or are disjoint, and the distinct left cosets form a partition of G.

When are the two partitions, into left cosets and into right cosets, the same? Evidently $Ha = aH$ for every $a \in G$ if and only if $a^{-1}Ha = H$ for every $a \in G$ or, since a may be replaced by a^{-1}, if and only if $a^{-1}ha \in H$ for every $h \in H$ and every $a \in G$. A subgroup H which satisfies this condition is said to be 'invariant' or *normal*.

Any group G obviously has two normal subgroups, namely G itself and the subset $\{e\}$ which contains only the identity element. A group G is said to be *simple* if it has no other normal subgroups and if these two are distinct (i.e., G contains more than one element).

We now show that if H is a normal subgroup of a group G, then the collection of all cosets of H can be given the structure of a group. Since $Ha = aH$ and $HH = H$, we have

$$(Ha)(Hb) \; = \; H(Ha)b \; = \; Hab.$$

Thus if we define the product $Ha \cdot Hb$ of the cosets Ha and Hb to be the coset Hab, the definition does not depend on the choice of coset representatives. Clearly multiplication of cosets is associative, the coset $H = He$ is an identity element and the coset Ha^{-1} is an inverse of the coset Ha. The new group thus constructed is called the *factor group* or *quotient group* of G by the normal subgroup H, and is denoted by G/H.

A mapping $f: G \to G'$ of a group G into a group G' is said to be a (group) *homomorphism* if

$$f(ab) \; = \; f(a)f(b) \quad \text{for all } a,b \in G.$$

By taking $a = b = e$, we see that this implies that $f(e) = e'$ is the identity element of G'. By taking $b = a^{-1}$, it now follows that $f(a^{-1})$ is the inverse of $f(a)$ in G'. Since the subset $f(G)$ of G' is closed under both multiplication and inversion, it is a subgroup of G'.

If $g: G' \to G''$ is a homomorphism of the group G' into a group G'', then the composite map $g \circ f: G \to G''$ is also a homomorphism.

The *kernel* of the homomorphism f is defined to be the set N of all $a \in G$ such that $f(a) = e'$ is the identity element of G'. The kernel is a subgroup of G, since if $a \in N$ and $b \in N$, then $ab \in N$ and $a^{-1} \in N$. Moreover, it is a normal subgroup, since $a \in N$ and $c \in G$ imply $c^{-1}ac \in N$.

For any $a \in G$, put $a' = f(a) \in G'$. The coset Na is the set of all $x \in G$ such that $f(x) = a'$, and the map $Na \to a'$ is a bijection from the collection of all cosets of N to $f(G)$. Since f is a homomorphism, Nab is mapped to $a'b'$. Hence the map $Na \to a'$ is a homomorphism of the factor group G/N to $f(G)$.

A mapping $f: G \to G'$ of a group G into a group G' is said to be a (group) *isomorphism* if it is both bijective and a homomorphism. The inverse mapping $f^{-1} : G' \to G$ is then also an isomorphism. (An *automorphism* of a group G is an isomorphism of G with itself.)

Thus we have shown that, if $f: G \to G'$ is a homomorphism of a group G into a group G', with kernel N, then the factor group G/N is isomorphic to $f(G)$.

Suppose now that G is an arbitrary group and a any element of G. We have already defined a^{-1}, the inverse of a. We now inductively define a^n, for any integer n, by putting

$$a^0 = e, \; a^1 = a,$$
$$a^n = a(a^{n-1}), \; a^{-n} = a^{-1}(a^{-1})^{n-1} \text{ if } n > 1.$$

It is readily verified that, for all $m,n \in \mathbb{Z}$,

$$a^m a^n = a^{m+n}, \quad (a^m)^n = a^{mn}.$$

The set $\langle a \rangle = \{a^n: n \in \mathbb{Z}\}$ is a commutative subgroup of G, the *cyclic subgroup generated by a*. Evidently $\langle a \rangle$ contains a and is contained in every subgroup of G which contains a.

If we regard $\mathbb{Z}$ as a group under addition, then the mapping $n \to a^n$ is a homomorphism of $\mathbb{Z}$ onto $\langle a \rangle$. Consequently $\langle a \rangle$ is isomorphic to the factor group $\mathbb{Z}/N$, where N is the subgroup of $\mathbb{Z}$ consisting of all integers n such that $a^n = e$. Evidently $0 \in N$, and $n \in N$ implies $-n \in N$. Thus either $N = \{0\}$ or N contains a positive integer. In the latter case, let s be the least positive integer in N. By Proposition 14, for any integer n there exist integers q,r such that

$$n = qs + r, \quad 0 \le r < s.$$

If $n \in N$, then also $r = n - qs \in N$ and hence $r = 0$, by the definition of s. It follows that $N = s\mathbb{Z}$ is the subgroup of $\mathbb{Z}$ consisting of all multiples of s. Thus either $\langle a \rangle$ is isomorphic to $\mathbb{Z}$, and is an infinite group, or $\langle a \rangle$ is isomorphic to the factor group $\mathbb{Z}/s\mathbb{Z}$, and is a finite group of order s. We say that the element a itself is of *infinite order* if $\langle a \rangle$ is infinite and of *order* s if $\langle a \rangle$ is of order s.

It is easily seen that in a *commutative* group the set of all elements of finite order is a subgroup, called its *torsion subgroup*.

If S is any nonempty subset of a group G, then the set $\langle S \rangle$ of all finite products $a_1^{\varepsilon_1} a_2^{\varepsilon_1} \dots a_n^{\varepsilon_n}$, where $n \in \mathbb{N}$, $a_j \in S$ and $\varepsilon_j = \pm 1$, is a subgroup of G, called the subgroup *generated* by S. Evidently $\langle S \rangle$ is contained in every subgroup of G which contains S.

Two elements a,b of a group G are said to be *conjugate* if $b = x^{-1}ax$ for some $x \in G$. It is easy to see that conjugacy is an equivalence relation. For $a = a^{-1}aa$, if $b = x^{-1}ax$ then $a = (x^{-1})^{-1}bx^{-1}$, and $b = x^{-1}ax$, $c = y^{-1}by$ together imply $c = (xy)^{-1}axy$. Consequently G may be partitioned into *conjugacy classes*, so that two elements of G are conjugate if and only if they belong to the same conjugacy class.

For any element a of a group G, the set N_a of all elements of G which commute with a,

$$N_a = \{x \in G : xa = ax\},$$

is closed under multiplication and inversion. Thus N_a is a subgroup of G, called the *centralizer* of a in G.

If y and z lie in the same right coset of N_a, so that $z = xy$ for some $x \in N_a$, then $zy^{-1}a = azy^{-1}$ and hence $y^{-1}ay = z^{-1}az$. Conversely, if $y^{-1}ay = z^{-1}az$, then y and z lie in the same right coset of N_a. If G is finite, it follows that the number of elements in the conjugacy class containing a is equal to the number of right cosets of the subgroup N_a, i.e. to the *index* of the subgroup N_a in G, and hence it divides the order of G.

To conclude, we mention a simple way of creating new groups from given ones. Let G, G' be groups and let $G \times G'$ be the set of all ordered pairs (a,a') with $a \in G$ and $a' \in G'$. Then $G \times G'$ acquires the structure of a group if we define the product $(a,a') \cdot (b,b')$ of (a,a') and (b,b') to be $(ab,a'b')$. Multiplication is clearly associative, (e,e') is an identity element and (a^{-1},a'^{-1}) is an inverse for (a,a'). The group thus constructed is called the *direct product* of G and G', and is again denoted by $G \times G'$.

8 Rings and fields

A nonempty set R is said is said to be a *ring* if two binary operations, + (addition) and ·
(multiplication), are defined with the properties

(i) R is a commutative group under addition, with 0 (*zero*) as identity element and $-a$ as
inverse of a;

(ii)   multiplication is associative: $(ab)c = a(bc)$ for all $a,b,c \in R$;

(iii) there exists an identity element 1 (*one*) for multiplication: $a1 = a = 1a$ for every $a \in R$;

(iv)   addition and multiplication are connected by the two distributive laws:

$$(a + b)c = (ac) + (bc), \quad c(a + b) = (ca) + (cb) \quad \text{for all } a,b,c \in R.$$

The elements 0 and 1 are necessarily uniquely determined. If, in addition, multiplication is
commutative:

$$ab = ba \quad \text{for all } a,b \in R,$$

then R is said to be a *commutative* ring. In a commutative ring either one of the two
distributive laws implies the other.

It may seem inconsistent to require that addition is commutative, but not multiplication.
However, the commutative law for addition is actually a consequence of the other axioms for a
ring. For, by the first distributive law we have

$$(a + b)(1 + 1) = a(1 + 1) + b(1 + 1) = a + a + b + b,$$

and by the second distributive law

$$(a + b)(1 + 1) = (a + b)1 + (a + b)1 = a + b + a + b.$$

Since a ring is a group under addition, by comparing these two relations we obtain first

$$a + a + b = a + b + a$$

and then $a + b = b + a$.

As examples, the set $\mathbb{Z}$ of all integers is a commutative ring, with the usual definitions of
addition and multiplication, whereas if $n > 1$, the set $M_n(\mathbb{Z})$ of all $n{\times}n$ matrices with entries
from $\mathbb{Z}$ is a noncommutative ring, with the usual definitions of matrix addition and
multiplication.

A very different example is the collection $\mathcal{P}(X)$ of all subsets of a given set X. If we define the sum $A + B$ of two subsets A,B of X to be their *symmetric difference*, i.e. the set of all elements of X which are in either A or B, but not in both:

$$A + B = (A \cup B) \setminus (A \cap B) = (A \cup B) \cap (A^c \cup B^c),$$

and the product AB to be the set of all elements of X which are in both A and B:

$$AB = A \cap B,$$

it is not difficult to verify that $\mathcal{P}(X)$ is a commutative ring, with the empty set $\varnothing$ as identity element for addition and the whole set X as identity element for multiplication. For every $A \in \mathcal{P}(X)$, we also have

$$A + A = \varnothing, \quad AA = A.$$

The set operations are in turn determined by the ring operations:

$$A \cup B = A + B + AB, \quad A \cap B = AB, \quad A^c = A + X.$$

A ring R is said to be a *Boolean ring* if $aa = a$ for every $a \in R$. It follows that $a + a = 0$ for every $a \in R$, since

$$a + a = (a + a)(a + a) = a + a + a + a.$$

Moreover, a Boolean ring is commutative, since

$$a + b = (a + b)(a + b) = a + b + ab + ba$$

and $ba = - ba$, by what we have already proved.

For an arbitrary set X, any nonempty subset of $\mathcal{P}(X)$ which is closed under union, intersection and complementation can be given the structure of a Boolean ring in the manner just described. It was proved by Stone (1936) that every Boolean ring may be obtained in this way. Thus the algebraic laws of set theory may be replaced by the more familiar laws of algebra and all such laws are consequences of a small number among them.

We now return to arbitrary rings. In the same way as for $\mathbb{Z}$, in any ring R we have

$$a0 = 0 = 0a \quad \text{for every } a$$

and

$$(-a)b = - (ab) = a(-b) \quad \text{for all } a,b.$$

It follows that R contains only one element if $1 = 0$. We will say that the ring R is 'trivial' in this case.

Suppose R is a nontrivial ring. Then, viewing R as a group under addition, the cyclic subgroup $\langle 1 \rangle$ is either infinite, and isomorphic to $\mathbb{Z}/0\mathbb{Z}$, or finite of order s, and isomorphic to $\mathbb{Z}/s\mathbb{Z}$ for some positive integer s. The ring R is said to have *characteristic* 0 in the first case and *characteristic s* in the second case.

For any positive integer n, write

$$na: = a + \ldots + a \quad (n \text{ summands}).$$

If R has characteristic $s > 0$, then $sa = 0$ for every $a \in R$, since

$$sa = (1 + \ldots + 1)a = 0a = 0.$$

On the other hand, $n1 \neq 0$ for every positive integer $n < s$, by the definition of characteristic.

An element a of a nontrivial ring R is said to be 'invertible' or a *unit* if there exists an element a^{-1} such that

$$a^{-1}a = 1 = aa^{-1}.$$

The element a^{-1} is then uniquely determined and is called the *inverse* of a. For example, 1 is a unit and is its own inverse. If a is a unit, then a^{-1} is also a unit and its inverse is a. If a and b are units, then ab is also a unit and its inverse is $b^{-1}a^{-1}$. It follows that the set $R^{\times}$ of all units is a group under multiplication.

A nontrivial ring R in which every nonzero element is invertible is said to be a *division ring*. Thus all nonzero elements of a division ring form a group under multiplication, the *multiplicative group* of the division ring. A *field* is a commutative division ring.

A nontrivial commutative ring R is said to be an *integral domain* if it has no 'divisors of zero', i.e. if $a \neq 0$ and $b \neq 0$ imply $ab \neq 0$. A division ring also has no divisors of zero, since if $a \neq 0$ and $b \neq 0$, then $a^{-1}ab = b \neq 0$, and hence $ab \neq 0$.

As examples, the set $\mathbb{Q}$ of rational numbers, the set $\mathbb{R}$ of real numbers and the set $\mathbb{C}$ of complex numbers are all fields, with the usual definitions of addition and multiplication. The set $\mathbb{H}$ of quaternions is a division ring, and the set $\mathbb{Z}$ of integers is a commutative integral domain, but neither is a field.

In an integral domain, the additive order of any nonzero element a is the same as the additive order of 1, since $ma = (m1)a = 0$ if and only if $m1 = 0$. Furthermore, the characteristic of an integral domain, and in particular of a division ring, is either 0 or a prime number. For assume $n = lm$, where l and m are positive integers less than n. If $n1 = 0$, then

$$(l1)(m1) = n1 = 0.$$

Since there are no divisors of zero, either $l1 = 0$ or $m1 = 0$, and hence the characteristic cannot be n.

A subset S of a ring R is said to be a (two-sided) *ideal* if it is a subgroup of R under addition and if, for every $a \in S$ and $c \in R$, both $ac \in S$ and $ca \in S$.

Any ring R has two obvious ideals, namely R itself and the subset $\{0\}$. It is said to be *simple* if it has no other ideals and is nontrivial.

Any division ring is simple. For if an ideal S of a division ring R contains $a \neq 0$, then for every $c \in R$ we have $c = (ca^{-1})a \in S$.

Conversely, if a *commutative* ring R is simple, then it is a field. For, if a is any nonzero element of R, the set

$$S_a = \{xa : x \in R\}$$

is an ideal (since R is commutative). Since S_a contains $1a = a \neq 0$, we must have $S_a = R$. Hence $1 = xa$ for some $x \in R$. Thus every nonzero element of R is invertible.

If R is a commutative ring and $a_1,...,a_m \in R$, then the set S consisting of all elements $x_1 a_1 + ... + x_m a_m$, where $x_j \in R$ ($1 \leq j \leq m$), is evidently an ideal of R, the ideal *generated* by $a_1,...,a_m$. An ideal of this type is said to be *finitely generated*.

We now show that if S is an ideal of the ring R, then the set $\mathcal{S}$ of all cosets $S + a$ of S can be given the structure of a ring. The ring R is a commutative group under addition. Hence, as we saw in §7, $\mathcal{S}$ acquires the structure of a (commutative) group under addition if we define the sum of $S + a$ and $S + b$ to be $S + (a + b)$. If $x = s + a$ and $x' = s' + b$ for some $s,s' \in S$, then $xx' = s'' + ab$, where $s'' = ss' + as' + sb$. Since S is an ideal, $s'' \in S$. Thus without ambiguity we may define the product of the cosets $S + a$ and $S + b$ to be the coset $S + ab$. Evidently multiplication is associative, $S + 1$ is an identity element for multiplication and both distributive laws hold. The new ring thus constructed is called the *quotient ring* of R by the ideal S, and is denoted by R/S.

A mapping $f: R \to R'$ of a ring R into a ring R' is said to be a (ring) *homomorphism* if, for all $a,b \in R$,

$$f(a + b) = f(a) + f(b), \quad f(ab) = f(a)f(b),$$

and if $f(1) = 1'$ is the identity element for multiplication in R'.

The *kernel* of the homomorphism f is the set N of all $a \in R$ such that $f(a) = 0'$ is the identity element for addition in R'. The kernel is an ideal of R, since it is a subgroup under addition and since $a \in N$, $c \in R$ imply $ac \in N$ and $ca \in N$.

For any $a \in R$, put $a' = f(a) \in R'$. The coset $N + a$ is the set of all $x \in R$ such that $f(x) = a'$, and the map $N + a \to a'$ is a bijection from the collection of all cosets of N to $f(R)$.

Since f is a homomorphism, $N + (a + b)$ is mapped to $a' + b'$ and $N + ab$ is mapped to $a'b'$. Hence the map $N + a \to a'$ is also a homomorphism of the quotient ring R/N into $f(R)$.

A mapping $f: R \to R'$ of a ring R into a ring R' is said to be a (ring) *isomorphism* if it is both bijective and a homomorphism. The inverse mapping $f^{-1}: R' \to R$ is then also an isomorphism. (An *automorphism* of a ring R is an isomorphism of R with itself.)

Thus we have shown that, if $f: R \to R'$ is a homomorphism of a ring R into a ring R', with kernel N, then the quotient ring R/N is isomorphic to $f(R)$.

An ideal M of a ring R is said to be *maximal* if $M \ne R$ and if there are no ideals S such that $M \subset S \subset R$.

Let M be an ideal of the ring R. If S is an ideal of R which contains M, then the set S' of all cosets $M + a$ with $a \in S$ is an ideal of R/M. Conversely, if S' is an ideal of R/M, then the set S of all $a \in R$ such that $M + a \in S'$ is an ideal of R which contains M. It follows that M is a maximal ideal of R if and only if R/M is simple. Hence an ideal M of a commutative ring R is maximal if and only if the quotient ring R/M is a field.

To conclude, we mention a simple way of creating new rings from given ones. Let R, R' be rings and let $R \times R'$ be the set of all ordered pairs (a, a') with $a \in R$ and $a' \in R'$. As we saw in the previous section, $R \times R'$ acquires the structure of a (commutative) group under addition if we define the sum $(a, a') + (b, b')$ of (a, a') and (b, b') to be $(a + b, a' + b')$. If we define their product $(a, a') \cdot (b, b')$ to be $(ab, a'b')$, then $R \times R'$ becomes a ring, with $(0, 0')$ as identity element for addition and $(1, 1')$ as identity element for multiplication. The ring thus constructed is called the *direct sum* of R and R', and is denoted by $R \oplus R'$.

9 Vector spaces and associative algebras

Although we assume some knowledge of linear algebra, it may be useful to place the basic definitions and results in the context of the preceding sections. A set V is said to be a *vector space* over a division ring D if it is a commutative group under an operation $+$ (addition) and there exists a map $\varphi: D \times V \to V$ (multiplication by a scalar) such that, if $\varphi(\alpha, v)$ is denoted by αv then, for all $\alpha, \beta \in D$ and all $v, w \in V$,

(i) $\alpha(v + w) = \alpha v + \alpha w$,

(ii) $(\alpha + \beta)v = \alpha v + \beta v$,

(iii) $(\alpha\beta)v = \alpha(\beta v)$,

(iv) $1v = v$,

where 1 is the identity element for multiplication in D. The elements of V will be called *vectors* and the elements of D *scalars*.

For example, for any positive integer n, the set D^n of all n-tuples of elements of the division ring D is a vector space over D if addition and multiplication by a scalar are defined by

$$(\alpha_1,...,\alpha_n) + (\beta_1,...,\beta_n) = (\alpha_1 + \beta_1,...,\alpha_n + \beta_n),$$
$$\alpha(\alpha_1,...,\alpha_n) = (\alpha\alpha_1,...,\alpha\alpha_n).$$

The special cases $D = \mathbb{R}$ and $D = \mathbb{C}$ have many applications.

As another example, the set $\mathscr{C}(I)$ of all continuous functions $f: I \to \mathbb{R}$, where I is an interval of the real line, is a vector space over the field $\mathbb{R}$ of real numbers if addition and multiplication by a scalar are defined, for every $t \in I$, by

$$(f + g)(t) = f(t) + g(t),$$
$$(\alpha f)(t) = \alpha f(t).$$

Let V be an arbitrary vector space over a division ring D. If O is the identity element of V with respect to addition, then

$$\alpha O = O \quad \text{for every } \alpha \in D,$$

since $\alpha O = \alpha(O + O) = \alpha O + \alpha O$. Similarly, if 0 is the identity element of D with respect to addition, then

$$0v = O \quad \text{for every } v \in V,$$

since $0v = (0 + 0)v = 0v + 0v$. Furthermore,

$$(-\alpha)v = -(\alpha v) \quad \text{for all } \alpha \in D \text{ and } v \in V,$$

since $O = 0v = (\alpha + (-\alpha))v = \alpha v + (-\alpha)v$, and

$$\alpha v \ne O \quad \text{if } \alpha \ne 0 \text{ and } v \ne O,$$

since $\alpha^{-1}(\alpha v) = (\alpha^{-1}\alpha)v = 1v = v$.

From now on we will denote the zero elements of D and V by the same symbol 0. This is easier on the eye and in practice is not confusing.

A subset U of a vector space V is said to be a *subspace* of V if it is a vector space under the same operations as V itself. It is easily seen that a nonempty subset U is a subspace of V if (and only if) it is closed under addition and multiplication by a scalar. For then, if $u \in U$, also $-u = (-1)u \in U$, and so U is an additive subgroup of V. The other requirements for a vector space are simply inherited from V.

For example, if $1 \leq m < n$, the set of all $(\alpha_1,...,\alpha_n) \in D^n$ with $\alpha_1 = ... = \alpha_m = 0$ is a subspace of D^n. Also, the set $\mathscr{C}'(I)$ of all continuously differentiable functions $f: I \to \mathbb{R}$ is a subspace of $\mathscr{C}(I)$. Two obvious subspaces of any vector space V are V itself and the subset $\{0\}$ which contains only the zero vector.

If U_1 and U_2 are subspaces of a vector space V, then their *intersection* $U_1 \cap U_2$, which necessarily contains 0, is again a subspace of V. The *sum* $U_1 + U_2$, consisting of all vectors $u_1 + u_2$ with $u_1 \in U_1$ and $u_2 \in U_2$, is also a subspace of V. Evidently $U_1 + U_2$ contains U_1 and U_2 and is contained in every subspace of V which contains both U_1 and U_2. If $U_1 \cap U_2 = \{0\}$, the sum $U_1 + U_2$ is said to be *direct*, and is denoted by $U_1 \oplus U_2$, since it may be identified with the set of all ordered pairs (u_1,u_2), where $u_1 \in U_1$ and $u_2 \in U_2$.

Let V be an arbitrary vector space over a division ring D and let $\{v_1,...,v_m\}$ be a finite subset of V. A vector v in V is said to be a *linear combination* of $v_1,...,v_m$ if

$$v = \alpha_1 v_1 + ... + \alpha_m v_m$$

for some $\alpha_1,...,\alpha_m \in D$. The *coefficients* $\alpha_1,...,\alpha_m$ need not be uniquely determined. Evidently a vector v is a linear combination of $v_1,...,v_m$ if it is a linear combination of some proper subset, since we can add the remaining vectors with zero coefficients.

If S is any nonempty subset of V, then the set $\langle S \rangle$ of all vectors in V which are linear combinations of finitely many elements of S is a subspace of V, the subspace 'spanned' or *generated* by S. Clearly $S \subseteq \langle S \rangle$ and $\langle S \rangle$ is contained in every subspace of V which contains S.

A finite subset $\{v_1,...,v_m\}$ of V is said to be *linearly dependent* (over D) if there exist $\alpha_1,...,\alpha_m \in D$, not all zero, such that

$$\alpha_1 v_1 + ... + \alpha_m v_m = 0,$$

and is said to be *linearly independent* otherwise.

For example, in $\mathbb{R}^3$ the vectors

$$v_1 = (1,0,1), \quad v_2 = (1,1,0), \quad v_3 = (1,1/2,1/2)$$

are linearly dependent, since $v_1 + v_2 - 2v_3 = 0$. On the other hand, the vectors

$$e_1 = (1,0,0), \quad e_2 = (0,1,0), \quad e_3 = (0,0,1)$$

are linearly independent, since $\alpha_1 e_1 + \alpha_2 e_2 + \alpha_3 e_3 = (\alpha_1,\alpha_2,\alpha_3)$, and this is 0 only if $\alpha_1 = \alpha_2 = \alpha_3 = 0$.

In any vector space V, the set $\{v\}$ containing the single vector v is linearly independent if $v \neq 0$ and linearly dependent if $v = 0$. If $v_1,\ldots,v_m$ are linearly independent, then any vector $v \in \langle v_1,\ldots,v_m \rangle$ has a unique representation as a linear combination of $v_1,\ldots,v_m$, since if

$$\alpha_1 v_1 + \ldots + \alpha_m v_m = \beta_1 v_1 + \ldots + \beta_m v_m,$$

then

$$(\alpha_1 - \beta_1)v_1 + \ldots + (\alpha_m - \beta_m)v_m = 0$$

and hence

$$\alpha_1 - \beta_1 = \ldots = \alpha_m - \beta_m = 0.$$

Evidently the vectors $v_1,\ldots,v_m$ are linearly dependent if some proper subset is linearly dependent. Hence any nonempty subset of a linearly independent set is again linearly independent.

A subset S of a vector space V is said to be a *basis* for V if S is linearly independent and $\langle S \rangle = V$. In the previous example, the vectors e_1,e_2,e_3 are a basis for $\mathbb{R}^3$, since they are not only linearly independent but also generate $\mathbb{R}^3$.

It may be shown that if a vector space V is generated by a finite subset T, and if S is a linearly independent subset of T, then V has a basis B such that $S \subseteq B \subseteq T$. Thus any nontrivial finitely generated vector space has a basis. Furthermore, for any linearly independent set S, there is a basis B with $S \subseteq B$. It follows also that any two bases contain the same number of elements.

If V has a basis containing n elements, we say V has *dimension n* and write $\dim V = n$. We say that V has infinite dimension if it is not finitely generated, and has dimension 0 if it contains only the vector 0.

For example, the field $\mathbb{C}$ of complex numbers may be regarded as a 2-dimensional vector space over the field $\mathbb{R}$ of real numbers, with basis $\{1,i\}$.

Again, D^n has dimension n as a vector space over the division ring D, since it has the basis

$$e_1 = (1,0,\ldots,0), \quad e_2 = (0,1,\ldots,0), \quad \ldots , \quad e_n = (0,0,\ldots,1).$$

On the other hand, the real vector space $\mathscr{C}(I)$ of all continuous functions $f\colon I \to \mathbb{R}$ has infinite dimension if the interval I contains more than one point since, for any positive integer n, the real polynomials of degree less than n form an n-dimensional subspace.

The first of these examples is readily generalized. If E and F are fields with $F \subseteq E$, we can regard E as a vector space over F. If this vector space is finite-dimensional, we say that E is a *finite extension* of F and define the *degree* of E over F to be the dimension $[E:F]$ of this vector space.

Any subspace U of a finite-dimensional vector space V is again finite-dimensional. Moreover, dim $U \le$ dim V, with equality only if $U = V$. If U_1 and U_2 are subspaces of V, then

$$\dim (U_1 + U_2) + \dim (U_1 \cap U_2) = \dim U_1 + \dim U_2.$$

Let V and W be vector spaces over the same division ring D. A map $T: V \to W$ is said to be *linear*, or a *linear transformation*, or a 'vector space homomorphism', if for all $v,v' \in V$ and every $\alpha \in D$,

$$T(v + v') = Tv + Tv', \quad T(\alpha v) = \alpha(Tv).$$

Since the first condition implies that T is a homomorphism of the additive group of V into the additive group of W, it follows that $T0 = 0$ and $T(-v) = -Tv$.

For example, if (τ_{jk}) is an $m \times n$ matrix with entries from the division ring D, then the map $T: D^m \to D^n$ defined by

$$T(\alpha_1,...,\alpha_m) = (\beta_1,...,\beta_n),$$

where

$$\beta_k = \alpha_1 \tau_{1k} + ... + \alpha_m \tau_{mk} \quad (1 \le k \le n),$$

is linear. It is easily seen that every linear map of D^m into D^n may be obtained in this way.

As another example, if $\mathscr{C}'(I)$ is the real vector space of all continuously differentiable functions $f: I \to \mathbb{R}$, then the map $T: \mathscr{C}'(I) \to \mathscr{C}(I)$ defined by $Tf = f'$ (the derivative of f) is linear.

Let U,V,W be vector spaces over the same division ring D. If $T: V \to W$ and $S: U \to V$ are linear maps, then the composite map $T \circ S: U \to W$ is again linear. For linear maps it is customary to write TS instead of $T \circ S$. The identity map $I: V \to V$ defined by $Iv = v$ for every $v \in V$ is clearly linear. If a linear map $T: V \to W$ is bijective, then its inverse map $T^{-1}: W \to V$ is again linear.

If $T: V \to W$ is a linear map, then the set N of all $v \in V$ such that $Tv = 0$ is a subspace of V, called the *nullspace* or *kernel* of T. Since $Tv = Tv'$ if and only if $T(v - v') = 0$, the map T is injective if and only if its kernel is $\{0\}$, i.e. when T is *nonsingular*.

For any subspace U of V, its image $TU = \{Tv: v \in U\}$ is a subspace of W. In particular, TV is a subspace of W, called the *range* of T. Thus the map T is surjective if and only if its range is W.

If V is finite-dimensional, then the range R of T is also finite-dimensional and

$$\dim R = \dim V - \dim N,$$

(since $R \approx V/N$). The dimensions of R and N are called respectively the *rank* and *nullity* of T. It follows that, if dim $V =$ dim W, then T is injective if and only if it is surjective.

Two vector spaces V,W over the same division ring D are said to be *isomorphic* if there exists a bijective linear map $T: V \to W$. As an example, if V is an n-dimensional vector space over the division ring D, then V is isomorphic to D^n. For if $v_1,...,v_n$ is a basis for V and if $v = \alpha_1 v_1 + ... + \alpha_n v_n$ is an arbitrary element of V, the map $v \to (\alpha_1,...,\alpha_n)$ is linear and bijective.

Thus there is essentially only one vector space of given finite dimension over a given division ring. However, vector spaces do not always present themselves in the concrete form D^n. An example is the set of solutions of a system of homogeneous linear equations with real coefficients. Hence, even if one is only interested in the finite-dimensional case, it is still desirable to be acquainted with the abstract definition of a vector space.

Let V and W be vector spaces over the same division ring D. We can define the *sum $S + T$* of two linear maps $S: V \to W$ and $T: V \to W$ by

$$(S + T)v = Sv + Tv.$$

This is again a linear map, and it is easily seen that with this definition of addition the set of all linear maps of V into W is a commutative group. If D is a field, i.e. if multiplication in D is commutative, then for any $\alpha \in D$ the map αT defined by

$$(\alpha T)v = \alpha(Tv)$$

is again linear, and with these definitions of addition and multiplication by a scalar the set of all linear maps of V into W is a vector space over D. (If the division ring D is not a field, it is necessary to consider 'right' vector spaces over D, as well as 'left' ones.)

If $V = W$, then the *product TS* is also defined and it is easily verified that the set of all linear maps of V into itself is a ring, with the identity map I as identity element for multiplication. The bijective linear maps of V to itself are the units of this ring and thus form a group under multiplication, the *general linear group GL(V)*.

The *centre* of a ring R is the set of all $c \in R$ such that $ac = ca$ for every $a \in R$. An *associative algebra A* over a field F is a ring containing F in its centre. On account of the ring structure, we can regard A as a vector space over F. The associative algebra is said to be *finite-dimensional* if it is finite-dimensional as a vector space over F.

For example, the set $M_n(F)$ of all $n \times n$ matrices with entries from the field F is a finite-dimensional associative algebra, with the usual definitions of addition and multiplication, and with $\alpha \in F$ identified with the matrix αI.

More generally, if D is a division ring containing F in its centre, then the set $M_n(D)$ of all $n{\times}n$ matrices with entries from D is an associative algebra over F. It is finite-dimensional if D itself is finite dimensional over F.

By the definition for rings, an associative algebra A is *simple* if $A \neq \{0\}$ and A has no ideals except $\{0\}$ and A. It is not difficult to show that, for any division ring D containing F in its centre, the associative algebra $M_n(D)$ is simple. It was proved by Wedderburn (1908) that any finite-dimensional simple associative algebra has the form $M_n(D)$, where D is a division ring containing F in its centre and of finite dimension over F.

If $F = \mathbb{C}$, the fundamental theorem of algebra implies that $\mathbb{C}$ is the only such D. If $F = \mathbb{R}$, there are three choices for D, by the following theorem of Frobenius (1878):

PROPOSITION 31 *If a division ring D contains the real field $\mathbb{R}$ in its centre and is of finite dimension as a vector space over $\mathbb{R}$, then D is isomorphic to $\mathbb{R}, \mathbb{C}$ or $\mathbb{H}$.*

Proof Suppose first that D is a field and $D \neq \mathbb{R}$. If $a \in D \setminus \mathbb{R}$ then, since D is finite-dimensional over $\mathbb{R}$, a is a root of a monic polynomial with real coefficients, which we may assume to be of minimal degree. Since $a \notin \mathbb{R}$, the degree is not 1 and the fundamental theorem of algebra implies that it must be 2. Thus

$$a^2 - 2\lambda a + \mu = 0$$

for some $\lambda,\mu \in \mathbb{R}$ with $\lambda^2 < \mu$. Then $\mu - \lambda^2 = \rho^2$ for some nonzero $\rho \in \mathbb{R}$ and $i = (a - \lambda)/\rho$ satisfies $i^2 = -1$. Thus D contains the field $\mathbb{R}(i) = \mathbb{R} + i\mathbb{R}$. But, since D is a field, the only $x \in D$ such that $x^2 = -1$ are i and $-i$. Hence the preceding argument shows that actually $D = \mathbb{R}(i)$. Thus D is isomorphic to the field $\mathbb{C}$ of complex numbers.

Suppose now that D is not commutative. Let a be an element of D which is not in the centre of D, and let M be an $\mathbb{R}$-subspace of D of maximal dimension which is commutative and which contains both a and the centre of D. If $x \in D$ commutes with every element of M, then $x \in M$. Hence M is a maximal commutative subset of D. It follows that if $x \in M$ and $x \neq 0$ then also $x^{-1} \in M$, since $xy = yx$ for all $y \in M$ implies $yx^{-1} = x^{-1}y$ for all $y \in M$. Similarly $x,x' \in M$ implies $xx' \in M$. Thus M is a field which properly contains $\mathbb{R}$. Hence, by the first part of the proof, M is isomorphic to $\mathbb{C}$. Thus $M = \mathbb{R}(i)$, where $i^2 = -1$, $[M:\mathbb{R}] = 2$ and $\mathbb{R}$ is the centre of D.

If $x \in D \setminus M$, then $b = (x + ixi)/2$ satisfies

$$bi = (xi - ix)/2 = -ib \neq 0.$$

Hence $b \in D \setminus M$ and $b^2 i = ib^2$. But, in the same way as before, $N = \mathbb{R} + \mathbb{R}b$ is a maximal subfield of D containing b and $\mathbb{R}$, and $N = \mathbb{R}(j)$, where $j^2 = -1$. Thus $b^2 = \alpha + \beta b$, where $\alpha, \beta \in \mathbb{R}$. In fact, since $b^2 i = ib^2$, we must have $\beta = 0$. Similarly $j = \gamma + \delta b$, where $\gamma, \delta \in \mathbb{R}$ and $\delta \neq 0$. Since $j^2 = \gamma^2 + 2\gamma\delta b + \delta^2\alpha = -1$, we must have $\gamma = 0$. Thus $j = \delta b$ and $ji = -ij$.

If we put $k = ij$, it now follows that

$$k^2 = -1, \quad jk = i = -kj, \quad ki = j = -ik.$$

Since no $\mathbb{R}$-linear combination of $1,i,j$ has these properties, the elements $1,i,j,k$ are $\mathbb{R}$-linearly independent. But, by Proposition 32 below, $[D:M] = [M:\mathbb{R}] = 2$. Hence $[D:\mathbb{R}] = 4$ and $1,i,j,k$ are a basis for D over $\mathbb{R}$. Thus D is isomorphic to the division ring $\mathbb{H}$ of quaternions. $\square$

To complete the proof of Proposition 31 we now prove

PROPOSITION 32 *Let D be a division ring which, as a vector space over its centre C, has finite dimension $[D:C]$. If M is a maximal subfield of D, then $[D:M] = [M:C]$.*

Proof Put $n = [D:C]$ and let $e_1,\ldots,e_n$ be a basis for D as a vector space over C. Obviously we may suppose $n > 1$. We show first that if $a_1,\ldots,a_n$ are elements of D such that

$$a_1 x e_1 + \ldots + a_n x e_n = 0 \quad \text{for every } x \in D,$$

then $a_1 = \ldots = a_n = 0$. Assume that there exists such a set $\{a_1,\ldots,a_n\}$ with not all elements zero and choose one with the minimal number of nonzero elements. We may suppose the notation chosen so that $a_i \neq 0$ for $i \leq r$ and $a_i = 0$ for $i > r$ and, by multiplying on the left by a_1^{-1}, we may further suppose that $a_1 = 1$. For any $y \in D$ we have

$$a_1 y x e_1 + \ldots + a_n y x e_n = 0 = y(a_1 x e_1 + \ldots + a_n x e_n)$$

and hence

$$(a_1 y - y a_1)x e_1 + \ldots + (a_n y - y a_n)x e_n = 0.$$

Since $a_i y = y a_i$ for $i = 1$ and for $i > r$, our choice of $\{a_1,\ldots,a_n\}$ implies that $a_i y = y a_i$ for all i. Since this holds for every $y \in D$, it follows that $a_i \in C$ for all i. But this is a contradiction, since $e_1,\ldots,e_n$ is a basis for D over C and $a_1 e_1 + \ldots + a_n e_n = 0$.

The map $T_{jk}: D \to D$ defined by $T_{jk}x = e_j x e_k$ is a linear transformation of D as a vector space over C. By what we have just proved, the n^2 linear maps T_{jk} $(j,k = 1,\ldots,n)$ are linearly independent over C. Consequently every linear transformation of D as a vector space over C is a C-linear combination of the maps T_{jk}.

Suppose now that $T: D \to D$ is a linear transformation of D as a vector space over M. Since $C \subseteq M$, T is also a linear transformation of D as a vector space over C and hence has the form

$$Tx \;=\; a_1 x e_1 + \dots + a_n x e_n$$

for some $a_1,\dots,a_n \in D$. But $T(bx) = b(Tx)$ for all $b \in M$ and $x \in D$. Hence

$$(a_1 b - b a_1) x e_1 + \dots + (a_n b - b a_n) x e_n \;=\; 0 \quad \text{for every } x \in D,$$

which implies $a_i b = b a_i$ $(i = 1,\dots,n)$. Since this holds for all $b \in M$ and M is a maximal subfield of D, it follows that $a_i \in M$ $(i = 1,\dots,n)$.

Let $\mathcal{T}$ denote the set of all linear transformations of D as a vector space over M. By what we have already proved, every $T \in \mathcal{T}$ is an M-linear combination of the maps $T_1,\dots,T_n$, where $T_i x = x e_i$ $(i = 1,\dots,n)$, and the maps $T_1,\dots,T_n$ are linearly independent over M. Consequently the dimension of $\mathcal{T}$ as a vector space over M is n. But $\mathcal{T}$ has dimension $[D:M]^2$ as a vector space over M. Hence $[D:M]^2 = n$. Since $n = [D:M]\,[M:C]$, it follows that $[D:M] = [M:C]$. $\quad\square$

10 Inner product spaces

Let F denote either the real field $\mathbb{R}$ or the complex field $\mathbb{C}$. A vector space V over F is said to be an *inner product space* if there exists a map $(u,v) \to \langle u,v \rangle$ of $V \times V$ into F such that, for every $\alpha \in F$ and all $u,u',v \in V$,

(i) $\langle \alpha u,v \rangle \;=\; \alpha\,\langle u,v \rangle$,

(ii) $\langle u + u',v \rangle \;=\; \langle u,v \rangle + \langle u',v \rangle$,

(iii) $\langle v,u \rangle \;=\; \overline{\langle u,v \rangle}$,

(iv) $\langle u,u \rangle \;>\; 0$ if $u \neq O$.

If $F = \mathbb{R}$, then (iii) simply says that $\langle v,u \rangle = \langle u,v \rangle$, since a real number is its own complex conjugate. The restriction $u \neq O$ is necessary in (iv), since (i) and (iii) imply that

$$\langle u,O \rangle \;=\; \langle O,v \rangle \;=\; 0 \quad \text{for all } u,v \in V.$$

It follows from (ii) and (iii) that

$$\langle u, v + v' \rangle \;=\; \langle u,v \rangle + \langle u,v' \rangle \quad \text{for all } u,v,v' \in V,$$

and from (i) and (iii) that

$$\langle u, \alpha v \rangle = \overline{\alpha} \langle u, v \rangle \quad \text{for every } \alpha \in F \text{ and all } u, v \in V.$$

The standard example of an inner product space is the vector space F^n, with the inner product of $x = (\xi_1, \ldots, \xi_n)$ and $y = (\eta_1, \ldots, \eta_n)$ defined by

$$\langle x, y \rangle = \xi_1 \overline{\eta}_1 + \ldots + \xi_n \overline{\eta}_n.$$

Another example is the vector space $\mathscr{C}(I)$ of all continuous functions $f: I \to F$, where $I = [a, b]$ is a compact subinterval of $\mathbb{R}$, with the inner product of f and g defined by

$$\langle f, g \rangle = \int_a^b f(t) \overline{g(t)} \, dt.$$

In an arbitrary inner product space V we define the *norm* $\|v\|$ of a vector $v \in V$ by

$$\|v\| = \langle v, v \rangle^{1/2}.$$

Thus $\|v\| \geq 0$, with equality if and only if $v = O$. Evidently

$$\|\alpha v\| = |\alpha| \|v\| \quad \text{for all } \alpha \in F \text{ and } v \in V.$$

Inner products and norms are connected by *Schwarz's inequality*:

$$|\langle u, v \rangle| \leq \|u\| \|v\| \quad \text{for all } u, v \in V,$$

with equality if and only if u and v are linearly dependent. For the proof we may suppose that u and v are linearly independent, since it is easily seen that equality holds if $u = \lambda v$ or $v = \lambda u$ for some $\lambda \in F$. Then, for all $\alpha, \beta \in F$, not both 0,

$$0 < \langle \alpha u + \beta v, \alpha u + \beta v \rangle = |\alpha|^2 \langle u, u \rangle + \alpha \overline{\beta} \langle u, v \rangle + \overline{\alpha} \beta \overline{\langle u, v \rangle} + |\beta|^2 \langle v, v \rangle.$$

If we choose $\alpha = \langle v, v \rangle$ and $\beta = -\langle u, v \rangle$, this takes the form

$$0 < \|u\|^2 \|v\|^4 - 2\|v\|^2 |\langle u, v \rangle|^2 + |\langle u, v \rangle|^2 \|v\|^2 = \{\|u\|^2 \|v\|^2 - |\langle u, v \rangle|^2\} \|v\|^2.$$

Hence

$$|\langle u, v \rangle|^2 < \|u\|^2 \|v\|^2,$$

as we wished to show. We follow common practice by naming the inequality after Schwarz (1885), but (cf. §4) it had already been proved for $\mathbb{R}^n$ by Cauchy (1821) and for $\mathscr{C}(I)$ by Bunyakovskii (1859).

It follows from Schwarz's inequality that

$$\|u+v\|^2 \; = \; \|u\|^2 + 2\Re\langle u,v\rangle + \|v\|^2$$

$$\leq \; \|u\|^2 + 2|\langle u,v\rangle| + \|v\|^2 \; \leq \; \{\|u\| + \|v\|\}^2.$$

Thus

$$\|u+v\| \; \leq \; \|u\| + \|v\| \quad \text{for all } u,v \in V,$$

with strict inequality if u and v are linearly independent.

It now follows that V acquires the structure of a metric space if we define the distance between u and v by

$$d(u,v) \; = \; \|u-v\|.$$

In the case $V = \mathbb{R}^n$ this is the *Euclidean distance*

$$d(x,y) \; = \; \left(\textstyle\sum_{j=1}^{n} |\xi_j - \eta_j|^2\right)^{1/2},$$

and in the case $V = \mathscr{C}(I)$ it is is the *L^2-norm*

$$d(f,g) \; = \; \left(\textstyle\int_a^b |f(t) - g(t)|^2\, dt\right)^{1/2}.$$

The norm in any inner product space V satisfies the *parallelogram law*:

$$\|u+v\|^2 + \|u-v\|^2 \; = \; 2\|u\|^2 + 2\|v\|^2 \quad \text{for all } u,v \in V.$$

This may be immediately verified by substituting $\|w\|^2 = \langle w,w\rangle$ throughout and using the linearity of the inner product. The geometrical interpretation is that in any parallelogram the sum of the squares of the lengths of the two diagonals is equal to the sum of the squares of the lengths of all four sides.

It may be shown that any normed vector space which satisfies the parallelogram law can be given the structure of an inner product space by defining

$$\langle u,v\rangle \; = \; \{\|u+v\|^2 - \|u-v\|^2\}/4 \quad \text{if } F = \mathbb{R},$$

$$= \; \{\|u+v\|^2 - \|u-v\|^2 + i\|u+iv\|^2 - i\|u-iv\|^2\}/4 \quad \text{if } F = \mathbb{C}.$$

(Cf. the argument for $F = \mathbb{Q}$ in §4 of Chapter XIII.)

In an arbitrary inner product space V a vector u is said to be 'perpendicular' or *orthogonal* to a vector v if $\langle u,v\rangle = 0$. The relation is symmetric, since $\langle u,v\rangle = 0$ implies $\langle v,u\rangle = 0$. For orthogonal vectors u,v, the *law of Pythagoras* holds:

$$\|u+v\|^2 \; = \; \|u\|^2 + \|v\|^2.$$

More generally, a subset E of V is said to be *orthogonal* if $\langle u,v \rangle = 0$ for all $u,v \in E$ with $u \neq v$. It is said to be *orthonormal* if, in addition, $\langle u,u \rangle = 1$ for every $u \in E$. An orthogonal set which does not contain O may be converted into an orthonormal set by replacing each $u \in E$ by $u/\|u\|$.

For example, if $V = F^n$, then the basis vectors

$$e_1 = (1,0,\ldots,0), \quad e_2 = (0,1,\ldots,0), \quad \ldots, \quad e_n = (0,0,\ldots,1)$$

form an orthonormal set. It is easily verified also that, if $I = [0,1]$, then in $\mathscr{C}(I)$ the functions $e_n(t) = e^{2\pi i n t}$ ($n \in \mathbb{Z}$) form an orthonormal set.

Let $\{e_1,\ldots,e_m\}$ be *any* orthonormal set in the inner product space V and let U be the vector subspace generated by $e_1,\ldots,e_m$. The norm of a vector $u = \alpha_1 e_1 + \ldots + \alpha_m e_m \in U$ is given by

$$\|u\|^2 = |\alpha_1|^2 + \ldots + |\alpha_m|^2,$$

which shows that $e_1,\ldots,e_m$ are linearly independent.

To find the *best approximation* in U to a given vector $v \in V$, put

$$w = \gamma_1 e_1 + \ldots + \gamma_m e_m,$$

where

$$\gamma_j = \langle v,e_j \rangle \quad (j = 1,\ldots,m).$$

Then $\langle w,e_j \rangle = \langle v,e_j \rangle$ ($j = 1,\ldots,m$) and hence $\langle v - w, w \rangle = 0$. Consequently, by the law of Pythagoras,

$$\|v\|^2 = \|v - w\|^2 + \|w\|^2.$$

Since $\|w\|^2 = |\gamma_1|^2 + \ldots + |\gamma_m|^2$, this yields *Bessel's inequality*:

$$|\langle v,e_1 \rangle|^2 + \ldots + |\langle v,e_m \rangle|^2 \leq \|v\|^2,$$

with strict inequality if $v \notin U$. For any $u \in U$, we also have $\langle v - w, w - u \rangle = 0$ and so, by Pythagoras again,

$$\|v - u\|^2 = \|v - w\|^2 + \|w - u\|^2.$$

This shows that w is the unique nearest point of U to v.

From any linearly independent set of vectors $v_1,\ldots,v_m$ we can inductively construct an orthonormal set $e_1,\ldots,e_m$ such that $e_1,\ldots,e_k$ span the same vector subspace as $v_1,\ldots,v_k$ for $1 \leq k \leq m$. We begin by taking $e_1 = v_1/\|v_1\|$. Now suppose $e_1,\ldots,e_k$ have been determined. If

$$w = v_{k+1} - \langle v_{k+1},e_1 \rangle e_1 - \ldots - \langle v_{k+1},e_k \rangle e_k,$$

then $\langle w, e_j \rangle = 0$ $(j = 1,\ldots,k)$. Moreover $w \neq O$, since w is a linear combination of $v_1,\ldots,v_{k+1}$ in which the coefficient of v_{k+1} is 1. By taking $e_{k+1} = w/\|w\|$, we obtain an orthonormal set $e_1,\ldots,e_{k+1}$ spanning the same linear subspace as $v_1,\ldots,v_{k+1}$. This construction is known as *Schmidt's orthogonalization process*, because of its use by E. Schmidt (1907) in his treatment of linear integral equations. The (normalized) Legendre polynomials are obtained by applying the process to the linearly independent functions $1, t, t^2,\ldots$ in the space $\mathscr{C}(I)$, where $I = [-1,1]$.

It follows that any finite-dimensional inner product space V has an orthonormal basis $e_1,\ldots,e_n$ and that

$$\|v\|^2 = \sum_{j=1}^{n} |\langle v, e_j \rangle|^2 \quad \text{for every } v \in V.$$

In an infinite-dimensional inner product space V an orthonormal set E may even be uncountably infinite. However, for a given $v \in V$, there are at most countably many vectors $e \in E$ for which $\langle v, e \rangle \neq 0$. For if $\{e_1,\ldots,e_m\}$ is any finite subset of E then, by Bessel's inequality,

$$\sum_{j=1}^{m} |\langle v, e_j \rangle|^2 \leq \|v\|^2$$

and so, for each $n \in \mathbb{N}$, there are at most $n^2 - 1$ vectors $e \in E$ for which $|\langle v, e \rangle| > \|v\|/n$.

If the vector subspace U of all finite linear combinations of elements of E is dense in V then, by the best approximation property of finite orthonormal sets, *Parseval's equality* holds:

$$\sum_{e \in E} |\langle v, e \rangle|^2 = \|v\|^2 \quad \text{for every } v \in V.$$

Parseval's equality holds for the inner product space $\mathscr{C}(I)$, where $I = [0,1]$, and the orthonormal set $E = \{e^{2\pi i n t}: n \in \mathbb{Z}\}$ since, by *Weierstrass's approximation theorem* (see the references in §6 of Chapter XI), every $f \in \mathscr{C}(I)$ is the uniform limit of a sequence of *trigonometric polynomials*. The result in this case was formally derived by Parseval (1805).

An *almost periodic function*, in the sense of Bohr (1925), is a function $f: \mathbb{R} \to \mathbb{C}$ which can be uniformly approximated on $\mathbb{R}$ by *generalized trigonometric polynomials*

$$\sum_{j=1}^{m} c_j e^{i\lambda_j t},$$

where $c_j \in \mathbb{C}$ and $\lambda_j \in \mathbb{R}$ $(j = 1,\ldots,m)$. For any almost periodic functions f, g, the limit

$$\langle f, g \rangle = \lim_{T \to \infty} (1/2T) \int_{-T}^{T} f(t)\overline{g(t)}\, dt$$

exists. The set $\mathscr{B}$ of all almost periodic functions acquires in this way the structure of an inner product space. The set $E = \{e^{i\lambda t}: \lambda \in \mathbb{R}\}$ is an uncountable orthonormal set and Parseval's equality holds for this set.

A finite-dimensional inner product space is necessarily complete as a metric space, i.e. every fundamental sequence converges. However, an infinite-dimensional inner product space need not be complete, as $\mathscr{C}(I)$ already illustrates. An inner product space which is complete is said to be a *Hilbert space*.

The case considered by Hilbert (1906) was the vector space ℓ^2 of all infinite sequences $x = (\xi_1,\xi_2,\dots)$ of complex numbers such that $\sum_{k \geq 1} |\xi_k|^2 < \infty$, with

$$\langle x,y \rangle = \sum_{k \geq 1} \xi_k \overline{\eta}_k.$$

Another example is the vector space $L^2(I)$, where $I = [0,1]$, of all (equivalence classes of) Lebesgue measurable functions $f: I \to \mathbb{C}$ such that $\int_0^1 |f(t)|^2 dt < \infty$, with

$$\langle f,g \rangle = \int_0^1 f(t)\overline{g(t)} \, dt.$$

With any $f \in L^2(I)$ we can associate a sequence $\hat{f} \in \ell^2$, consisting of the inner products $\langle f,e_n \rangle$, where $e_n(t) = e^{2\pi i n t}$ $(n \in \mathbb{Z})$, in some fixed order. The map $\mathscr{F}: L^2(I) \to \ell^2$ thus defined is linear and, by Parseval's equality,

$$\|\mathscr{F}f\| = \|f\|.$$

In fact $\mathscr{F}$ is an *isometry* since, by the *theorem of Riesz–Fischer* (1907), it is bijective.

11 Further remarks

A fund of information about numbers in different cultures is contained in Menninger [52]. A good popular book is Dantzig [18].

The algebra of sets was created by Boole (1847), who used the symbols + and · instead of $\cup$ and $\cap$, as is now customary. His ideas were further developed, with applications to logic and probability theory, in Boole [10]. A simple system of axioms for Boolean algebra was given by Huntingdon [39]. For an introduction to Stone's representation theorem, referred to in §8, see Stone [69]; there are proofs in Halmos [30] and Sikorski [66]. For applications of Boolean algebras to switching circuits see, for example, Rudeanu [62]. Boolean algebra is studied in the more general context of lattice theory in Birkhoff [6].

Dedekind's axioms for $\mathbb{N}$ may be found on p. 67 of [19], which contains also his earlier construction of the real numbers from the rationals by means of cuts. Some interesting comments on the axioms **(N1)-(N3)** are contained in Henkin [34]. Starting from these axioms, Landau [47] gives a detailed derivation of the basic properties of $\mathbb{N}, \mathbb{Q}, \mathbb{R}$ and $\mathbb{C}$.

 I. The expanding universe of numbers

The argument used to extend $\mathbb{N}$ to $\mathbb{Z}$ shows that any commutative *semigroup* satisfying the cancellation law may be embedded in a commutative *group*. The argument used to extend $\mathbb{Z}$ to $\mathbb{Q}$ shows that any commutative *ring* without divisors of zero may be embedded in a *field*.

An example of an ordered field which does not have the Archimedean property, although every fundamental sequence is (trivially) convergent, is the field $^*\mathbb{R}$ of hyperreal numbers, constructed by Abraham Robinson (1961). Hyperreal numbers are studied in Stroyan and Luxemburg [70].

The 'arithmetization of analysis' had a gradual evolution, which is traced in Chapitre VI (by Dugac) of Dieudonné *et al.* [22]. A modern text on real analysis is Rudin [63]. In Lemma 7 of Chapter VI we will show that all norms on $\mathbb{R}^n$ are equivalent.

The contraction principle (Proposition 26) has been used to prove the *central limit theorem* of probability theory by Hamedani and Walter [32]. Bessaga (1959) has proved a *converse* of the contraction principle: Let E be an arbitrary set, $f: E \to E$ a map of E to itself and θ a real number such that $0 < \theta < 1$. If each iterate f^n ($n \in \mathbb{N}$) has at most one fixed point and if some iterate has a fixed point, then a complete metric d can be defined on E such that $\mathrm{d}(f(x'),f(x'')) \le \theta \, \mathrm{d}(x',x'')$ for all $x',x'' \in E$. A short proof is given by Jachymski [40].

There are other important fixed point theorems besides Proposition 26. *Brouwer's fixed point theorem* states that, if $B = \{x \in \mathbb{R}^n: |x| \le 1\}$ is the n-dimensional closed unit ball, every continuous map $f: B \to B$ has a fixed point. For an elementary proof, see Kulpa [44]. The *Lefschetz fixed point theorem* requires a knowledge of algebraic topology, even for its statement. Fixed point theorems are extensively treated in Dugundji and Granas [23] (and in A. Granas and J. Dugundji, *Fixed Point Theory*, Springer-Verlag, New York, 2003).

For a more detailed discussion of differentiability for functions of several variables see, for example, Fleming [26] and Dieudonné [21]. The inverse function theorem (Proposition 27) is a local result. Some global results are given by Atkinson [5] and Chichilnisky [14]. For a holomorphic version of Proposition 28 and for the simple way in which higher-order equations may be replaced by systems of first-order equations see, e.g., Coddington and Levinson [16].

The formula for the roots of a cubic was first published by Cardano [12], but it was discovered by del Ferro and again by Tartaglia, who accused Cardano of breaking a pledge of secrecy. Cardano is judged less harshly by historians today than previously. His book, which contained developments of his own and also the formula for the roots of a quartic discovered by his pupil Ferrari, was the most significant Western contribution to mathematics for more than a thousand years.

Proposition 29 still holds, but is more difficult to prove, if in its statement "has a nonzero derivative" is replaced by "which is not constant". Read [57] shows that the basic results of complex analysis may be deduced from this stronger form of Proposition 29 without the use of complex integration.

A field F is said to be *algebraically closed* if every polynomial of positive degree with coefficients from F has a root in F. Thus the 'fundamental theorem of algebra' says that the field $\mathbb{C}$ of complex numbers is algebraically closed. The proofs of this theorem due to Argand–Cauchy and Euler–Lagrange–Laplace are given in Chapter 4 (by Remmert) of Ebbinghaus *et al.* [24]. As shown on p. 77 of [24], the latter method provides, in particular, a simple proof for the existence of n-th roots.

Wall [72] gives a proof of the fundamental theorem of algebra, based on the notion of topological degree, and Ahlfors [1] gives the most common complex analysis proof, based on Liouville's theorem that a function holomorphic in the whole complex plane is bounded only if it is a constant. A form of Liouville's theorem is easily deduced from Proposition 29: if the power series

$$p(z) = a_0 + a_1 z + a_2 z^2 + \dots$$

converges and $|p(z)|$ is bounded for all $z \in \mathbb{C}$, then $a_n = 0$ for every $n \geq 1$.

The representation of trigonometric functions by complex exponentials appears in §138 of Euler [25]. The various algebraic formulas involving trigonometric functions, such as

$$\cos 3x = 4 \cos^3 x - 3 \cos x,$$

are easily established by means of this representation and the addition theorem for the exponential function.

Some texts on complex analysis are Ahlfors [1], Caratheodory [11] and Narasimhan [56].

The 19th century literature on quaternions is surveyed in Rothe [59]. Although Hamilton hoped that quaternions would prove as useful as complex numbers, a quaternionic analysis analogous to complex analysis was first developed by Fueter (1935). A good account is given by Sudbery [71].

One significant contribution of quaternions was indirect. After Hamilton had shown the way, other 'hypercomplex' number systems were constructed, which led eventually to the structure theory of associative algebras discussed below.

It is not difficult to show that any *automorphism* of $\mathbb{H}$, i.e. any bijective map $T: \mathbb{H} \to \mathbb{H}$ such that

$$T(x + y) = Tx + Ty, \quad T(xy) = (Tx)(Ty) \quad \text{for all } x,y \in \mathbb{H},$$

has the form $Tx = uxu^{-1}$ for some quaternion u with norm 1.

For octonions and their uses, see van der Blij [8] and Springer and Veldkamp [67]. The group of all automorphisms of the algebra $\mathbb{O}$ is the exceptional simple Lie group G_2. The other four exceptional simple Lie groups are also all related to $\mathbb{O}$ in some way.

Of wider significance are the associative algebras introduced in 1878 by Clifford [15] (pp. 266-276) as a common generalization of quaternions and Grassmann algebra. *Clifford algebras* were used by Lipschitz (1886) to represent orthogonal transformations in n-dimensional space. There is an extensive discussion of Clifford algebras in Deheuvels [20]. For their applications in physics, see Salingaros and Wene [64].

Proposition 32 has many uses. The proof given here is extracted from Nagahara and Tominaga [55].

It was proved by both Kervaire (1958) and Milnor (1958) that if a division algebra A (not necessarily associative) contains the real field $\mathbb{R}$ in its centre and is of finite dimension as a vector space over $\mathbb{R}$, then this dimension must be 1,2,4 or 8 (but the algebra need not be isomorphic to $\mathbb{R},\mathbb{C},\mathbb{H}$ or $\mathbb{O}$). All known proofs use deep results from algebraic topology, which was first applied to the problem by H. Hopf (1940). For more information about the proof, see Chapter 11 (by Hirzebruch) of Ebbinghaus *et al.* [24].

When is the product of two sums of squares again a sum of squares? To make the question precise, call a triple (r,s,t) of positive integers 'admissible' if there exist real numbers ρ_{ijk} $(1 \le i \le t, 1 \le j \le r, 1 \le k \le s)$ such that, for every $x = (\xi_1,...,\xi_r) \in \mathbb{R}^r$ and every $y = (\eta_1,...,\eta_s) \in \mathbb{R}^s$,

$$(\xi_1^2 + ... + \xi_r^2)(\eta_1^2 + ... + \eta_s^2) = \zeta_1^2 + ... + \zeta_t^2,$$

where

$$\zeta_i = \sum_{j=1}^{r} \sum_{k=1}^{s} \rho_{ijk}\xi_j\eta_k \, .$$

The question then becomes, which triples (r,s,t) are admissible? It is obvious that $(1,1,1)$ is admissible and the relation $n(x)n(y) = n(xy)$ for the norms of complex numbers, quaternions and octonions shows that (t,t,t) is admissible also for $t = 2,4,8$. It was proved by Hurwitz (1898) that (t,t,t) is admissible for no other values of t. A survey of the general problem is given by Shapiro [65].

General introductions to algebra are provided by Birkhoff and MacLane [7] and Herstein [35]. More extended treatments are given in Jacobson [41] and Lang [48].

The theory of groups is treated in M. Hall [29] and Rotman [60]. An especially significant class of groups is studied in Humphreys [38].

If H is a subgroup of a finite group G, then it is possible to choose a system of left coset representatives of H which is also a system of right coset representatives. This interesting, but not very useful, fact belongs to combinatorics rather than to group theory. We mention it because it was the motivation for the theorem of P. Hall (1935) on *systems of distinct representatives*, also known as the 'marriage theorem'. Further developments are described in Mirsky [53]. For quantitative versions, with applications to operations research, see Ford and Fulkerson [27].

The theory of rings separates into two parts. Noncommutative ring theory, which now incorporates the structure theory of associative algebras, is studied in Herstein [36], Kasch [42] and Lam [46]. Commutative ring theory, which grew out of algebraic number theory and algebraic geometry, is studied in Atiyah and Macdonald [4] and Kunz [45].

Field theory was established as an independent subject of study in 1910 by Steinitz [68]. The books of Jacobson [41] and Lang [48] treat also the more recent theory of ordered fields, due to Artin and Schreier (1927).

Fields and groups are connected with one another by *Galois theory*. This subject has its origin in attempts to solve polynomial equations 'by radicals'. The founder of the subject is really Lagrange (1770/1). By developing his ideas, Ruffini (1799) and Abel (1826) showed that polynomial equations of degree greater than 4 cannot, in general, be solved by radicals. Abel (1829) later showed that polynomial equations *can* be solved by radicals if their 'Galois group' is commutative. In honour of this result, commutative groups are often called *abelian*.

Galois (1831, published posthumously in 1846) introduced the concept of normal subgroup and stated a necessary and sufficient condition for a polynomial equation to be solvable by radicals. The significance of Galois theory today lies not in this result, despite its historical importance, but in the much broader 'fundamental theorem of Galois theory'. In the form given it by Dedekind (1894) and Artin (1944), this establishes a correspondence between extension fields and groups of automorphisms, and provides a framework for the solution of a number of algebraic problems.

Morandi [54] and Rotman [61] give modern accounts of Galois theory. The historical development is traced in Kiernan [43]. In recent years attention has focussed on the problem of determining which finite groups occur as Galois groups over a given field; for an introductory account, see Matzat [51].

Some texts on linear algebra and matrix theory are Halmos [31], Horn and Johnson [37], Mal'cev [50] and Gantmacher [28].

The older literature on associative algebras is surveyed in Cartan [13]. The texts on noncommutative rings cited above give modern introductions.

A vast number of characterizations of inner product spaces, in addition to the parallelogram law, is given in Amir [3]. The theory of Hilbert space is treated in the books of Riesz and

Sz.-Nagy [58] and Akhiezer and Glazman [2]. For its roots in the theory of integral equations, see Hellinger and Toeplitz [33]. Almost periodic functions are discussed from different points of view in Bohr [9], Corduneanu [17] and Maak [49]. The convergence of Fourier series is treated in Zygmund [73], for example.

12 Selected references

[1] L.V. Ahlfors, *Complex analysis*, 3rd ed., McGraw-Hill, New York, 1978.

[2] N.I. Akhiezer and I.M. Glazman, *Theory of linear operators in Hilbert space*, English transl. by E.R. Dawson based on 3rd Russian ed., Pitman, London, 1981.

[3] D. Amir, *Characterizations of inner product spaces*, Birkhäuser, Basel, 1986.

[4] M.F. Atiyah and I.G. Macdonald, *Introduction to commutative algebra*, Addison-Wesley, Reading, Mass., 1969.

[5] F.V. Atkinson, The reversibility of a differentiable mapping, *Canad. Math. Bull.* **4** (1961), 161-181.

[6] G. Birkhoff, *Lattice theory*, corrected reprint of 3rd ed., American Mathematical Society, Providence, R.I., 1979.

[7] G. Birkhoff and S. MacLane, *A survey of modern algebra*, 3rd ed., Macmillan, New York, 1965.

[8] F. van der Blij, History of the octaves, *Simon Stevin* **34** (1961), 106-125.

[9] H. Bohr, *Almost periodic functions*, English transl. by H. Cohn and F. Steinhardt, Chelsea, New York, 1947.

[10] G. Boole, *An investigation of the laws of thought, on which are founded the mathematical theories of logic and probability*, reprinted, Dover, New York, 1957. [Original edition, 1854]

[11] C. Caratheodory, *Theory of functions of a complex variable*, English transl. by F. Steinhardt, 2 vols., 2nd ed., Chelsea, New York, 1958/1960.

[12] G. Cardano, *The great art or the rules of algebra*, English transl. by T.R. Witmer, M.I.T. Press, Cambridge, Mass., 1968. [Latin original, 1545]

[13] E. Cartan, Nombres complexes, *Encyclopédie des sciences mathématiques, Tome* I, *Fasc.* 4, *Art.* I.5, Gauthier-Villars, Paris, 1908. [Reprinted in *Oeuvres complètes, Partie* II, *Vol.* 1, pp. 107-246.]

[14] G. Chichilnisky, Topology and invertible maps, *Adv. in Appl. Math.* **21** (1998), 113-123.

[15] W.K. Clifford, *Mathematical Papers*, reprinted, Chelsea, New York, 1968.

[16] E.A. Coddington and N. Levinson, *Theory of ordinary differential equations*, McGraw-Hill, New York, 1955.

[17] C. Corduneanu, *Almost periodic functions*, English transl. by G. Berstein and E. Tomer, Interscience, New York, 1968.

[18] T. Dantzig, *Number: The language of science*, 3rd ed., Allen & Unwin, London, 1947.

[19] R. Dedekind, *Essays on the theory of numbers*, English transl. by W.W. Beman, reprinted, Dover, New York, 1963.

[20] R. Deheuvels, *Formes quadratiques et groupes classiques*, Presses Universitaires de France, Paris, 1981.

[21] J. Dieudonné, *Foundations of modern analysis*, enlarged reprint, Academic Press, New York, 1969.

[22] J. Dieudonné *et al.*, *Abrégé d'histoire des mathématiques 1700-1900*, reprinted, Hermann, Paris, 1996.

[23] J. Dugundji and A. Granas, *Fixed point theory* I, PWN, Warsaw, 1982.

[24] H.-D. Ebbinghaus *et al.*, *Numbers*, English transl. of 2nd German ed. by H.L.S. Orde, Springer-Verlag, New York, 1990.

[25] L. Euler, *Introduction to analysis of the infinite, Book* I, English transl. by J.D. Blanton, Springer-Verlag, New York, 1988.

[26] W. Fleming, *Functions of several variables*, 2nd ed., Springer-Verlag, New York, 1977.

[27] L.R. Ford Jr. and D.R. Fulkerson, *Flows in networks*, Princeton University Press, Princeton, N.J., 1962.

[28] F.R. Gantmacher, *The theory of matrices*, English transl. by K.A. Hirsch, 2 vols., Chelsea, New York, 1959.

[29] M. Hall, *The theory of groups*, reprinted, Chelsea, New York, 1976.

[30] P.R. Halmos, *Lectures on Boolean algebras*, Van Nostrand, Princeton, N.J., 1963.

[31] P.R. Halmos, *Finite-dimensional vector spaces*, 2nd ed., reprinted, Springer-Verlag, New York, 1974.

[32] G.G. Hamedani and G.G. Walter, A fixed point theorem and its application to the central limit theorem, *Arch. Math.* **43** (1984), 258-264.

[33] E. Hellinger and O. Toeplitz, *Integralgleichungen und Gleichungen mit unendlichvielen Unbekannten*, reprinted, Chelsea, New York, 1953. [Original edition, 1928]

[34] L. Henkin, On mathematical induction, *Amer. Math. Monthly* **67** (1960), 323-338.

[35] I.N. Herstein, *Topics in algebra*, reprinted, Wiley, London, 1976.

[36] I.N Herstein, *Noncommutative rings*, reprinted, Mathematical Association of America, Washington, D.C., 1994.

[37] R.A. Horn and C.A. Johnson, *Matrix analysis*, corrected reprint, Cambridge University Press, 1990.

[38] J.E. Humphreys, *Reflection groups and Coxeter groups*, Cambridge University Press, 1990.

[39] E.V. Huntingdon, Boolean algebra: A correction, *Trans. Amer. Math. Soc.* **35** (1933), 557-558.

[40] J. Jachymski, A short proof of the converse to the contraction principle and some related results, *Topol. Methods Nonlinear Anal.* **15** (2000), 179-186.

[41] N. Jacobson, *Basic Algebra I,II*, 2nd ed., Freeman, New York, 1985/1989.

[42] F. Kasch, *Modules and rings*, English transl. by D.A.R. Wallace, Academic Press, London, 1982.

[43] B.M. Kiernan, The development of Galois theory from Lagrange to Artin, *Arch. Hist. Exact Sci.* **8** (1971), 40-154.

[44] W. Kulpa, The Poincaré–Miranda theorem, *Amer. Math. Monthly* **104** (1997), 545-550.

[45] E. Kunz, *Introduction to commutative algebra and algebraic geometry*, English transl. by M. Ackerman, Birkhäuser, Boston, Mass., 1985.

[46] T.Y. Lam, *A first course in noncommutative rings*, Springer-Verlag, New York, 1991.

[47] E. Landau, *Foundations of analysis*, English transl. by F. Steinhardt, 3rd ed., Chelsea, New York, 1966. [German original, 1930]

[48] S. Lang, *Algebra*, corrected reprint of 3rd ed., Addison-Wesley, Reading, Mass., 1994.

[49] W. Maak, *Fastperiodische Funktionen*, Springer-Verlag, Berlin, 1950.

[50] A.I. Mal'cev, *Foundations of linear algebra*, English transl. by T.C. Brown, Freeman, San Francisco, 1963.

[51] B.H. Matzat, Über das Umkehrproblem der Galoisschen Theorie, *Jahresber. Deutsch. Math.-Verein.* **90** (1988), 155-183.

[52] K. Menninger, *Number words and number symbols*, English transl. by P. Broneer, M.I.T. Press, Cambridge, Mass., 1969.

[53] L. Mirsky, *Transversal theory*, Academic Press, London, 1971.

[54] P. Morandi, *Field and Galois theory*, Springer, New York, 1996.

[55] T. Nagahara and H. Tominaga, Elementary proofs of a theorem of Wedderburn and a theorem of Jacobson, *Abh. Math. Sem. Univ. Hamburg* **41** (1974), 72-74.

[56] R. Narasimhan, *Complex analysis in one variable*, Birkhäuser, Boston, Mass., 1985.

[57] A.H. Read, Higher derivatives of analytic functions from the standpoint of functional analysis, *J. London Math. Soc.* **36** (1961), 345-352.

[58] F. Riesz and B. Sz.-Nagy, *Functional analysis*, English transl. by L.F. Boron of 2nd French ed., F. Ungar, New York, 1955.

[59] H. Rothe, Systeme geometrischer Analyse, *Encyklopädie der Mathematischen Wissenschaften* III 1.2, pp. 1277-1423, Teubner, Leipzig, 1914-1931.

[60] J.J. Rotman, *An introduction to the theory of groups*, 4th ed., Springer-Verlag, New York, 1995.

[61] J. Rotman, *Galois theory*, 2nd ed., Springer-Verlag, New York, 1998.

[62] S. Rudeanu, *Boolean functions and equations*, North-Holland, Amsterdam, 1974.

[63] W. Rudin, *Principles of mathematical analysis*, 3rd ed., McGraw-Hill, New York, 1976.

[64] N.A. Salingaros and G.P. Wene, The Clifford algebra of differential forms, *Acta Appl. Math.* **4** (1985), 271-292.

[65] D.B. Shapiro, Products of sums of squares, *Exposition. Math.* **2** (1984), 235-261.

[66] R. Sikorski, *Boolean algebras*, 3rd ed., Springer-Verlag, New York, 1969.

[67] T.A. Springer and F.D. Veldkamp, *Octonions, Jordan algebras, and exceptional groups*, Springer, Berlin, 2000.

[68] E. Steinitz, *Algebraische Theorie der Körper*, reprinted, Chelsea, New York, 1950.

[69] M.H. Stone, The representation of Boolean algebras, *Bull. Amer. Math. Soc.* **44** (1938), 807-816.

[70] K.D. Stroyan and W.A.J. Luxemburg, *Introduction to the theory of infinitesimals*, Academic Press, New York, 1976.

[71] A. Sudbery, Quaternionic analysis, *Math. Proc. Cambridge Philos. Soc.* **85** (1979), 199-225.

[72] C.T.C. Wall, *A geometric introduction to topology*, reprinted, Dover, New York, 1993.

[73] A. Zygmund, *Trigonometric series*, 3rd ed., Cambridge University Press, 2003.

II
Divisibility

1 Greatest common divisors

In the set $\mathbb{N}$ of all positive integers we can perform two basic operations: addition and multiplication. In this chapter we will be primarily concerned with the second operation.

Multiplication has the following properties:

(M1) *if $ab = ac$, then $b = c$;* (cancellation law)
(M2) *$ab = ba$ for all a,b;* (commutative law)
(M3) *$(ab)c = a(bc)$ for all a,b,c;* (associative law)
(M4) *$1a = a$ for all a.* (identity element)

For any $a,b \in \mathbb{N}$ we say that b *divides* a, or that b is a *factor* of a, or that a is a *multiple* of b if $a = ba'$ for some $a' \in \mathbb{N}$. We write $b|a$ if b divides a and $b \nmid a$ if b does not divide a. For example, $2|6$, since $6 = 2{\times}3$, but $4 \nmid 6$. (We sometimes use $\times$ instead of $\cdot$ for the product of positive integers.) The following properties of divisibility follow at once from the definition:

(i)   *$a|a$ and $1|a$ for every a;*
(ii)   *if $b|a$ and $c|b$, then $c|a$;*
(iii) *if $b|a$, then $b|ac$ for every c;*
(iv)   *$bc|ac$ if and only if $b|a$;*
(v)   *if $b|a$ and $a|b$, then $b = a$.*

For any $a,b \in \mathbb{N}$ we say that d is a *common divisor* of a and b if $d|a$ and $d|b$. We say that a common divisor d of a and b is a *greatest common divisor* if every common divisor of a and b divides d. The greatest common divisor of a and b is uniquely determined, if it exists, and will be denoted by (a,b).

The greatest common divisor of a and b is indeed the *numerically* greatest common divisor. However, it is preferable not to define greatest common divisors in this way, since the concept is then available for algebraic structures in which there is no relation of magnitude and only the operation of multiplication is defined.

PROPOSITION 1 *Any $a,b \in \mathbb{N}$ have a greatest common divisor (a,b).*

Proof   Without loss of generality we may suppose $a \geq b$. If b divides a, then $(a,b) = b$. Assume that there exists a pair a,b without greatest common divisor and choose one for which a is a minimum. Then $1 < b < a$, since b does not divide a. Since also $1 \leq a - b < a$, the pair $a - b, b$ has a greatest common divisor d. Since any common divisor of a and b divides $a - b$, and since d divides $(a - b) + b = a$, it follows that d is a greatest common divisor of a and b. But this is a contradiction. $\square$

The proof of Proposition 1 uses not only the multiplicative structure of the set $\mathbb{N}$, but also its ordering and additive structure. To see that there is a reason for this, consider the set S of all positive integers of the form $4k + 1$. The set S is closed under multiplication, since

$$(4j + 1)(4k + 1) = 4(4jk + j + k) + 1,$$

and we can define divisibility and greatest common divisors in S by simply replacing $\mathbb{N}$ by S in our previous definitions. However, although the elements 693 and 189 of S have the common divisors 9 and 21, they have no greatest common divisor according to this definition.

In the following discussion we use the result of Proposition 1, but make no further appeal to either addition or order.

For any $a,b \in \mathbb{N}$ we say that h is a *common multiple* of a and b if $a|h$ and $b|h$. We say that a common multiple h of a and b is a *least common multiple* if h divides every common multiple of a and b. The least common multiple of a and b is uniquely determined, if it exists, and will be denoted by $[a,b]$.

It is evident that, for every a,

$$(a,1) = 1, \quad [a,1] = a,$$
$$(a,a) = a = [a,a].$$

PROPOSITION 2 *Any $a,b \in \mathbb{N}$ have a least common multiple $[a,b]$. Moreover,*

$$(a,b)\,[a,b] = ab.$$

Furthermore, for all $a,b,c \in \mathbb{N}$,

$$(ac,bc) = (a,b)c, \quad [ac,bc] = [a,b]c,$$

$$([a,b],[a,c]) = [a,(b,c)], \quad [(a,b),(a,c)] = (a,[b,c]).$$

Proof We show first that $(ac,bc) = (a,b)c$. Put $d = (a,b)$. Clearly cd is a common divisor of ac and bc, and so $(ac,bc) = qcd$ for some $q \in \mathbb{N}$. Thus $ac = qcda'$, $bc = qcdb'$ for some $a',b' \in \mathbb{N}$. It follows that $a = qda'$, $b = qdb'$. Thus qd is a common divisor of a and b. Hence qd divides d, which implies $q = 1$.

If g is any common multiple of a and b, then ab divides ga and gb, and hence ab also divides (ga,gb). But, by what we have just proved,

$$(ga,gb) = (a,b)g = dg.$$

Hence $h: = ab/d$ divides g. Since h is clearly a common multiple of a and b, it follows that $h = [a,b]$. Replacing a,b by ac,bc, we now obtain

$$[ac,bc] = acbc/(ac,bc) = abc/(a,b) = hc.$$

If we put

$$A = ([a,b],[a,c]), \quad B = [a,(b,c)],$$

then by what we have already proved,

$$A = (ab/(a,b),\ ac/(a,c)),$$
$$B = a(b,c)/(a,(b,c)) = (ab/(a,(b,c)),\ ac/(a,(b,c))).$$

Since any common divisor of $ab/(a,b)$ and $ac/(a,c)$ is also a common divisor of $ab/(a,(b,c))$ and $ac/(a,(b,c))$, it follows that A divides B. On the other hand, a divides A, since a divides $[a,b]$ and $[a,c]$, and similarly (b,c) divides A. Hence B divides A. Thus $B = A$.

The remaining statement of the proposition is proved in the same way, with greatest common divisors and least common multiples interchanged. $\square$

The last two statements of Proposition 2 are referred to as the distributive laws, since if the greatest common divisor and least common multiple of a and b are denoted by $a \wedge b$ and $a \vee b$ respectively, they take the form

$$(a \vee b) \wedge (a \vee c) = a \vee (b \wedge c), \quad (a \wedge b) \vee (a \wedge c) = a \wedge (b \vee c).$$

Properties (i), (ii) and (v) at the beginning of the section say that divisibility is a *partial ordering* of the set $\mathbb{N}$ with 1 as least element. The existence of greatest common divisors and least common multiples says that $\mathbb{N}$ is a *lattice* with respect to this partial ordering. The distributive laws say that $\mathbb{N}$ is actually a *distributive* lattice.

We say that $a,b \in \mathbb{N}$ are *relatively prime*, or *coprime*, if $(a,b) = 1$. Divisibility properties in this case are much simpler:

PROPOSITION 3 *For any $a,b,c \in \mathbb{N}$ with $(a,b) = 1$,*

(i)   *if $a|c$ and $b|c$, then $ab|c$;*
(ii)   *if $a|bc$, then $a|c$;*
(iii) *$(a,bc) = (a,c)$;*
(iv)   *if also $(a,c) = 1$, then $(a,bc) = 1$;*
(v)   *$(a^m,b^n) = 1$ for all $m,n \geq 1$.*

Proof To prove (i), note that $[a,b]$ divides c and $[a,b] = ab$. To prove (ii), note that a divides $(ac,bc) = (a,b)c = c$. To prove (iii), note that any common divisor of a and bc divides c, by (ii). Obviously (iii) implies (iv), and (v) follows by induction. $\square$

PROPOSITION 4 *If $a,b \in \mathbb{N}$ and $(a,b) = 1$, then any divisor of ab can be uniquely expressed in the form de, where $d|a$ and $e|b$. Conversely, any product of this form is a divisor of ab.*

Proof The proof is based on Proposition 3. Suppose c divides ab and put $d = (a,c)$, $e = (b,c)$. Then $(d,e) = 1$ and hence de divides c. If $a = da'$ and $c = dc'$, then $(a',c') = 1$ and $e|c'$. On the other hand, $c'|a'b$ and hence $c'|b$. Since $e = (b,c)$, it follows that $c' = e$ and $c = de$.

Suppose $de = d'e'$, where d,d' divide a and e,e' divide b. Then $d|d'$, since $(d,e') = 1$, and similarly $d'|d$, since $(d',e) = 1$. Hence $d' = d$ and $e' = e$.

The final statement of the proposition is obvious. $\square$

It follows from Proposition 4 that if $c^n = ab$, where $(a,b) = 1$, then $a = d^n$ and $b = e^n$ for some $d,e \in \mathbb{N}$.

The greatest common divisor and least common multiple of any finite set of elements of $\mathbb{N}$ may be defined in the same way as for sets of two elements. By induction we easily obtain:

PROPOSITION 5 *Any $a_1,\ldots,a_n \in \mathbb{N}$ have a greatest common divisor $(a_1,\ldots,a_n)$ and a least common multiple $[a_1,\ldots,a_n]$. Moreover,*

(i)   *$(a_1,a_2,\ldots,a_n) = (a_1,(a_2,\ldots,a_n))$, $[a_1,a_2,\ldots,a_n] = [a_1,[a_2,\ldots,a_n]]$;*
(ii)   *$(a_1c,\ldots,a_nc) = (a_1,\ldots,a_n)c$, $[a_1c,\ldots,a_nc] = [a_1,\ldots,a_n]c$;*
(iii) *$(a_1,\ldots,a_n) = a/[a/a_1,\ldots,a/a_n]$, $[a_1,\ldots,a_n] = a/(a/a_1,\ldots,a/a_n)$, where $a = a_1\cdots a_n$.* $\square$

We can use the distributive laws to show that

$$([a,b],[a,c],[b,c]) = [(a,b),(a,c),(b,c)].$$

In fact the left side is equal to $\{a \vee (b \wedge c)\} \wedge (b \vee c)$, whereas the right side is equal to

$$(\check{a} \wedge c) \vee \{a \wedge (b \vee c)\} = \{(b \wedge c) \vee a\} \wedge \{(b \wedge c) \vee (b \vee c)\}$$
$$= \{a \vee (b \wedge c)\} \wedge (b \vee c).$$

If

$$a = (a_1,...,a_m), \quad b = (b_1,...,b_n),$$

then ab is the greatest common divisor of all products $a_j b_k$, since $(a_j b_1,...,a_j b_n) = a_j b$ and $(a_1 b,...,a_m b) = ab$.

Similarly, if

$$a = [a_1,...,a_m], \quad b = [b_1,...,b_n],$$

then ab is the least common multiple of all products $a_j b_k$.

It is easily shown by induction that if $(a_i,a_j) = 1$ for $1 \le i < j \le m$, then

$$(a_1 \cdots a_m, c) = (a_1,c) \cdots (a_m,c), \quad [a_1 \cdots a_m, c] = [a_1,...,a_m,c].$$

PROPOSITION 6 *If $a \in \mathbb{N}$ has two factorizations*

$$a = b_1 \cdots b_m = c_1 \cdots c_n,$$

then these factorizations have a common refinement, i.e. there exist $d_{jk} \in \mathbb{N}$ ($1 \le j \le m$, $1 \le k \le n$) such that

$$b_j = \prod_{k=1}^{n} d_{jk}, \quad c_k = \prod_{j=1}^{m} d_{jk}.$$

Proof We show first that if $a = a_1 \cdots a_n$ and $d|a$, then $d = d_1 \cdots d_n$, where $d_i|a_i$ ($1 \le i \le n$). We may suppose that $n > 1$ and that the assertion holds for products of less than n elements of $\mathbb{N}$. Put $a' = a_1 \cdots a_{n-1}$ and $d' = (a',d)$. Then $d' = d_1 \cdots d_{n-1}$, where $d_i|a_i$ ($1 \le i < n$). Moreover $a'' = a'/d'$ and $d'' = d/d'$ are coprime. Since $d'' = d/d'$ divides $a'' a_n = a/d'$, the greatest common divisor $a_n = (a_n a'', a_n d'')$ is divisible by d''. Thus we can take $d_n = d''$.

We return now to the proposition. Since $c_1 | \prod_j b_j$, we can write $c_1 = \prod_j d_{j1}$, where $d_{j1}|b_j$. Put $b_j' = b_j/d_{j1}$. Then

$$\prod_j b_j' = a/c_1 = c_2 \cdots c_n.$$

Hence we can write $c_2 = \prod_j d_{j2}$, where $d_{j2}|b_j'$. Proceeding in this way, we obtain factorizations $c_k = \prod_j d_{jk}$ such that $\prod_k d_{jk}$ divides b_j. In fact, since

$$\prod_{j,k} d_{jk} = a = \prod_j b_j,$$

we must have $b_j = \prod_k d_{jk}$. $\square$

Instead of defining divisibility and greatest common divisors in the set $\mathbb{N}$ of all positive integers, we can define them in the set $\mathbb{Z}$ of all integers by simply replacing $\mathbb{N}$ by $\mathbb{Z}$ in the previous definitions. The properties (i)-(v) at the beginning of this section continue to hold, provided that in (iv) we require $c \neq 0$ and in (v) we alter the conclusion to $b = \pm a$. We now list some additional properties:

(i)′ *$a|0$ for every a;*
(ii)′ *if $0|a$, then $a = 0$;*
(iii)′ *if $c|a$ and $c|b$, then $c|ax + by$ for all x,y.*

Greatest common divisors and least common multiples still exist, but uniqueness holds only up to sign. With this understanding, Propositions 2–4 continue to hold, and so also do Propositions 5 and 6 if we require $a \neq 0$. It is evident that, for every a,

$$(a,0) = a, \quad [a,0] = 0.$$

More generally, we can define divisibility in any *integral domain*, i.e. a commutative ring in which $a \neq 0$ and $b \neq 0$ together imply $ab \neq 0$. The properties (i)-(v) at the beginning of the section continue to hold, provided that in (iv) we require $c \neq 0$ and in (v) we alter the conclusion to $b = ua$, where u is a *unit*, i.e. $u|1$. The properties (i)′-(iii)′ above also remain valid.

We define a *GCD domain* to be an integral domain in which any pair of elements has a greatest common divisor. This implies that any pair of elements also has a least common multiple. Uniqueness now holds only up to unit multiples. With this understanding Propositions 2–6 continue to hold in any GCD domain in the same way as for $\mathbb{Z}$.

An important example, which we will consider in Section 3, of a GCD domain other than $\mathbb{Z}$ is the *polynomial ring $K[t]$*, consisting of all polynomials in t with coefficients from an arbitrary field K. The units in this case are the nonzero elements of K.

Another example, which we will meet in §4 of Chapter VI, is the valuation ring R of a non-archimedean valued field. In this case, for any $a,b \in R$, either $a|b$ or $b|a$ and so (a,b) is either a or b.

In the same way that the ring $\mathbb{Z}$ of integers may be embedded in the field $\mathbb{Q}$ of rational numbers, any integral domain R may be embedded in a field K, its *field of fractions*, so that any nonzero $c \in K$ has the form $c = ab^{-1}$, where $a,b \in R$ and $b \neq 0$. If R is a GCD domain we can further require $(a,b) = 1$, and a,b are then uniquely determined apart from a common unit multiple. The field of fractions of the polynomial ring $K[t]$ is the field $K(t)$ of *rational functions*.

In our discussion of divisibility so far we have avoided all mention of prime numbers. A positive integer $a \neq 1$ is said to be *prime* if 1 and a are its only positive divisors, and otherwise is said to be *composite*.

For example, 2,3 and 5 are primes, but $4 = 2{\times}2$ and $6 = 2{\times}3$ are composite. The significance of the primes is that, as far as multiplication is concerned, they are the 'atoms' and the composite integers are the 'molecules'. This is made precise in the following so-called *fundamental theorem of arithmetic*:

PROPOSITION 7 *If $a \in \mathbb{N}$ and $a \neq 1$, then a can be represented as a product of finitely many primes. Moreover, the representation is unique, except for the order of the factors.*

Proof Assume, on the contrary, that some composite $a_1 \in \mathbb{N}$ is not a product of finitely many primes. Since a_1 is composite, it has a factorization $a_1 = a_2 b_2$, where $a_2, b_2 \in \mathbb{N}$ and $a_2, b_2 \neq 1$. At least one of a_2, b_2 must be composite and not a product of finitely many primes, and we may choose the notation so that a_2 has these properties. The preceding argument can now be repeated with a_2 in place of a_1. Proceeding in this way, we obtain an infinite sequence (a_k) of positive integers such that a_{k+1} divides a_k and $a_{k+1} \neq a_k$ for each $k \geq 1$. But then the sequence (a_k) has no least element, which contradicts Proposition I.3.

Suppose now that

$$a = p_1 \cdots p_m = q_1 \cdots q_n$$

are two representations of a as a product of primes. Then, by Proposition 6, there exist $d_{jk} \in \mathbb{N}$ $(1 \leq j \leq m, 1 \leq k \leq n)$ such that

$$p_j = \prod_{k=1}^{n} d_{jk}, \quad q_k = \prod_{j=1}^{m} d_{jk}.$$

Since p_1 is a prime, we must have $d_{1k_1} = p_1$ for some $k_1 \in \{1,\ldots,n\}$, and since q_{k_1} is a prime, we must have $q_{k_1} = d_{1k_1} = p_1$. The same argument can now be applied to

$$a' = \prod_{j \neq 1} p_j = \prod_{k \neq k_1} q_k.$$

It follows that $m = n$ and $q_1,\ldots,q_n$ is a permutation of $p_1,\ldots,p_m$. $\square$

It should be noted that factorization into primes would not be unique if we admitted 1 as a prime. The fundamental theorem of arithmetic may be reformulated in the following way: any $a \in \mathbb{N}$ can be uniquely represented in the form

$$a = \prod_p p^{\alpha_p},$$

where p runs through the primes and the α_p are non-negative integers, only finitely many of which are nonzero. It is easily seen that if $b \in \mathbb{N}$ has the analogous representation

$$b = \prod_p p^{\beta_p},$$

then $b|a$ if and only if $\beta_p \le \alpha_p$ for all p. It follows that the greatest common divisor and least common multiple of a and b have the representations

$$(a,b) = \prod_p p^{\gamma_p}, \quad [a,b] = \prod_p p^{\delta_p},$$

where

$$\gamma_p = \min\{\alpha_p,\beta_p\}, \quad \delta_p = \max\{\alpha_p,\beta_p\}.$$

The fundamental theorem of arithmetic extends at once from $\mathbb{N}$ to $\mathbb{Q}$: any nonzero rational number a can be uniquely represented in the form

$$a = u\prod_p p^{\alpha_p},$$

where $u = \pm 1$ is a unit, p runs through the primes and the α_p are integers (not necessarily non-negative), only finitely many of which are nonzero.

The following property of primes was already established in Euclid's *Elements* (Book VII, Proposition 30):

PROPOSITION 8 *If p is a prime and $p|bc$, then $p|b$ or $p|c$.*

Proof If p does not divide b, we must have $(p,b) = 1$. But then p divides c, by Proposition 3(ii). $\square$

The property in Proposition 8 actually characterizes primes. For if a is composite, then $a = bc$, where $b,c \ne 1$. Thus $a|bc$, but $a \nmid b$ and $a \nmid c$.

We consider finally the extension of these notions to an arbitrary integral domain R. For any nonzero $a,b \in R$, we say that a divisor b of a is a *proper divisor* if a does not divide b (i.e., if a and b do not differ only by a unit factor). We say that $p \in R$ is *irreducible* if p is neither zero nor a unit and if every proper divisor of p is a unit. We say that $p \in R$ is *prime* if p is neither zero nor a unit and if $p|bc$ implies $p|b$ or $p|c$.

By what we have just said, the notions of 'prime' and 'irreducible' coincide if $R = \mathbb{Z}$, and the same argument applies if R is any GCD domain. However, in an arbitrary integral domain R, although any prime element is irreducible, an irreducible element need not be prime. (For example, in the integral domain R consisting of all complex numbers of the form $a + b\sqrt{-5}$, where $a,b \in \mathbb{Z}$, $6 = 2{\times}3 = (1 + \sqrt{-5})(1 - \sqrt{-5})$ has two essentially distinct factorizations into irreducibles, and thus none of these irreducibles is prime.)

The proof of Proposition 7 shows that, in an arbitrary integral domain R, every element which is neither zero nor a unit can be represented as a product of finitely many irreducible elements if and only if the following *chain condition* is satisfied:

(#) *there exists no infinite sequence (a_n) of elements of R such that a_{n+1} is a proper divisor of a_n for every n.*

Furthermore, the representation is *essentially unique* (i.e. unique except for the order of the factors and for multiplying them by units) if and only if R is also a GCD domain.

An integral domain R is said to be *factorial* (or a 'unique factorization domain') if the 'fundamental theorem of arithmetic' holds in R, i.e. if every element which is neither zero nor a unit has such an essentially unique representation as a product of finitely many irreducibles. By the above remarks, an integral domain R is factorial if and only if it is a GCD domain satisfying the chain condition (#).

For future use, we define an element of a factorial domain to be *square-free* if it is neither zero nor a unit and if, in its representation as a product of irreducibles, no factor is repeated. In particular, a positive integer is square-free if and only if it is a nonempty product of distinct primes.

2 The Bézout identity

If a,b are arbitrary integers with $a \neq 0$, then there exist unique integers q,r such that

$$b = qa + r, \ \ 0 \leq r < |a|.$$

In fact qa is the greatest multiple of a which does not exceed b. The integers q and r are called the *quotient* and *remainder* in the 'division' of b by a.

(For $a > 0$ this was proved in Proposition I.14. It follows that if a and n are positive integers, any positive integer b less than a^n has a unique representation 'to the base a':

$$b = b_0 + b_1 a + \ldots + b_{n-1} a^{n-1},$$

where $0 \leq b_j < a$ for all j. In fact b_{n-1} is the quotient in the division of b by a^{n-1}, b_{n-2} is the quotient in the division of the remainder by a^{n-2}, and so on.)

If a,b are arbitrary integers with $a \neq 0$, then there exist also integers q,r such that

$$b = qa + r, \quad |r| \leq |a|/2.$$

In fact qa is the nearest multiple of a to b. Thus q and r are not uniquely determined if b is midway between two consecutive multiples of a.

Both these *division algorithms* have their uses. We will be impartial and merely use the fact that

$$b = qa + r, \quad |r| < |a|.$$

An *ideal* in the commutative ring $\mathbb{Z}$ of all integers is defined to be a nonempty subset J such that if $a,b \in J$ and $x,y \in \mathbb{Z}$, then also $ax + by \in J$.

For example, if $a_1,...,a_n$ are given elements of $\mathbb{Z}$, then the set of all linear combinations $a_1 x_1 + ... + a_n x_n$ with $x_1,...,x_n \in \mathbb{Z}$ is an ideal, the ideal *generated* by $a_1,...,a_n$. An ideal generated by a single element, i.e. the set of all multiples of that element, is said to be a *principal ideal*.

LEMMA 9 *Any ideal J in the ring $\mathbb{Z}$ is a principal ideal.*

Proof If 0 is the only element of J, then 0 generates J. Otherwise there is a nonzero $a \in J$ with minimum absolute value. For any $b \in J$, we can write $b = qa + r$, for some $q,r \in \mathbb{Z}$ with $|r| < |a|$. By the definition of an ideal, $r \in J$ and so, by the definition of a, $r = 0$. Thus a generates J. $\square$

PROPOSITION 10 *Any $a,b \in \mathbb{Z}$ have a greatest common divisor $d = (a,b)$. Moreover, for any $c \in \mathbb{Z}$, there exist $x,y \in \mathbb{Z}$ such that*

$$ax + by = c$$

if and only if d divides c.

Proof Let J be the ideal generated by a and b. By Lemma 9, J is generated by a single element d. Since $a,b \in J$, d is a common divisor of a and b. On the other hand, since $d \in J$, there exist $u,v \in \mathbb{Z}$ such that $d = au + bv$. Hence any common divisor of a and b also divides d. Thus $d = (a,b)$. The final statement of the proposition follows immediately since, by definition, $c \in J$ if and only if there exist $x,y \in \mathbb{Z}$ such that $ax + by = c$. $\square$

It is readily shown that if the 'linear Diophantine' equation $ax + by = c$ has a solution $x_0, y_0 \in \mathbb{Z}$, then all solutions $x,y \in \mathbb{Z}$ are given by the formula

$$x = x_0 + kb/d, \quad y = y_0 - ka/d,$$

where $d = (a,b)$ and k is an arbitrary integer.

Proposition 10 provides a new proof for the existence of greatest common divisors and, in addition, it shows that the greatest common divisor of two integers can be represented as a linear combination of them. This representation is usually referred to as the *Bézout identity*, although it was already known to Bachet (1624) and even earlier to the Hindu mathematicians Aryabhata (499) and Brahmagupta (628).

In exactly the same way that we proved Proposition 10 – or, alternatively, by induction from Proposition 10 – we can prove

PROPOSITION 11 *Any finite set $a_1,...,a_n$ of elements of $\mathbb{Z}$ has a greatest common divisor $d = (a_1,...,a_n)$. Moreover, for any $c \in \mathbb{Z}$, there exist $x_1,...,x_n \in \mathbb{Z}$ such that*

$$a_1 x_1 + ... + a_n x_n = c$$

if and only if d divides c. $\square$

The proof which we gave for Proposition 10 is a pure existence proof – it does not help us to find the greatest common divisor. The following constructive proof was already given in Euclid's *Elements* (Book VII, Proposition 2). Let a,b be arbitrary integers. Since $(0,b) = b$, we may assume $a \neq 0$. Then there exist integers q,r such that

$$b = qa + r, \quad |r| < |a|.$$

Put $a_0 = b$, $a_1 = a$ and repeatedly apply this procedure:

$$a_0 = q_1 a_1 + a_2, \quad |a_2| < |a_1|,$$
$$a_1 = q_2 a_2 + a_3, \quad |a_3| < |a_2|,$$
$$\cdots\cdots$$
$$a_{N-2} = q_{N-1} a_{N-1} + a_N, \quad |a_N| < |a_{N-1}|,$$
$$a_{N-1} = q_N a_N.$$

The process must eventually terminate as shown, because otherwise we would obtain an infinite sequence of positive integers with no least element. We claim that a_N is a greatest common divisor of a and b. In fact, working forwards from the first equation we see that any common divisor c of a and b divides each a_k and so, in particular, a_N. On the other hand, working backwards from the last equation we see that a_N divides each a_k and so, in particular, a and b.

The Bézout identity can also be obtained in this way, although Euclid himself lacked the necessary algebraic notation. Define sequences (x_k), (y_k) by the recurrence relations

$$x_{k+1} = x_{k-1} - q_k x_k, \quad y_{k+1} = y_{k-1} - q_k y_k \quad (1 \le k < N),$$

with the starting values

$$x_0 = 0, \ x_1 = 1, \ \text{resp.} \ y_0 = 1, \ y_1 = 0.$$

It is easily shown by induction that $a_k = ax_k + by_k$ and so, in particular, $a_N = ax_N + by_N$.

The Euclidean algorithm is quite practical. For example, the reader may use it to verify that 13 is the greatest common divisor of 2171 and 5317, and that

$$49 \times 5317 - 120 \times 2171 \ = \ 13.$$

However, the first proof given for Proposition 10 also has its uses: there is some advantage in separating the conceptual from the computational and the proof actually rests on more general principles, since there are quadratic number fields whose ring of integers is a 'principal ideal domain' that does not possess any Euclidean algorithm.

It is not visibly obvious that the binomial coefficients

$$^{m+n}C_n \ = \ (m + 1)...(m + n)/1{\cdot}2{\cdot}...{\cdot}n$$

are integers for all positive integers m,n, although it is apparent from their combinatorial interpretation. However, the property is readily proved by induction, using the relation

$$^{m+n}C_n \ = \ ^{m+n-1}C_n + \ ^{m+n-1}C_{n-1}.$$

Binomial coefficients have other arithmetic properties. Hermite observed that $^{m+n}C_n$ is divisible by the integers $(m + n)/(m,n)$ and $(m + 1)/(m + 1,n)$. In particular, the *Catalan numbers* $(n + 1)^{-1} \, ^{2n}C_n$ are integers. The following proposition is a substantial generalization of these results and illustrates the application of Proposition 10.

PROPOSITION 12 *Let (a_n) be a sequence of nonzero integers such that, for all $m,n \geq 1$, every common divisor of a_m and a_n divides a_{m+n}, and every common divisor of a_m and a_{m+n} divides a_n. Then, for all $m,n \geq 1$,*

(i) $(a_m,a_n) = a_{(m,n)}$;

(ii) $A_{m,n} := a_{m+1}{\cdots}a_{m+n}/a_1{\cdots}a_n \in \mathbb{Z}$;

(iii) $A_{m,n}$ *is divisible by* $a_{m+n}/(a_m,a_n)$, *by* $a_{m+1}/(a_{m+1},a_n)$ *and by* $a_{n+1}/(a_m,a_{n+1})$;

(iv) $(A_{m,n-1}, A_{m+1,n}, A_{m-1,n+1}) = (A_{m-1,n}, A_{m+1,n-1}, A_{m,n+1})$.

Proof The hypotheses imply that

$$(a_m,a_n) = (a_m,a_{m+n}) \ \text{for all} \ m,n \geq 1.$$

Since $a_m = (a_m, a_m)$, it follows by induction that $a_m | a_{km}$ for all $k \geq 1$. Moreover,

$$(a_{km}, a_{(k+1)m}) = a_m,$$

since every common divisor of a_{km} and $a_{(k+1)m}$ divides a_m.

Put $d = (m, n)$. Then $m = dm'$, $n = dn'$, where $(m', n') = 1$. Thus there exist integers u, v such that $m'u - n'v = 1$. By replacing u, v by $u + tn', v + tm'$ with any $t > \max\{|u|, |v|\}$, we may assume that u and v are both positive. Then

$$(a_{mu}, a_{nv}) = (a_{(n'v+1)d}, a_{n'vd}) = a_d.$$

Since a_d divides (a_m, a_n) and (a_m, a_n) divides (a_{mu}, a_{nv}), this implies $(a_m, a_n) = a_d$. This proves (i).

Since $a_1 | a_{m+1}$, it is evident that $A_{m,1} \in \mathbb{Z}$ for all $m \geq 1$. We assume that $n > 1$ and $A_{m,n} \in \mathbb{Z}$ for all smaller values of n and all $m \geq 1$. Since it is trivial that $A_{0,n} \in \mathbb{Z}$, we assume also that $m \geq 1$ and $A_{m,n} \in \mathbb{Z}$ for all smaller values of m. By Proposition 10, there exist $x, y \in \mathbb{Z}$ such that

$$a_m x + a_n y = a_{m+n},$$

since (a_m, a_n) divides a_{m+n}. Since

$$A_{m,n} = \frac{a_{m+1} \cdots a_{m+n}}{a_1 \cdots a_n} = \frac{a_m a_{m+1} \cdots a_{m+n-1}}{a_1 \cdots a_n} x + \frac{a_{m+1} \cdots a_{m+n-1}}{a_1 \cdots a_{n-1}} y,$$

our induction hypotheses imply that $A_{m,n} \in \mathbb{Z}$. This proves (ii).

Since

$$a_{m+n} A_{m,n-1} = a_n A_{m,n},$$

a_{m+n} divides $(a_n, a_{m+n}) A_{m,n}$ and, since $(a_n, a_{m+n}) = (a_m, a_n)$, this implies that $a_{m+n}/(a_m, a_n)$ divides $A_{m,n}$.

Similarly, since

$$a_{m+1} A_{m+1,n} = a_{m+n+1} A_{m,n}, \quad a_{m+1} A_{m+1,n-1} = a_n A_{m,n},$$

a_{m+1} divides $(a_n, a_{m+n+1}) A_{m,n}$ and, since $(a_n, a_{m+n+1}) = (a_{m+1}, a_n)$, it follows that $a_{m+1}/(a_{m+1}, a_n)$ divides $A_{m,n}$. In the same way, since

$$a_{n+1} A_{m,n+1} = a_{m+n+1} A_{m,n}, \quad a_{n+1} A_{m-1,n+1} = a_m A_{m,n},$$

a_{n+1} divides $(a_m, a_{m+n+1}) A_{m,n}$ and hence $a_{n+1}/(a_m, a_{n+1})$ divides $A_{m,n}$. This proves (iii).

By multiplying by $a_1 \cdots a_{n+1}/a_{m+2} \cdots a_{m+n-1}$, we see that (iv) is equivalent to

$$(a_n a_{n+1} a_{m+1}, \; a_{n+1} a_{m+n} a_{m+n+1}, \; a_m a_{m+1} a_{m+n}) = (a_{n+1} a_m a_{m+1}, \; a_n a_{n+1} a_{m+n}, \; a_{m+1} a_{m+n} a_{m+n+1}).$$

Since here the two sides are interchanged when m and n are interchanged, it is sufficient to show that any common divisor e of the three terms on the right is also a common divisor of the three terms on the left. We have

$$(a_{n+1} a_m a_{m+1}, \; a_n a_{n+1} a_{m+1}) = a_{n+1} a_{m+1}(a_m, a_n) = a_{n+1} a_{m+1}(a_m, a_{m+n})$$
$$= (a_{n+1} a_m a_{m+1}, \; a_{m+1} a_{n+1} a_{m+n}),$$

and similarly

$$(a_n a_{n+1} a_{m+n}, \; a_{n+1} a_{m+n} a_{m+n+1}) = (a_n a_{n+1} a_{m+n}, \; a_{m+1} a_{n+1} a_{m+n}),$$
$$(a_{m+1} a_{m+n} a_{m+n+1}, \; a_m a_{m+1} a_{m+n}) = (a_{m+1} a_{m+n} a_{m+n+1}, \; a_{m+1} a_{n+1} a_{m+n}).$$

Hence if we put $g = a_{m+1} a_{n+1} a_{m+n}$, then

$$(e,g) = (e, a_n a_{n+1} a_{m+1}) = (e, a_{n+1} a_{m+n} a_{m+n+1}) = (e, a_m a_{m+1} a_{m+n})$$

and if we put $f = (e,g)$, then

$$1 = (e/f, a_n a_{n+1} a_{m+1}/f) = (e/f, a_{n+1} a_{m+n} a_{m+n+1}/f) = (e/f, a_m a_{m+1} a_{m+n}/f).$$

Hence $(e/f, P/f^3) = 1$, where

$$P = a_n a_{n+1} a_{m+1} \cdot a_{n+1} a_{m+n} a_{m+n+1} \cdot a_m a_{m+1} a_{m+n}.$$

But P is divisible by e^3, since we can also write

$$P = a_{n+1} a_m a_{m+1} \cdot a_n a_{n+1} a_{m+n} \cdot a_{m+1} a_{m+n} a_{m+n+1}.$$

Hence the previous relation implies $e/f = 1$. Thus $e = f$ is a common divisor of $a_n a_{n+1} a_{m+1}$, $a_{n+1} a_{m+n} a_{m+n+1}$ and $a_m a_{m+1} a_{m+n}$, as we wished to show. $\square$

For the binomial coefficient case, i.e. $a_n = n$, the property (iv) of Proposition 12 was discovered empirically by Gould (1972) and then proved by Hillman and Hoggatt (1972). It states that if in the *Pascal triangle* one picks out the hexagon surrounding a particular element, then the greatest common divisor of three alternately chosen vertices is equal to the greatest common divisor of the remaining three vertices. Hillman and Hoggatt also gave generalizations along the lines of Proposition 12.

The hypotheses of Proposition 12 are also satisfied if $a_n = q^n - 1$, for some integer $q > 1$, since in this case $a_{m+n} = a_m a_n + a_m + a_n$. The corresponding *q-binomial coefficients* were studied by Gauss and, as mentioned in Chapter XIII, they play a role in the theory of partitions.

We may also take (a_n) to be the sequence defined recurrently by

$$a_1 = 1, \quad a_2 = c, \quad a_{n+2} = ca_{n+1} + ba_n \quad (n \geq 1),$$

where b and c are coprime positive integers. Indeed it is easily shown by induction that

$$(a_n, a_{n+1}) = (b, a_{n+1}) = 1 \text{ for all } n \geq 1.$$

By induction on m one may also show that

$$a_{m+n} = a_{m+1}a_n + ba_m a_{n-1} \text{ for all } m \geq 1, n > 1.$$

It follows that the hypotheses of Proposition 12 are satisfied. In particular, for $b = c = 1$, they are satisfied by the sequence of *Fibonacci numbers*.

We consider finally extensions of our results to more general algebraic structures. An integral domain R is said to be a *Bézout domain* if any $a, b \in R$ have a common divisor of the form $au + bv$ for some $u, v \in R$. Since such a common divisor is necessarily a greatest common divisor, any Bézout domain is a GCD domain. It is easily seen, by induction on the number of generators, that an integral domain is a Bézout domain if and only if every finitely generated ideal is a principal ideal. Thus Propositions 10 and 11 continue to hold if $\mathbb{Z}$ is replaced by any Bézout domain.

An integral domain R is said to be a *principal ideal domain* if every ideal is a principal ideal.

LEMMA 13 *An integral domain R is a principal ideal domain if and only if it is a Bézout domain satisfying the chain condition*

(#) *there exists no infinite sequence (a_n) of elements of R such that a_{n+1} is a proper divisor of a_n for every n.*

Proof It is obvious that any principal ideal domain is a Bézout domain. Suppose R is a Bézout domain, but not a principal ideal domain. Then R contains an ideal J which is not finitely generated. Hence there exists a sequence (b_n) of elements of J such that b_{n+1} is not in the ideal J_n generated by $b_1, \ldots, b_n$. But J_n is a principal ideal. If a_n generates J_n, then a_{n+1} is a proper divisor of a_n for every n. Thus the chain condition is violated.

Suppose now that R is a Bézout domain containing a sequence (a_n) such that a_{n+1} is a proper divisor of a_n for every n. Let J denote the set of all elements of R which are divisible by at least one term of this sequence. Then J is an ideal. For if $a_j|b$ and $a_k|c$, where $j \leq k$, then also $a_k|b$ and hence $a_k|bx + cy$ for all $x, y \in R$. If J were generated by a single element a, we would have $a|a_n$ for every n. On the other hand, since $a \in J$, $a_N|a$ for some N. Hence $a_N|a_{N+1}$. Since a_{N+1} is a proper divisor of a_N, this is a contradiction. Thus R is not a principal ideal domain. $\square$

It follows from the remarks at the end of Section 1 that a principal ideal domain is factorial, i.e. any element which is neither zero nor a unit can be represented as a product of finitely many irreducibles and the representation is essentially unique.

In the next section we will show that the ring $K[t]$ of all polynomials in one indeterminate t with coefficients from an arbitrary field K is a principal ideal domain.

It may be shown that the ring of all algebraic integers is a Bézout domain, and likewise the ring of all functions which are holomorphic in a nonempty connected open subset G of the complex plane $\mathbb{C}$. However, neither is a principal ideal domain. In the former case there are no irreducibles, since any algebraic integer a has the factorization $a = \sqrt{a} \cdot \sqrt{a}$. In the latter case $z - \zeta$ is an irreducible for any $\zeta \in G$, but the chain condition is violated. For example, take

$$a_n(z) = f(z)/(z - \zeta_1)...(z - \zeta_n),$$

where $f(z)$ is a non-identically vanishing function which is holomorphic in G and has infinitely many zeros $\zeta_1, \zeta_2, ...$ in G.

3 Polynomials

In this section we study the most important example of a principal ideal domain other than $\mathbb{Z}$, namely the ring $K[t]$ of all polynomials in t with coefficients from an arbitrary field K (e.g., $K = \mathbb{Q}$ or $\mathbb{C}$).

The attitude adopted towards polynomials in algebra is different from that adopted in analysis. In analysis we regard 't' as a variable which can take different values; in algebra we regard 't' simply as a symbol, an 'indeterminate', on which we can perform various algebraic operations. Since the concept of function is so pervasive, the algebraic approach often seems mysterious at first sight and it seems worthwhile taking the time to give a precise meaning to an 'indeterminate'.

Let R be an integral domain (e.g., $R = \mathbb{Z}$ or $\mathbb{Q}$). A *polynomial* with coefficients from R is defined to be a sequence $f = (a_0, a_1, a_2, ...)$ of elements of R in which at most finitely many terms are nonzero. The sum and product of two polynomials

$$f = (a_0, a_1, a_2, \dots), \quad g = (b_0, b_1, b_2, \dots)$$

are defined by

$$f + g = (a_0+b_0, a_1+b_1, a_2+b_2, \dots),$$
$$fg = (a_0b_0, a_0b_1+a_1b_0, a_0b_2+a_1b_1+a_2b_0, \dots).$$

It is easily verified that these are again polynomials and that the set $R[t]$ of all polynomials with coefficients from R is a commutative ring with $O = (0,0,0, \dots)$ as zero element. (By dropping the requirement that at most finitely many terms are nonzero, we obtain the ring $R[[t]]$ of all *formal power series* with coefficients from R.)

We define the *degree* $\partial(f)$ of a polynomial $f = (a_0, a_1, a_2, \dots) \neq O$ to be the greatest integer n for which $a_n \neq 0$ and we put

$$|f| = 2^{\partial(f)}, \quad |O| = 0.$$

It is easily verified that, for all polynomials f, g,

$$|f + g| \leq \max\{|f|, |g|\}, \quad |fg| = |f||g|.$$

Since $|f| \geq 0$, with equality if and only if $f = O$, the last property implies that $R[t]$ *is an integral domain*. Thus we can define divisibility in $R[t]$, as explained in Section 1.

The set of all polynomials of the form $(a_0, 0, 0, \dots)$ is a subdomain isomorphic to R. By identifying this set with R, we may regard R as embedded in $R[t]$. The only units in $R[t]$ are the units in R, since $1 = ef$ implies $1 = |e||f|$ and hence $|e| = 1$.

If we put $t = (0, 1, 0, 0, \dots)$, then

$$t^2 = tt = (0,0,1,0, \dots), \quad t^3 = tt^2 = (0,0,0,1, \dots), \quad \dots \ .$$

Hence if the polynomial $f = (a_0, a_1, a_2, \dots)$ has degree n, it can be uniquely expressed in the form

$$f = a_0 + a_1 t + \dots + a_n t^n \quad (a_n \neq 0).$$

We refer to the elements $a_0, a_1, \dots, a_n$ of R as the *coefficients* of f. In particular, a_0 is the *constant* coefficient and a_n the *highest* coefficient. We say that f is *monic* if its highest coefficient $a_n = 1$.

If also

$$g = b_0 + b_1 t + \dots + b_m t^m \quad (b_m \neq 0),$$

then the sum and product assume their familiar forms:

$$f + g = (a_0 + b_0) + (a_1 + b_1)t + (a_2 + b_2)t^2 + \ldots,$$
$$fg = a_0 b_0 + (a_0 b_1 + a_1 b_0)t + (a_0 b_2 + a_1 b_1 + a_2 b_0)t^2 + \ldots.$$

Suppose now that $R = K$ is a field, and let

$$f = a_0 + a_1 t + \ldots + a_n t^n \quad (a_n \neq 0),$$
$$g = b_0 + b_1 t + \ldots + b_m t^m \quad (b_m \neq 0)$$

be any two nonzero elements of $K[t]$. If $|g| < |f|$, i.e. if $m < n$, then $g = qf + r$, with $q = O$ and $r = g$. Suppose on the other hand that $|f| \le |g|$. Then

$$g = a_n^{-1} b_m t^{m-n} f + g^{\dagger},$$

where $g^{\dagger} \in K[t]$ and $|g^{\dagger}| < |g|$. If $|f| \le |g^{\dagger}|$, the process can be repeated with $g^{\dagger}$ in place of g. Continuing in this way, we obtain $q, r \in K[t]$ such that

$$g = qf + r, \quad |r| < |f|.$$

Moreover, q and r are uniquely determined, since if also

$$g = q_1 f + r_1, \quad |r_1| < |f|,$$

then

$$(q - q_1)f = r_1 - r, \quad |r_1 - r| < |f|,$$

which is only possible if $q = q_1$.

Ideals in $K[t]$ can be defined in the same way as for $\mathbb{Z}$ and the proof of Lemma 9 remains valid. Thus $K[t]$ is a principal ideal domain and, *a fortiori*, a GCD domain.

The Euclidean algorithm can also be applied in $K[t]$ in the same way as for $\mathbb{Z}$ and again, from the sequence of polynomials $f_0, f_1, \ldots, f_N$ which it provides to determine the greatest common divisor f_N of f_0 and f_1 we can obtain polynomials u_k, v_k such that

$$f_k = f_1 u_k + f_0 v_k \quad (0 \le k \le N).$$

We can actually say more for polynomials than for integers, since if

$$f_{k-1} = q_k f_k + f_{k+1}, \quad |f_{k+1}| < |f_k|,$$

then $|f_{k-1}| = |q_k||f_k|$ and hence, by induction,

$$|f_{k-1}||u_k| = |f_0|, \quad |f_{k-1}||v_k| = |f_1| \quad (1 < k \le N).$$

It may be noted in passing that the Euclidean algorithm can also be applied in the ring $K[t,t^{-1}]$ of *Laurent polynomials*. A Laurent polynomial $f \neq O$, with coefficients from the field K, has the form

$$f = a_m t^m + a_{m+1} t^{m+1} + \dots + a_n t^n,$$

where $m,n \in \mathbb{Z}$ with $m \leq n$ and $a_j \in K$ with $a_m a_n \neq 0$. Thus $f = t^m f_0$, where $f_0 \in K[t]$. If we put

$$|f| = 2^{n-m}, \quad |O| = 0,$$

then the division algorithm for ordinary polynomials implies a corresponding algorithm for Laurent polynomials: for any $f,g \in K[t,t^{-1}]$ with $f \neq O$, there exist $q,r \in K[t,t^{-1}]$ such that

$$g = qf + r, \quad |r| < |f|.$$

We return now to ordinary polynomials. The general definition for integral domains in Section 1 means, in the present case, that a polynomial $p \in K[t]$ is *irreducible* if it has positive degree and if every proper divisor has degree zero.

It follows that any polynomial of degree 1 is irreducible. However, there may exist also irreducible polynomials of higher degree. For example, we will show shortly that the polynomial $t^2 - 2$ is irreducible in $\mathbb{Q}[t]$. For $K = \mathbb{C}$, however, every irreducible polynomial has degree 1, by the fundamental theorem of algebra (Theorem I.30) and Proposition 14 below. It follows that, for $K = \mathbb{R}$, every irreducible polynomial has degree 1 or 2.

It is obvious that the chain condition (#) of Section 1 holds in the integral domain $K[t]$, since if g is a proper divisor of f, then $|g| < |f|$. It follows that any polynomial of positive degree can be represented as a product of finitely many irreducible polynomials and that the representation is essentially unique.

We now consider the connection between polynomials in the sense of algebra (polynomial forms) and polynomials in the sense of analysis (polynomial functions). Let K be a field and $f \in K[t]$:

$$f = a_0 + a_1 t + \dots + a_n t^n.$$

If we replace 't' by $c \in K$ we obtain an element of K, which we denote by $f(c)$:

$$f(c) = a_0 + a_1 c + \dots + a_n c^n.$$

A rapid procedure ('Horner's rule') for calculating $f(c)$ is to use the recurrence relations

$$f_0 = a_n, \quad f_j = f_{j-1} c + a_{n-j} \quad (j = 1,\dots,n).$$

It is readily shown by induction that

$$f_j = a_n c^j + a_{n-1} c^{j-1} + \ldots + a_{n-j},$$

and hence $f(c) = f_n$ is obtained with just n multiplications and n additions.

It is easily seen that $f = g + h$ implies $f(c) = g(c) + h(c)$, and $f = gh$ implies $f(c) = g(c)h(c)$. Thus the mapping $f \to f(c)$ is a 'homomorphism' of $K[t]$ into K. A simple consequence is the so-called *remainder theorem*:

PROPOSITION 14 *Let K be a field and $c \in K$. If $f \in K[t]$, then*

$$f = (t - c)g + f(c),$$

for some $g \in K[t]$.

   In particular, f is divisible by $t - c$ if and only if $f(c) = 0$.

Proof We already know that there exist $q, r \in K[t]$ such that

$$f = (t - c)q + r, \quad |r| \le 1.$$

Thus $r \in K$ and the homomorphism properties imply that $f(c) = r$. $\quad\square$

   We say that $c \in K$ is a *root* of the polynomial $f \in K[t]$ if $f(c) = 0$.

PROPOSITION 15 *Let K be a field. If $f \in K[t]$ is a polynomial of degree $n \ge 0$, then f has at most n distinct roots in K.*

Proof If f is of degree 0, then $f = c$ is a nonzero element of K and f has no roots. Suppose now that $n \ge 1$ and the result holds for polynomials of degree less than n. If c is a root of f then, by Proposition 14, $f = (t - c)g$ for some $g \in K[t]$. Since g has degree $n - 1$, it has at most $n - 1$ roots. But every root of f distinct from c is a root of g. Hence f has at most n roots. $\quad\square$

   We consider next properties of the integral domain $R[t]$, when R is an integral domain rather than a field (e.g., $R = \mathbb{Z}$). The famous Pythagorean proof that $\sqrt{2}$ is irrational is considerably generalized by the following result:

PROPOSITION 16 *Let R be a GCD domain and K its field of fractions. Let*

$$f = a_0 + a_1 t + \ldots + a_n t^n$$

be a polynomial of degree $n > 0$ with coefficients $a_j \in R$ $(0 \le j \le n)$. If $c \in K$ is a root of f and $c = ab^{-1}$, where $a, b \in R$ and $(a,b) = 1$, then $b | a_n$ and $a | a_0$.

   In particular, if f is monic, then $c \in R$.

Proof We have

$$a_0 b^n + a_1 a b^{n-1} + \ldots + a_{n-1} a^{n-1} b + a_n a^n = 0.$$

Hence $b | a_n a^n$ and $a | a_0 b^n$. Since $(a^n, b) = (a, b^n) = 1$, by Proposition 3(v), the result follows from Proposition 3(ii). $\square$

The polynomial $t^2 - 2$ has no integer roots, since $0,1,-1$ are not roots and if $c \in \mathbb{Z}$ and $c \neq 0,1,-1$, then $c^2 \geq 4$. Consequently, by Proposition 16, the polynomial $t^2 - 2$ also has no rational roots. It now follows from Proposition 14 that $t^2 - 2$ is irreducible in $\mathbb{Q}[t]$, since it has no divisors of degree 1.

Proposition 16 was known to Euler (1774) for the case $R = \mathbb{Z}$. In this case it shows that to obtain all rational roots of a polynomial with rational coefficients we need test only a finite number of possibilities, which can be explicitly enumerated. For example, if $z \in \mathbb{Z}$, the cubic polynomial $t^3 + zt + 1$ has no rational roots unless $z = 0$ or $z = -2$.

It was shown by Gauss (1801), again for the case $R = \mathbb{Z}$, that Proposition 16 may itself be considerably generalized. His result may be formulated in the following way:

PROPOSITION 17 *Let $f, g \in R[t]$, where R is a GCD domain with field of fractions K. Then g divides f in $R[t]$ if and only if g divides f in $K[t]$ and the greatest common divisor of the coefficients of g divides the greatest common divisor of the coefficients of f.*

Proof For any polynomial $f \in R[t]$, let $c(f)$ denote the greatest common divisor of its coefficients. We say that f is *primitive* if $c(f) = 1$. We show first that the product $f = gh$ of two primitive polynomials g, h is again primitive.

Let

$$g = b_0 + b_1 t + \ldots, \quad h = c_0 + c_1 t + \ldots, \quad f = a_0 + a_1 t + \ldots,$$

and assume on the contrary that the coefficients a_i have a common divisor d which is not a unit. Then d does not divide all the coefficients b_j, nor all the coefficients c_k. Let b_m, c_n be the first coefficients of g, h which are not divisible by d. Then

$$a_{m+n} = \sum_{j+k=m+n} b_j c_k$$

and d divides every term on the right, except possibly $b_m c_n$. In fact, since $d | a_{m+n}$, d must also divide $b_m c_n$. Hence we cannot have both $(d, b_m) = 1$ and $(d, c_n) = 1$. Consequently we can replace d by a proper divisor d', again not a unit, for which $m' + n' > m + n$. Since there exists a divisor d for which $m + n$ is a maximum, this yields a contradiction.

Now let f,g be polynomials in $R[t]$ such that g divides f in $K[t]$. Thus $f = gH$, where $H \in K[t]$. We can write $H = ab^{-1}h_0$, where a,b are coprime elements of R and h_0 is a primitive polynomial in $R[t]$. Also

$$f = c(f)f_0, \quad g = c(g)g_0,$$

where f_0,g_0 are primitive polynomials in $R[t]$. Hence

$$bc(f)f_0 = ac(g)g_0h_0.$$

Since g_0h_0 is primitive, it follows that

$$bc(f) = ac(g).$$

If $H \in R[t]$, then $b = 1$ and so $c(g)|c(f)$. Conversely, if $c(g)|c(f)$, then $bc(f)/c(g) = a$. Since $(a,b) = 1$, this implies that $b = 1$ and $H \in R[t]$. $\square$

COROLLARY 18 *If R is a GCD domain, then $R[t]$ is also a GCD domain. If, moreover, R is a factorial domain, then $R[t]$ is also a factorial domain.*

Proof Let K denote the field of fractions of R. Since $K[t]$ is a GCD domain and $R[t] \subseteq K[t]$, $R[t]$ is certainly an integral domain. If $f,g \in R[t]$, then there exists a primitive polynomial $h_0 \in R[t]$ which is a greatest common divisor of f and g in $K[t]$. It follows from Proposition 17 that

$$h = (c(f),c(g))h_0$$

is a greatest common divisor of f and g in $R[t]$.

This proves the first statement of the corollary. It remains to show that if R also satisfies the chain condition (#), then $R[t]$ does likewise. But if $f_n \in R[t]$ and $f_{n+1}|f_n$ for every n, then f_n must be of constant degree for all large n. The second statement of the corollary now also follows from Proposition 17 and the chain condition in R. $\square$

It follows by induction that in the statement of Corollary 18 we may replace $R[t]$ by the ring $R[t_1,...,t_m]$ of all polynomials in finitely many indeterminates $t_1,...,t_m$ with coefficients from R. In particular, if K is a field, then any polynomial $f \in K[t_1,...,t_m]$ such that $f \notin K$ can be represented as a product of finitely many irreducible polynomials and the representation is essentially unique.

It is now easy to give examples of GCD domains which are not Bézout domains. Let R be a GCD domain which is not a field (e.g., $R = \mathbb{Z}$). Then some $a_0 \in R$ is neither zero nor a unit. By Corollary 18, $R[t]$ is a GCD domain and, by Proposition 17, the greatest common divisor in $R[t]$ of the polynomials a_0 and t is 1. If there existed $g,h \in R[t]$ such that

$$a_0 g + th = 1,$$

where $g = b_0 + b_1 t + \ldots$, then by equating constant coefficients we would obtain $a_0 b_0 = 1$, which is a contradiction. Thus $R[t]$ is not a Bézout domain.

As an application of the preceding results we show that if $a_1,\ldots,a_n$ are distinct integers, then the polynomial

$$f = \prod_{j=1}^{n}(t - a_j) - 1$$

is irreducible in $\mathbb{Q}[t]$. Assume, on the contrary, that $f = gh$, where $g,h \in \mathbb{Q}[t]$ and have positive degree. We may suppose without loss of generality that $g \in \mathbb{Z}[t]$ and that the greatest common divisor of the coefficients of g is 1. Since $f \in \mathbb{Z}[t]$, it then follows from Proposition 17 that also $h \in \mathbb{Z}[t]$. Thus $g(a_j)$ and $h(a_j)$ are integers for every j. Since $g(a_j)h(a_j) = -1$, it follows that $g(a_j) = -h(a_j)$. Thus the polynomial $g + h$ has the distinct roots $a_1,\ldots,a_n$. Since $g + h$ has degree less than n, this is possible only if $g + h = O$. Hence $f = -g^2$. But, since the highest coefficient of f is 1, this is a contradiction.

In general, it is not an easy matter to determine if a polynomial with rational coefficients is irreducible in $\mathbb{Q}[t]$. However, the following *irreducibility criterion*, due to Eisenstein (1850), is sometimes useful:

PROPOSITION 19 *If*

$$f(t) = a_0 + a_1 t + \ldots + a_{n-1}t^{n-1} + t^n$$

is a monic polynomial of degree n with integer coefficients such that $a_0,a_1,\ldots,a_{n-1}$ are all divisible by some prime p, but a_0 is not divisible by p^2, then f is irreducible in $\mathbb{Q}[t]$.

Proof Assume on the contrary that f is reducible. Then there exist polynomials $g(t),h(t)$ of positive degrees l,m with *integer* coefficients such that $f = gh$. If

$$g(t) = b_0 + b_1 t + \ldots + b_l t^l, \quad h(t) = c_0 + c_1 t + \ldots + c_m t^m,$$

then $a_0 = b_0 c_0$. The hypotheses imply that exactly one of b_0,c_0 is divisible by p. Without loss of generality, assume it to be b_0. Since p divides $a_1 = b_0 c_1 + b_1 c_0$, it follows that $p|b_1$. Since p divides $a_2 = b_0 c_2 + b_1 c_1 + b_2 c_0$, it now follows that $p|b_2$. Proceeding in this way, we see that p divides b_j for every $j \le l$. But, since $b_l c_m = 1$, this yields a contradiction. $\square$

It follows from Proposition 19 that, for any prime p, the p-th *cyclotomic polynomial*

$$\Phi_p(x) = x^{p-1} + x^{p-2} + \ldots + 1$$

is irreducible in $\mathbb{Q}[x]$. For $\Phi_p(x) = (x^p - 1)/(x - 1)$ and, if we put $x = 1 + t$, the transformed polynomial

$$\{(1 + t)^p - 1\}/t \;=\; t^{p-1} + {}^P C_{p-1}\, t^{p-2} + \ldots + {}^P C_2\, t + p$$

satisfies the hypotheses of Proposition 19.

For any field K, we define the *formal derivative* of a polynomial $f \in K[t]$,

$$f = a_0 + a_1 t + \ldots + a_n t^n,$$

to be the polynomial

$$f' = a_1 + 2a_2 t + \ldots + na_n t^{n-1}.$$

If the field K is of *characteristic* 0 (see Chapter I, §8), then $\partial(f') = \partial(f) - 1$.

Formal derivatives share the following properties with the derivatives of real analysis:

(i) $(f + g)' = f' + g'$;
(ii) $(cf)' = cf'$ *for any* $c \in K$;
(iii) $(fg)' = f'g + fg'$;
(iv) $(f^k)' = kf^{k-1}f'$ *for any* $k \in \mathbb{N}$.

The first two properties are easily established and the last two properties then need only be verified for monomials $f = t^m$, $g = t^n$.

We can use formal derivatives to determine when a polynomial is *square-free*:

PROPOSITION 20 *Let f be a polynomial of positive degree with coefficients from a field K. If f is relatively prime to its formal derivative f', then f is a product of irreducible polynomials, no two of which differ by a constant factor. Conversely, if f is such a product and if K has characteristic 0, then f is relatively prime to f'.*

Proof If $f = g^2 h$ for some polynomials $g, h \in K[t]$ with $\partial(g) > 0$ then, by the rules above,

$$f' = 2gg'h + g^2 h'.$$

Hence $g|f'$ and f, f' are not relatively prime.

On the other hand, if $f = p_1 \cdots p_m$ is a product of essentially distinct irreducible polynomials p_j, then

$$f' = p_1'p_2 \cdots p_m + p_1 p_2' p_3 \cdots p_m + \ldots + p_1 \cdots p_{m-1}p_m'.$$

If the field K has characteristic 0, then p_1' is of lower degree than p_1 and is not the zero polynomial. Thus the first term on the right is not divisible by p_1, but all the other terms are.

Therefore $p_1 \nmid f'$, and hence $(f',p_1) = 1$. Similarly, $(f',p_j) = 1$ for $1 < j \le m$. Since essentially distinct irreducible polynomials are relatively prime, it follows that $(f',f) = 1$. $\square$

For example, it follows from Proposition 20 that the polynomial $t^n - 1 \in K[t]$ is square-free if the characteristic of the field K does not divide the positive integer n.

4 Euclidean domains

An integral domain R is said to be *Euclidean* if it possesses a Euclidean algorithm, i.e. if there exists a map $\delta: R \to \mathbb{N} \cup \{0\}$ such that, for any $a,b \in R$ with $a \ne 0$, there exist $q,r \in R$ with the properties

$$b = qa + r, \quad \delta(r) < \delta(a).$$

It follows that $\delta(a) > \delta(0)$ for any $a \ne 0$. For there exist $q_1,a_1 \in R$ such that

$$0 = q_1 a + a_1, \quad \delta(a_1) < \delta(a),$$

and if $a_n \ne 0$ there exist $q_{n+1}, a_{n+1} \in R$ such that

$$0 = q_{n+1} a_n + a_{n+1}, \quad \delta(a_{n+1}) < \delta(a_n).$$

Repeatedly applying this process, we must arrive at $a_N = 0$ for some N, since the sequence $\{\delta(a_n)\}$ cannot decrease forever, and we then have $\delta(0) = \delta(a_N) < ... < \delta(a_1) < \delta(a)$.

By replacing δ by $\delta - \delta(0)$ *we may, and will, assume that* $\delta(0) = 0$, $\delta(a) > 0$ if $a \ne 0$.

Since the proof of Lemma 9 remains valid if $\mathbb{Z}$ is replaced by R and $|a|$ by $\delta(a)$, *any Euclidean domain is a principal ideal domain.*

The *polynomial ring* $K[t]$ is a Euclidean domain with $\delta(a) = |a| = 2^{\partial(a)}$. Polynomial rings are characterized among all Euclidean domains by the following result:

PROPOSITION 21 *For a Euclidean domain R, the following conditions are equivalent:*

(i) *for any $a,b \in R$ with $a \ne 0$, there exist* unique $q,r \in R$ *such that $b = qa + r$, $\delta(r) < \delta(a)$;*
(ii) *for any $a,b,c \in R$ with $c \ne 0$,*

$$\delta(a + b) \le \max\{\delta(a),\delta(b)\}, \quad \delta(a) \le \delta(ac).$$

Moreover, if one or other of these two conditions holds, then either R is a field and $\delta(a) = \delta(1)$ for every $a \ne 0$, or $R = K[t]$ for some field K and δ is an increasing function of $||$.

Proof Suppose first that (i) holds. If $a \neq 0$, $c \neq 0$, then from $0 = 0a - 0 = ca - ac$, we obtain $\delta(ac) \geq \delta(a)$, and this holds also if $a = 0$. If we take $c = -1$ and replace a by $-a$, we get $\delta(-a) = \delta(a)$. Since $b = 0(a + b) + b = 1(a + b) + (-a)$, it follows that either $\delta(b) \geq \delta(a + b)$ or $\delta(a) \geq \delta(a + b)$. Thus (i) $\Rightarrow$ (ii).

Suppose next that (ii) holds. Assume that, for some $a,b \in R$ with $a \neq 0$, there exist pairs q,r and q',r' such that

$$b = qa + r = q'a + r', \quad \max\{\delta(r),\delta(r')\} < \delta(a).$$

From (ii) we obtain first $\delta(-r) = \delta(r)$ and then $\delta(r' - r) \leq \max\{\delta(r),\delta(r')\} < \delta(a)$. Since $r' - r = a(q - q')$, this implies $q - q' = 0$ and hence $r' - r = 0$. Thus (ii) $\Rightarrow$ (i).

Suppose now that (i) and (ii) both hold. Then $\delta(1) \leq \delta(a)$ for any $a \neq 0$, since $a = 1a$. Furthermore, $\delta(a) = \delta(ae)$ for any unit e, since

$$\delta(a) \leq \delta(ae) \leq \delta(aee^{-1}) = \delta(a).$$

On the other hand, $\delta(a) = \delta(ae)$ for some $a \neq 0$ implies that e is a unit. For from

$$a = qae + r, \quad \delta(r) < \delta(ae),$$

we obtain $r = (1 - qe)a$, $\delta(r) < \delta(a)$, and hence $1 - qe = 0$. In particular, $\delta(e) = \delta(1)$ if and only if e is a unit.

The set K of all $a \in R$ such that $\delta(a) \leq \delta(1)$ thus consists of 0 and all units of R. Since $a,b \in K$ implies $a - b \in K$, it follows that K is a field. We assume that $K \neq R$, since otherwise we have the first alternative of the proposition.

Choose $x \in R \setminus K$ so that

$$\delta(x) = \min_{a \in R\setminus K} \delta(a).$$

For any $a \in R \setminus K$, there exist $q_0,r_0 \in R$ such that

$$a = q_0 x + r_0, \quad \delta(r_0) < \delta(x),$$

i.e. $r_0 \in K$. Then $\delta(q_0) < \delta(q_0 x) = \delta(a - r_0) \leq \delta(a)$. If $\delta(q_0) \geq \delta(x)$, i.e. if $q_0 \in R \setminus K$, then in the same way there exist $q_1,r_1 \in R$ such that

$$q_0 = q_1 x + r_1, \, r_1 \in K, \, \delta(q_1) < \delta(q_0).$$

After finitely many repetitions of this process we must arrive at some $q_{n-1} \in K$. Putting $r_n = q_{n-1}$, we obtain

$$a = r_n x^n + r_{n-1} x^{n-1} + \dots + r_0,$$

where $r_0,\dots,r_n \in K$ and $r_n \neq 0$. Since $\delta(r_j x^j) = \delta(x^j)$ if $r_j \neq 0$ and $\delta(x^j) < \delta(x^{j+1})$ for every j, it follows that $\delta(a) = \delta(x^n)$. Since the representation $a = qx^n + r$ with $\delta(r) < \delta(x^n)$ is unique, it follows that $r_0,\dots,r_n$ are uniquely determined by a. Define a map $\psi: R \to K[t]$ by

$$\psi(r_n x^n + r_{n-1} x^{n-1} + \dots + r_0) = r_n t^n + r_{n-1} t^{n-1} + \dots + r_0.$$

Then ψ is a bijection and actually an isomorphism, since it preserves sums and products. Furthermore $\delta(a) >, =,$ or $< \delta(b)$ according as $|\psi(a)| >, =,$ or $< |\psi(b)|$. $\square$

Some significant examples of principal ideal domains are provided by quadratic fields, which will be studied in Chapter III. Any quadratic number field has the form $\mathbb{Q}(\sqrt{d})$, where $d \in \mathbb{Z}$ is square-free and $d \neq 1$. The set $\mathcal{O}_d$ of all algebraic integers in $\mathbb{Q}(\sqrt{d})$ is an integral domain. In the equivalent language of binary quadratic forms, it was known to Gauss that $\mathcal{O}_d$ is a principal ideal domain for nine negative values of d, namely

$$d = -1,-2,-3,-7,-11,-19,-43,-67,-163.$$

Heilbronn and Linfoot (1934) showed that there was at most one additional negative value of d for which $\mathcal{O}_d$ is a principal ideal domain. Stark (1967) proved that this additional value does not in fact exist, and soon afterwards it was observed that a gap in a previous proof by Heegner (1952) could be filled without difficulty. It is conjectured that $\mathcal{O}_d$ is a principal ideal domain for infinitely many positive values of d, but this remains unproved.

Much work has been done on determining for which quadratic number fields $\mathbb{Q}(\sqrt{d})$ the ring of integers $\mathcal{O}_d$ is a Euclidean domain. Although we regard being Euclidean more as a useful property than as an important concept, we report here the results which have been obtained for their intrinsic interest.

The ring $\mathcal{O}_d$ is said to be *norm-Euclidean* if it is Euclidean when one takes $\delta(a)$ to be the absolute value of the *norm* of a. It has been shown that $\mathcal{O}_d$ is norm-Euclidean for precisely the following values of d:

$$d = -11,-7,-3,-2,-1,2,3,5,6,7,11,13,17,19,21,29,33,37,41,57,73.$$

It is known that, for $d < 0$, $\mathcal{O}_d$ is Euclidean only if it is norm-Euclidean. Comparing the two lists, we see that for $d = -19,-43,-67,-163$, $\mathcal{O}_d$ is a principal ideal domain, but not a Euclidean domain. On the other hand it is also known that, for $d = 69$, $\mathcal{O}_d$ is Euclidean but not norm-Euclidean.

5 Congruences

The invention of a new notation often enables one to replace a long, involved argument by simple and mechanical algebraic operations. This is well illustrated by the congruence notation.

Two integers a and b are said to be *congruent modulo* a third integer m if m divides $a - b$, and this is denoted by $a \equiv b \bmod m$. For example,

$$13 \equiv 4 \bmod 3, \quad 13 \equiv -7 \bmod 5, \quad 19 \equiv 7 \bmod 4.$$

The notation is a modification by Gauss of the notation $a = b \bmod m$ used by Legendre, as Gauss explicitly acknowledged (*D.A.*, §2). (If a and b are not congruent modulo m, we write $a \not\equiv b \bmod m$.) Congruence has, in fact, many properties in common with equality:

(C1) $a \equiv a \bmod m$ *for all a,m;*   (reflexive law)

(C2) *if* $a \equiv b \bmod m$, *then* $b \equiv a \bmod m$;   (symmetric law)

(C3) *if* $a \equiv b$ *and* $b \equiv c \bmod m$, *then* $a \equiv c \bmod m$;   (transitive law)

(C4) *if* $a \equiv a'$ *and* $b \equiv b' \bmod m$, *then* $a + b \equiv a' + b'$ *and* $ab \equiv a'b' \bmod m$.

(replacement laws)

The proofs of these properties are very simple. For any a,m we have $a - a = 0 = m \cdot 0$. If m divides $a - b$, then it also divides $b - a = -(a - b)$. If m divides both $a - b$ and $b - c$, then it also divides $(a - b) + (b - c) = a - c$. Finally, if m divides both $a - a'$ and $b - b'$, then it also divides $(a - a') + (b - b') = (a + b) - (a' + b')$ and $(a - a')b + a'(b - b') = ab - a'b'$.

The properties **(C1)–(C3)** state that congruence mod m is an *equivalence relation*. Since $a = b$ implies $a \equiv b \bmod m$, it is a coarsening of the equivalence relation of equality (but coincides with it if $m = 0$). The corresponding equivalence classes are called *residue classes*. The set $\mathbb{Z}$ with equality replaced by congruence mod m will be denoted by $\mathbb{Z}_{(m)}$. If $m > 0$, $\mathbb{Z}_{(m)}$ has cardinality m, since an arbitrary integer a can be uniquely represented in the form $a = qm + r$, where $r \in \{0,1,\ldots,m - 1\}$ and $q \in \mathbb{Z}$. The particular r which represents a given $a \in \mathbb{Z}$ is referred to as the *least non-negative residue* of a mod m.

The replacement laws imply that the associative, commutative and distributive laws for addition and multiplication are inherited from $\mathbb{Z}$ by $\mathbb{Z}_{(m)}$. Hence $\mathbb{Z}_{(m)}$ is a commutative ring, with 0 as an identity element for addition and 1 as an identity element for multiplication. However, $\mathbb{Z}_{(m)}$ is not an integral domain if m is composite, since if $m = m'm''$ with $1 < m' < m$, then

$$m'm'' \equiv 0, \quad \text{but } m' \not\equiv 0, \, m'' \not\equiv 0 \bmod m.$$

On the other hand, if $ab \equiv ac \bmod m$ and $(a,m) = 1$, then $b \equiv c \bmod m$, by Proposition 3(ii). Thus factors which are relatively prime to the modulus can be cancelled.

In algebraic terms, $\mathbb{Z}_{(m)}$ is the *quotient ring* $\mathbb{Z}/m\mathbb{Z}$ of $\mathbb{Z}$ with respect to the ideal $m\mathbb{Z}$ generated by m, and the elements of $\mathbb{Z}_{(m)}$ are the *cosets* of this ideal. For convenience, rather than necessity, we suppose from now on that $m > 1$.

Congruences enter implicitly into many everyday problems. For example, the ring $\mathbb{Z}_{(2)}$ contains two distinct elements, 0 and 1, with the addition and multiplication tables

$$0 + 0 = 1 + 1 = 0, \quad 0 + 1 = 1 + 0 = 1,$$
$$0 \cdot 0 = 0 \cdot 1 = 1 \cdot 0 = 0, \quad 1 \cdot 1 = 1.$$

This is the arithmetic of *odds* (1) and *evens* (0), which is used by electronic computers.

Again, to determine the day of the week on which one was born, from the date and day of the week today, is an easy calculation in the arithmetic of $\mathbb{Z}_{(7)}$ (remembering that $366 \equiv 2 \bmod 7$).

The well-known tests for divisibility of an integer by 3 or 9 are easily derived by means of congruences. Let the positive integer a have the decimal representation

$$a = a_0 + a_1 10 + \dots + a_n 10^n,$$

where $a_0, a_1, \dots, a_n \in \{0, 1, \dots, 9\}$. Since $10 \equiv 1 \bmod m$, where $m = 3$ or 9, the replacement laws imply that $10^k \equiv 1 \bmod m$ for any positive integer k and hence

$$a \equiv a_0 + a_1 + \dots + a_n \bmod m.$$

Thus a is divisible by 3 or 9 if and only if the sum of its digits is so divisible.

This can be used to check the accuracy of arithmetical calculations. Any equation involving only additions and multiplications must remain valid when equality is replaced by congruence mod m. For example, suppose we wish to check if

$$7714 \times 3036 = 23{,}419{,}804.$$

Taking congruences mod 9, we have on the left side $19 \times 12 \equiv 1 \times 3 \equiv 3$ and on the right side $5 + 14 + 12 \equiv 5 + 5 + 3 \equiv 4$. Since $4 \not\equiv 3 \bmod 9$, the original equation is incorrect (the 8 should be a 7).

Since the distinct squares in $\mathbb{Z}_{(4)}$ are 0 and 1, it follows that an integer $a \equiv 3 \bmod 4$ cannot be represented as the sum of two squares of integers. Similarly, since the distinct squares in $\mathbb{Z}_{(8)}$ are $0, 1, 4$, an integer $a \equiv 7 \bmod 8$ cannot be represented as the sum of three squares of integers.

The oldest known work on number theory is a Babylonian cuneiform text, from at least as early as 1600 B.C., which contains a list of right-angled triangles whose side lengths are all exact multiples of the unit length. By Pythagoras' theorem, the problem is to find positive integers x,y,z such that

$$x^2 + y^2 = z^2.$$

For example, 3,4,5 and 5,12,13 are solutions. The number of solutions listed suggests that the Babylonians not only knew the theorem of Pythagoras, but also had some rule for finding such *Pythagorean triples*. There are in fact infinitely many, and a rule for finding them all is given by Euclid in his *Elements* (Book X, Lemma 1 following Proposition 28). In the derivation of this rule we will use the preceding remark about sums of two squares.

We may assume that x and y are relatively prime since, if x,y,z is a Pythagorean triple for which x and y have greatest common divisor d, then $d^2|z^2$ and hence $d|z$, so that $x/d,y/d,z/d$ is also a Pythagorean triple. If x and y are relatively prime, then they are not both even and without loss of generality we may assume that x is odd. If y were also odd, we would have

$$z^2 = x^2 + y^2 \equiv 1 + 1 = 2 \ \ \mathrm{mod} \ 4,$$

which is impossible. Hence y is even and z is odd. Then 2 is a common divisor of $z + x$ and $z - x$, and is actually their greatest common divisor, since $(x,y) = 1$ implies $(x,z) = 1$. Since

$$(y/2)^2 \ = \ (z + x)/2 \cdot (z - x)/2$$

and the two factors on the right are relatively prime, they are also squares:

$$(z + x)/2 = a^2, \ \ (z - x)/2 = b^2,$$

where $a > b > 0$ and $(a,b) = 1$. Then

$$x = a^2 - b^2, \ y = 2ab, \ z = a^2 + b^2.$$

Moreover a and b cannot both be odd, since z is odd.

Conversely, if x,y,z are defined by these formulas, where a and b are relatively prime positive integers with $a > b$ and either a or b even, then x,y,z is a Pythagorean triple. Moreover x is odd, since z is odd and y even, and it is easily verified that $(x,y) = 1$. For given x and z, a^2 and b^2 are uniquely determined, and hence a and b are also. Thus different couples a,b give different solutions x,y,z.

To return to congruences, we now consider the structure of the ring $\mathbb{Z}_{(m)}$. If $a \equiv a' \ \mathrm{mod} \ m$ and $(a,m) = 1$, then also $(a',m) = 1$. Hence we may speak of an element of $\mathbb{Z}_{(m)}$ as being relatively prime to m. The set of all elements of $\mathbb{Z}_{(m)}$ which are relatively prime to m will be denoted by $\mathbb{Z}_{(m)}^{\times}$. If a is a *unit* of the ring $\mathbb{Z}_{(m)}$, then clearly $a \in \mathbb{Z}_{(m)}^{\times}$. The following proposition shows that, conversely, if $a \in \mathbb{Z}_{(m)}^{\times}$, then a is a unit of the ring $\mathbb{Z}_{(m)}$.

PROPOSITION 22 *The set $\mathbb{Z}_{(m)}^{\times}$ is a commutative group under multiplication.*

Proof By Proposition 3(iv), $\mathbb{Z}_{(m)}^{\times}$ is closed under multiplication. Since multiplication is associative and commutative, it only remains to show that any $a \in \mathbb{Z}_{(m)}^{\times}$ has an inverse $a^{-1} \in \mathbb{Z}_{(m)}^{\times}$.

The elements of $\mathbb{Z}_{(m)}^{\times}$ may be taken to be the positive integers $c_1,...,c_h$ which are less than m and relatively prime to m, and we may choose the notation so that $c_1 = 1$. Since $ac_j \equiv ac_k$ mod m implies $c_j \equiv c_k$ mod m, the elements $ac_1,...,ac_h$ are distinct elements of $\mathbb{Z}_{(m)}^{\times}$ and hence are a permutation of $c_1,...,c_h$. In particular, $ac_i \equiv c_1$ mod m for one and only one value of i. (The existence of inverses also follows from the Bézout identity $au + mv = 1$, since this implies $au \equiv 1$ mod m. Hence the Euclidean algorithm provides a way of calculating a^{-1}.) $\square$

COROLLARY 23 *If p is a prime, then $\mathbb{Z}_{(p)}$ is a finite field with p elements.*

Proof We already know that $\mathbb{Z}_{(p)}$ is a commutative ring, whose distinct elements are represented by the integers $0,1,...,p - 1$. Since p is a prime, $\mathbb{Z}_{(p)}^{\times}$ consists of all nonzero elements of $\mathbb{Z}_{(p)}$. Since $\mathbb{Z}_{(p)}^{\times}$ is a multiplicative group, by Proposition 22, it follows that $\mathbb{Z}_{(p)}$ is a field. $\square$

The finite field $\mathbb{Z}_{(p)}$ will be denoted from now on by the more usual notation $\mathbb{F}_p$. Corollary 23, in conjunction with Proposition 15, implies that if p is a prime and f a polynomial of degree $n \geq 1$, then the congruence

$$f(x) \equiv 0 \ \bmod p$$

has at most n mutually incongruent solutions mod p. This is no longer true if the modulus is not a prime. For example, the congruence $x^2 - 1 \equiv 0$ mod 8 has the distinct solutions $x \equiv 1,3,5,7$ mod 8.

The *order* of the group $\mathbb{Z}_{(m)}^{\times}$, i.e. the number of positive integers less than m and relatively prime to m, is traditionally denoted by $\varphi(m)$, with the convention that $\varphi(1) = 1$. For example, if p is a prime, then $\varphi(p) = p - 1$. More generally, for any positive integer k,

$$\varphi(p^k) = p^k - p^{k-1},$$

since the elements of $\mathbb{Z}_{(p^k)}$ which are not in $\mathbb{Z}_{(p^k)}^\times$ are the multiples jp with $0 \le j < p^{k-1}$. By Proposition 4, if $m = m'm''$, where $(m',m'') = 1$, then $\varphi(m) = \varphi(m')\varphi(m'')$. Together with what we have just proved, this implies that if an arbitrary positive integer m has the factorization

$$m = p_1^{k_1} \cdots p_s^{k_s}$$

as a product of positive powers of distinct primes, then

$$\varphi(m) = p_1^{k_1-1}(p_1 - 1) \cdots p_s^{k_s-1}(p_s - 1).$$

In other words,

$$\varphi(m) = m \prod_{p \mid m} (1 - 1/p).$$

The function $\varphi(m)$ was first studied by Euler and is known as Euler's *phi*-function (or 'totient' function), although it was Gauss who decided on the letter φ. Gauss (*D.A.*, §39) also established the following property:

PROPOSITION 24 *For any positive integer n,*

$$\sum_{d \mid n} \varphi(d) = n,$$

where the summation is over all positive divisors d of n.

Proof Let d be a positive divisor of n and let S_d denote the set of all positive integers $m \le n$ such that $(m,n) = d$. Since $(m,n) = d$ if and only if $(m/d,n/d) = 1$, the cardinality of S_d is $\varphi(n/d)$. Moreover every positive integer $m \le n$ belongs to exactly one such set S_d. Hence

$$n = \sum_{d \mid n} \varphi(n/d) = \sum_{d \mid n} \varphi(d),$$

since n/d runs through the positive divisors of n at the same time as d. $\square$

Much of the significance of Euler's function stems from the following property:

PROPOSITION 25 *If m is a positive integer and a an integer relatively prime to m, then*

$$a^{\varphi(m)} \equiv 1 \mod m.$$

Proof Let $c_1,\ldots,c_h$, where $h = \varphi(m)$, be the distinct elements of $\mathbb{Z}_{(m)}^\times$. As we saw in the proof of Proposition 22, the elements $ac_1,\ldots,ac_h$ of $\mathbb{Z}_{(m)}^\times$ are just a permutation of $c_1,\ldots,c_h$. Forming their product, we obtain $a^h c_1 \cdots c_h \equiv c_1 \cdots c_h \mod m$. Since the c's are relatively prime to m, they can be cancelled and we are left with $a^h \equiv 1 \mod m$. $\square$

COROLLARY 26 *If p is a prime and a an integer not divisible by p, then $a^{p-1} \equiv 1 \bmod p$.*
□

Corollary 26 was stated without proof by Fermat (1640) and is commonly known as 'Fermat's little theorem'. The first published proof was given by Euler (1736), who later (1760) proved the general Proposition 25.

Proposition 25 is actually a very special case of Lagrange's theorem that the order of a subgroup of a finite group divides the order of the whole group. In the present case the whole group is $\mathbb{Z}_{(m)}^{\times}$ and the subgroup is the cyclic group generated by a.

Euler gave also another proof of Corollary 26, which has its own interest. For any two integers a,b and any prime p we have, by the binomial theorem,

$$(a + b)^p = \sum_{k=0}^{p} {}^{p}C_k \, a^k b^{p-k},$$

where the binomial coefficients

$$ {}^{p}C_k = (p - k + 1)\cdots p/1 \cdot 2 \cdot \ldots \cdot k $$

are integers. Moreover p divides ${}^{p}C_k$ for $0 < k < p$, since p divides ${}^{p}C_k \cdot k!$ and is relatively prime to $k!$ It follows that

$$(a + b)^p \equiv a^p + b^p \bmod p.$$

In particular, $(a + 1)^p \equiv a^p + 1 \bmod p$, from which we obtain by induction $a^p \equiv a \bmod p$ for every integer a. If p does not divide a, the factor a can be cancelled to give $a^{p-1} \equiv 1 \bmod p$.

The first part of this alternative proof generalizes to the statement that *in any commutative ring R, of prime characteristic p, the map $a \to a^p$ is a homomorphism*:

$$(a + b)^p = a^p + b^p, \quad (ab)^p = a^p b^p.$$

(As defined in §8 of Chapter I, R has *characteristic* k if k is the least positive integer such that the sum of k 1's is 0, and has *characteristic zero* if there is no such positive integer.) By way of illustration, we give one important application of this result.

We showed in §3 that, for any prime p, the polynomial

$$\Phi_p(x) = x^{p-1} + x^{p-2} + \ldots + 1$$

is irreducible in $\mathbb{Q}[x]$. The roots in $\mathbb{C}$ of $\Phi_p(x)$ are the p-th roots of unity, other than 1. By a quite different argument we now show that, for any positive integer n, the 'primitive' n-th roots of unity are the roots of a monic polynomial $\Phi_n(x)$ with integer coefficients which is irreducible in $\mathbb{Q}[x]$. The uniquely determined polynomial $\Phi_n(x)$ is called the n-th *cyclotomic polynomial*.

Let ζ be a *primitive n-th root of unity*, i.e. $\zeta^n = 1$ but $\zeta^k \neq 1$ for $0 < k < n$. It follows from Corollary 18 that ζ is a root of some monic irreducible polynomial $f(x) \in \mathbb{Z}[x]$ which divides $x^n - 1$. If p is a prime which does not divide n, then ζ^p is also a primitive n-th root of unity and, for the same reason, ζ^p is a root of some monic irreducible polynomial $g(x) \in \mathbb{Z}[x]$ which divides $x^n - 1$. We show first that $g(x) = f(x)$.

Assume on the contrary that $g(x) \neq f(x)$. Then

$$x^n - 1 \; = \; f(x)g(x)h(x)$$

for some $h(x) \in \mathbb{Z}[x]$. Since ζ is a root of $g(x^p)$, we also have

$$g(x^p) \; = \; f(x)k(x)$$

for some $k(x) \in \mathbb{Z}[x]$. If $\bar{f}(x),...$ denotes the polynomial in $\mathbb{F}_p[x]$ obtained from $f(x),...$ by reducing the coefficients mod p, then

$$x^n - 1 \; = \; \bar{f}(x)\,\bar{g}(x)\,\bar{h}(x), \quad \bar{g}(x^p) = \bar{f}(x)\,\bar{k}(x).$$

But $\bar{g}(x^p) = \bar{g}(x)^p$, since $\mathbb{F}_p[x]$ is a ring of characteristic p and $a^p = a$ for every $a \in \mathbb{F}_p$. Hence any irreducible factor $\bar{e}(x)$ of $\bar{f}(x)$ in $\mathbb{F}_p[x]$ also divides $\bar{g}(x)$. Consequently $\bar{e}(x)^2$ divides $x^n - 1$ in $\mathbb{F}_p[x]$. But $x^n - 1$ is relatively prime to its formal derivative nx^{n-1}, since $p \nmid n$, and so is square-free. This is the desired contradiction.

By applying this repeatedly for the same or different primes p, we see that ζ^m is a root of $f(x)$ for any positive integer m less than n and relatively prime to n. If ω is any n-th root of unity, then $\omega = \zeta^k$ for a unique k such that $0 \leq k < n$. If $(k,n) \neq 1$, then $\omega^d = 1$ for some proper divisor d of n (cf. Lemma 31 below). If such an ω were a root of $f(x)$, then $f(x)$ would divide $x^d - 1$, which is impossible since ζ is not a root of $x^d - 1$. Hence $f(x)$ does not depend on the original choice of primitive n-th root of unity, its roots being all the primitive n-th roots of unity. The polynomial $f(x)$ will now be denoted by $\Phi_n(x)$. Since $x^n - 1$ is square-free, we have

$$x^n - 1 \; = \; \prod_{d|n} \Phi_d(x).$$

This yields a new proof of Proposition 24, since $\Phi_d(x)$ has degree $\varphi(d)$.

As an application of Fermat's little theorem (Corollary 26) we now prove

PROPOSITION 27 *If p is a prime, then $(p-1)! + 1$ is divisible by p.*

Proof Since $1! + 1 = 2$, we may suppose that the prime p is odd. By Corollary 26, the polynomial $f(t) = t^{p-1} - 1$ has the distinct roots $1,2,...,p-1$ in the field $\mathbb{F}_p$. But the polynomial $g(t) = (t-1)(t-2)\cdots(t-p+1)$ has the same roots. Since $f(t) - g(t)$ is a polynomial of degree less than $p - 1$, it follows from Proposition 15 that $f(t) - g(t)$ is the zero polynomial. In particular, $f(t)$ and $g(t)$ have the same constant coefficient. Since $(-1)^{p-1} = 1$, this yields the result. $\square$

Proposition 27 is known as *Wilson's theorem*, although the first published proof was given by Lagrange (1773). Lagrange observed also that $(n - 1)! + 1$ is divisible by n *only if* n is prime. For suppose $n = n'n''$, where $1 < n',n'' < n$. If $n' \neq n''$, then both n' and n'' occur as factors in $(n - 1)!$ and hence n divides $(n - 1)!$ If $n' = n'' > 2$ then, since $n > 2n'$, both n' and $2n'$ occur as factors in $(n - 1)!$ and again n divides $(n - 1)!$ Finally, if $n = 4$, then n divides $(n - 1)! + 2$.

As another application of Fermat's little theorem, we prove *Euler's criterion for quadratic residues*. If p is a prime and a an integer not divisible by p, we say that a is a *quadratic residue*, or *quadratic nonresidue*, of p according as there exists, or does not exist, an integer c such that $c^2 \equiv a \bmod p$. Thus a is a quadratic residue of p if and only if it is a square in $\mathbb{F}_p^\times$. Euler's criterion is the first statement of the following proposition:

PROPOSITION 28 *If p is an odd prime and a an integer not divisible by p, then*

$$a^{(p-1)/2} \equiv 1 \ or -1 \ \bmod p,$$

according as a is a quadratic residue or nonresidue of p.

Moreover, exactly half of the integers $1,2,...,p - 1$ are quadratic residues of p.

Proof If a is a quadratic residue of p, then $a \equiv c^2 \bmod p$ for some integer c and hence, by Fermat's little theorem,

$$a^{(p-1)/2} \equiv c^{p-1} \equiv 1 \ \bmod p.$$

Since the polynomial $t^{(p-1)/2} - 1$ has at most $(p - 1)/2$ roots in the field $\mathbb{F}_p$, it follows that there are at most $r := (p - 1)/2$ distinct quadratic residues of p. On the other hand, no two of the integers $1^2, 2^2,...,r^2$ are congruent mod p, since $u^2 \equiv v^2 \bmod p$ implies $u \equiv v$ or $u \equiv -v \bmod p$. Hence there are exactly $(p - 1)/2$ distinct quadratic residues of p and, if b is a quadratic nonresidue of p, then $b^{(p-1)/2} \not\equiv 1 \bmod p$. Since $b^{p-1} \equiv 1 \bmod p$, and

$$b^{p-1} - 1 = (b^{(p-1)/2} - 1)(b^{(p-1)/2} + 1),$$

we must have $b^{(p-1)/2} \equiv -1 \bmod p$. $\square$

COROLLARY 29 *If p is an odd prime, then -1 is a quadratic residue of p if $p \equiv 1 \bmod 4$ and a quadratic nonresidue of p if $p \equiv 3 \bmod 4$.* $\square$

Euler's criterion may also be used to determine for what primes 2 is a quadratic residue:

PROPOSITION 30 *For any odd prime p, 2 is a quadratic residue of p if* $p \equiv \pm 1$ mod 8 *and a quadratic nonresidue if* $p \equiv \pm 3$ mod 8.

Proof Let A denote the set of all even integers a such that $p/2 < a < p$, and let B denote the set of all even integers b such that $0 < b < p/2$. Since $A \cup B$ is the set of all positive even integers less than p, it has cardinality $r := (p - 1)/2$. Evidently $a \in A$ if and only if $p - a$ is odd and $0 < p - a < p/2$. Hence the integers $1,2,...,r$ are just the elements of B, together with the integers $p - a$ ($a \in A$). If we denote the cardinality of A by $\#A$, it follows that

$$r! = \prod_{a \in A} (p - a) \prod_{b \in B} b$$

$$\equiv (-1)^{\#A} \prod_{a \in A} a \prod_{b \in B} b \mod p$$

$$= (-1)^{\#A} \, 2^r \, r!$$

Thus $2^r \equiv (-1)^{\#A}$ mod p and hence, by Proposition 28, 2 is a quadratic residue or nonresidue of p according as $\#A$ is even or odd. But $\#A = k$ if $p = 4k + 1$ and $\#A = k + 1$ if $p = 4k + 3$. The result follows. $\square$

We now introduce some simple group-theoretical concepts. Let G be a finite group and $a \in G$. Then there exist $j,k \in \mathbb{N}$ with $j < k$ such that $a^j = a^k$. Thus $a^{k-j} = 1$, where 1 is the identity element of G. The *order* of a is the least positive integer d such that $a^d = 1$.

LEMMA 31 *Let G be a finite group and a an element of G of order d. Then*

(i)   *for any* $k \in \mathbb{N}$, $a^k = 1$ *if and only if d divides k*;
(ii) *for any* $k \in \mathbb{N}$, a^k *has order* $d/(k,d)$;
(iii) $H = \{1,a,...,a^{d-1}\}$ *is a subgroup of G.*

Proof Any $k \in \mathbb{N}$ can be written in the form $k = qd + r$, where $q \geq 0$ and $0 \leq r < d$. Since $a^{qd} = (a^d)^q = 1$, we have $a^k = 1$ if and only if $a^r = 1$, i.e. if and only if $r = 0$, by the definition of d.

It follows that if a^k has order e, then $ke = [k,d]$. Since $[k,d] = kd/(k,d)$, this implies $e = d/(k,d)$. In particular, a^k again has order d if and only if $(k,d) = 1$.

If $0 \leq j,k < d$, put $i = j + k$ if $j + k < d$ and $i = j + k - d$ if $j + k \geq d$. Then $a^j a^k = a^i$, and so H contains the product of any two of its elements. If $0 < k < d$, then $a^k a^{d-k} = 1$, and so H contains also the inverse of any one of its elements. $\square$

The subgroup H in Lemma 31 is the *cyclic subgroup generated by a*. If G has order n, then d divides n, by Lagrange's theorem that the order of a subgroup divides the order of the whole group. We will be interested in the case $G = \mathbb{Z}_{(m)}^{\times}$. In this case there is no need to appeal to Lagrange's theorem, since $\mathbb{Z}_{(m)}^{\times}$ has order $\varphi(m)$, and it follows from Proposition 25 and Lemma 31(i) that d divides $\varphi(m)$.

A group G is *cyclic* if it coincides with the cyclic subgroup generated by one of its elements. For example, the n-th roots of unity in $\mathbb{C}$ form a cyclic group generated by $e^{2\pi i/n}$. In fact the generators of this group are just the primitive n-th roots of unity.

Our next result provides a sufficient condition for a finite group to be cyclic.

LEMMA 32 *A finite group G of order n is cyclic if, for each positive divisor d of n, there are at most d elements of G whose order divides d.*

Proof If H is a cyclic subgroup of G, then its order d divides n. Since all its elements are of order dividing d, the hypothesis of the lemma implies that any element of G whose order divides d must be in H. Furthermore, H contains exactly $\varphi(d)$ elements of order d since, if a generates H, a^k has order d if and only if $(k,d) = 1$.

For each divisor d of n, let $\psi(d)$ denote the number of elements of G of order d. Then, by what we have just proved, either $\psi(d) = 0$ or $\psi(d) = \varphi(d)$. But $\sum_{d\,|\,n} \psi(d) = n$, since the order of each element is a divisor of n, and $\sum_{d\,|\,n} \varphi(d) = n$, by Proposition 24. Hence we must have $\psi(d) = \varphi(d)$ for every $d|n$. In particular, the group G has $\psi(n) = \varphi(n)$ elements of order n. $\square$

The condition of Lemma 32 is also necessary. For let G be a finite cyclic group of order n, generated by the element a, and let d be a divisor of n. An element $x \in G$ has order dividing d if and only if $x^d = 1$. Thus the elements a^k of G of order dividing d are given by $k = jn/d$, with $j = 0,1,...,d - 1$.

We now return from group theory to number theory.

PROPOSITION 33 *For any prime p, the multiplicative group $\mathbb{F}_p^{\times}$ of the field $\mathbb{F}_p$ is cyclic.*

Proof Put $G = \mathbb{F}_p^{\times}$ and denote the order of G by n. For any divisor d of n, the polynomial $t^d - 1$ has at most d roots in $\mathbb{F}_p$. Hence there are at most d elements of G whose order divides d. The result now follows from Lemma 32. $\square$

The same argument shows that, for an arbitrary field K, any finite subgroup of the multiplicative group of K is cyclic.

In the terminology of number theory, an integer which generates $\mathbb{Z}_{(m)}^{\times}$ is said to be a *primitive root* of m. Primitive roots may be used to replace multiplications mod m by additions mod $\varphi(m)$ in the same way that logarithms were once used in analysis. If g is a primitive root of m, then the elements of $\mathbb{Z}_{(m)}^\times$ are precisely $1,g,g^2,\ldots,g^{n-1}$, where $n = \varphi(m)$. Thus for each $a \in \mathbb{Z}_{(m)}^\times$ we have $a \equiv g^\alpha \bmod m$ for a unique index α ($0 \le \alpha < n$). We can construct a table of these indices once and for all. If $a \equiv g^\alpha$ and $b \equiv g^\beta$, then $ab \equiv g^{\alpha+\beta}$. By replacing $\alpha + \beta$ by its least non-negative residue $\gamma \bmod n$ and going backwards in our table we can determine c such that $ab \equiv c \bmod m$.

For any prime p, an essentially complete proof for the existence of primitive roots of p was given by Euler (1774). Jacobi (1839) constructed tables of indices for all primes less than 1000.

We now use primitive roots to prove a general property of polynomials with coefficients from a finite field:

PROPOSITION 34 *If $f(x_1,\ldots,x_n)$ is a polynomial of degree less than n in n variables with coefficients from the finite field $\mathbb{F}_p$, then the number of zeros of f in $\mathbb{F}_p{}^n$ is divisible by the characteristic p. In particular, $(0,\ldots,0)$ is not the only zero of f if f has no constant term.*

Proof Put $K = \mathbb{F}_p$ and $g = 1 - f^{p-1}$. If $\alpha = (a_1,\ldots,a_n)$ is a zero of f, then $g(\alpha) = 1$. If α is not a zero of f, then $f(\alpha)^{p-1} = 1$ and $g(\alpha) = 0$. Hence the number N of zeros of f satisfies

$$N \equiv \sum_{\alpha \in K^n} g(\alpha) \bmod p.$$

We will complete the proof by showing that

$$\sum_{\alpha \in K^n} g(\alpha) = 0.$$

Since g has degree less than $n(p-1)$, it is a constant linear combination of polynomials of the form $x_1^{k_1}\cdots x_n^{k_n}$, where $k_1 + \ldots + k_n < n(p-1)$. Thus $k_j < p-1$ for at least one j. Since

$$\sum_{\alpha \in K^n} a_1^{k_1}\cdots a_n^{k_n} = \left(\sum_{a_1 \in K} a_1^{k_1}\right)\cdots\left(\sum_{a_n \in K} a_n^{k_n}\right),$$

it is enough to show that $S_k := \sum_{a \in K} a^k$ is zero for $0 \le k < p-1$. If $k = 0$, then $a^k = 1$ and $S_0 = p\cdot 1 = 0$. Suppose $1 \le k < p-1$ and let b be a generator for the multiplicative group $K^\times$ of K. Then $c := b^k \neq 1$ and

$$S_k = \sum_{j=1}^{p-1} c^j = c(c^{p-1} - 1)/(c - 1) = 0. \quad \square$$

The general case of Proposition 34 was first proved by Warning (1936), after the particular case had been proved by Chevalley (1936). As an illustration, the particular case implies that, for any integers a,b,c and any prime p, the congruence $ax^2 + by^2 + cz^2 \equiv 0 \bmod p$ has a solution in integers x,y,z not all divisible by p.

If m is not a prime, then $\mathbb{Z}_{(m)}$ is not a field. However, we now show that the group $\mathbb{Z}_{(m)}^{\times}$ is cyclic also if $m = p^2$ is the square of a prime.

Let g be a primitive root of p. It follows from the binomial theorem that

$$(g + p)^p \equiv g^p \mod p^2.$$

Hence, if $g^p \equiv g \mod p^2$, then $(g + p)^p \not\equiv g + p \mod p^2$. Thus, by replacing g by $g + p$ if necessary, we may assume that $g^{p-1} \not\equiv 1 \mod p^2$. If the order of g in $\mathbb{Z}_{(p^2)}^{\times}$ is d, then d divides $\varphi(p^2) = p(p - 1)$. But $\varphi(p) = p - 1$ divides d, since $g^d \equiv 1 \mod p^2$ implies $g^d \equiv 1 \mod p$ and g is a primitive root of p. Since p is prime and $d \neq p - 1$, it follows that $d = p(p - 1)$, i.e. $\mathbb{Z}_{(p^2)}^{\times}$ is cyclic with g as generator.

We briefly state some further results about primitive roots, although we will not use them. Gauss (*D.A.*, §89-92) showed that *the group* $\mathbb{Z}_{(m)}^{\times}$ *is cyclic if and only if* $m \in \{2, 4, p^k, 2p^k\}$, where p is an odd prime and $k \in \mathbb{N}$. Evidently 1 is a primitive root of 2 and 3 is a primitive root of 4. *If g is a primitive root of p^2, where p is an odd prime, then g is a primitive root of p^k for every $k \in \mathbb{N}$; and if $g' = g$ or $g + p^k$, according as g is odd or even, then g' is a primitive root of $2p^k$.*

By Fermat's little theorem, if p is prime, then $a^{p-1} \equiv 1 \mod p$ for every $a \in \mathbb{Z}$ such that $(a,p) = 1$. With the aid of primitive roots we will now show that there exist also composite integers n such that $a^{n-1} \equiv 1 \mod n$ for every $a \in \mathbb{Z}$ such that $(a,n) = 1$.

PROPOSITION 35 *For any integer $n > 1$, the following two statements are equivalent:*

(i) $a^{n-1} \equiv 1 \mod n$ *for every integer a such that $(a,n) = 1$;*

(ii)   *n is a product of distinct primes and, for each prime $p|n$, $p - 1$ divides $n - 1$.*

Proof Suppose first that (i) holds and assume that, for some prime p, $p^2|n$. As we have just proved, there exists a primitive root g of p^2. Evidently $p \nmid g$. It is easily seen that there exists $c \in \mathbb{N}$ such that $a = g + cp^2$ is relatively prime to n; in fact we can take c to be the product of the distinct prime factors of n, other than p, which do not divide g. Since n divides $a^{n-1} - 1$, also p^2 divides $a^{n-1} - 1$. But a, like g, is a primitive root of p^2, and so its order in $\mathbb{Z}_{(p^2)}^{\times}$ is $\varphi(p^2) = p(p - 1)$. Hence $p(p - 1)$ divides $n - 1$. But this contradicts $p|n$.

Now let p be any prime divisor of n and let g be a primitive root of p. In the same way as before, there exists $c \in \mathbb{N}$ such that $a = g + cp$ is relatively prime to n. Arguing as before, we see that $\varphi(p) = p - 1$ divides $n - 1$. This proves that (i) implies (ii).

Suppose next that (ii) holds and let a be any integer relatively prime to n. If p is a prime factor of n, then $p \nmid a$ and hence $a^{p-1} \equiv 1 \mod p$. Since $p - 1$ divides $n - 1$, it follows that

$a^{n-1} \equiv 1 \bmod p$. Thus $a^{n-1} - 1$ is divisible by each prime factor of n and hence, since n is squarefree, also by n itself. $\square$

Proposition 35 was proved by Carmichael (1910), and a composite integer n with the equivalent properties stated in the proposition is said to be a *Carmichael number*. Any Carmichael number n must be odd, since it has an odd prime factor p such that $p - 1$ divides $n - 1$. Furthermore a Carmichael number must have more than two prime factors. For assume $n = pq$, where $1 < p < q < n$ and $q - 1$ divides $n - 1$. Since $q \equiv 1 \bmod (q - 1)$, it follows that

$$0 \equiv pq - 1 \equiv p - 1 \bmod (q - 1),$$

which contradicts $p < q$.

The composite integer $561 = 3{\times}11{\times}17$ is a Carmichael number, since 560 is divisible by 2,10 and 16, and it is in fact the smallest Carmichael number. The taxicab number 1729, which Hardy reckoned to Ramanujan was uninteresting, is also a Carmichael number, since $1729 = 7{\times}13{\times}19$. Indeed it is not difficult to show that if p, $2p - 1$ and $3p - 2$ are all primes, with $p > 3$, then their product is a Carmichael number. Recently Alford, Granville and Pomerance (1994) confirmed a long-standing conjecture by proving that there are infinitely many Carmichael numbers.

Our next topic is of greater importance. Many arithmetical problems require for their solution the determination of an integer which is congruent to several given integers according to various given moduli. We consider first a simple, but important, special case.

PROPOSITION 36 *Let $m = m'm''$, where m' and m'' are relatively prime integers. Then, for any integers a',a'', there exists an integer a, which is uniquely determined $\bmod m$, such that*

$$a \equiv a' \bmod m', \quad a \equiv a'' \bmod m''.$$

Moreover, a is relatively prime to m if and only if a' is relatively prime to m' and a'' is relatively prime to m''.

Proof By Proposition 22, there exist integers c',c'' such that

$$c'm'' \equiv 1 \bmod m', \quad c''m' \equiv 1 \bmod m''.$$

Thus $e' := c'm''$ is congruent to 1 mod m' and congruent to 0 mod m''. Similarly $e'' := c''m'$ is congruent to 0 mod m' and congruent to 1 mod m''. It follows that $a = a'e' + a''e''$ is congruent to a' mod m' and congruent to a'' mod m''.

It is evident that if $b \equiv a \bmod m$, then also $b \equiv a' \bmod m'$ and $b \equiv a'' \bmod m''$. Conversely, if b satisfies these two congruences, then $b - a \equiv 0 \bmod m'$ and $b - a \equiv 0 \bmod m''$. Hence $b - a \equiv 0 \bmod m$, by Proposition 3(i).

Since m' and m'' are relatively prime, it follows from Proposition 3(iv) that $(a,m) = 1$ if and only if $(a,m') = (a,m'') = 1$. Since $a \equiv a' \bmod m'$ implies $(a,m') = (a',m')$, and $a \equiv a''$ $\bmod m''$ implies $(a,m'') = (a'',m'')$, this proves the last statement of the proposition. $\square$

In algebraic terms, Proposition 36 says that if $m = m'm''$, where m' and m'' are relatively prime integers, then the ring $\mathbb{Z}_{(m)}$ is (isomorphic to) the direct sum of the rings $\mathbb{Z}_{(m')}$ and $\mathbb{Z}_{(m'')}$. Furthermore, the group $\mathbb{Z}_{(m)}^{\times}$ is (isomorphic to) the direct product of the groups $\mathbb{Z}_{(m')}^{\times}$ and $\mathbb{Z}_{(m'')}^{\times}$.

Proposition 36 can be considerably generalized:

PROPOSITION 37 *For any integers $m_1,\dots,m_n$ and $a_1,\dots,a_n$, the simultaneous congruences*

$$x \equiv a_1 \bmod m_1, \dots , x \equiv a_n \bmod m_n$$

have a solution x if and only if

$$a_j \equiv a_k \bmod (m_j,m_k) \; for \; 1 \le j < k \le n.$$

Moreover, y is also a solution if and only if

$$y \equiv x \bmod [m_1,\dots,m_n].$$

Proof The necessity of the conditions is trivial. For if x is a solution and if $d_{jk} = (m_j,m_k)$ is the greatest common divisor of m_j and m_k, then $a_j \equiv x \equiv a_k \bmod d_{jk}$. Also, if y is another solution, then $y - x$ is divisible by $m_1,\dots,m_n$ and hence also by their least common multiple $[m_1,\dots,m_n]$.

We prove the sufficiency of the conditions by induction on n. Suppose first that $n = 2$ and $a_1 \equiv a_2 \bmod d$, where $d = (m_1,m_2)$. By the Bézout identity,

$$d = x_1 m_1 - x_2 m_2$$

for some $x_1,x_2 \in \mathbb{Z}$. Since $a_1 - a_2 = kd$ for some $k \in \mathbb{Z}$, it follows that

$$x := a_1 - kx_1 m_1 = a_2 - kx_2 m_2$$

is a solution.

Suppose next that $n > 2$ and the result holds for all smaller values of n. Then there exists $x' \in \mathbb{Z}$ such that

$$x' \equiv a_i \bmod m_i \text{ for } 1 \le i < n,$$

and x' is uniquely determined mod m', where $m' = [m_1,\ldots,m_{n-1}]$. Since any solution of the two congruences

$$x \equiv x' \bmod m', \ \ x \equiv a_n \bmod m_n$$

is also a solution of the given congruences, we need only show that $x' \equiv a_n \bmod (m',m_n)$. But, by the distributive law connecting greatest common divisors and least common multiples,

$$(m',m_n) \ = \ [(m_1,m_n),\ldots,(m_{n-1},m_n)].$$

Since $x' \equiv a_i \equiv a_n \bmod (m_i,m_n)$ for $1 \le i < n$, it follows that $x' \equiv a_n \bmod (m',m_n)$. $\square$

COROLLARY 38 *Let $m_1,\ldots,m_n$ be integers, any two of which are relatively prime, and let $m = m_1 \cdots m_n$ be their product. Then, for any given integers $a_1,\ldots,a_n$, there is a unique integer x mod m such that*

$$x \equiv a_1 \bmod m_1, \ \ldots \ , x \equiv a_n \bmod m_n.$$

Moreover, x is relatively prime to m if and only if a_i is relatively prime to m_i for $1 \le i \le n$. $\square$

Corollary 38 can also be proved by an extension of the argument used to prove Proposition 36. Both Proposition 37 and Corollary 38 are referred to as the *Chinese remainder theorem*. Sunzi (4th century A.D.) gave a procedure for obtaining the solution $x = 23$ of the simultaneous congruences

$$x \equiv 2 \bmod 3, \ x \equiv 3 \bmod 5, \ x \equiv 2 \bmod 7.$$

Qin Jiushao (1247) gave a general procedure for solving simultaneous congruences, the moduli of which need not be pairwise relatively prime, although he did not state the necessary condition for the existence of a solution. The problem appears to have its origin in the construction of calendars.

6 Sums of squares

Which positive integers n can be represented as a sum of two squares of integers? The question is answered completely by the following proposition, which was stated by Girard

(1625). Fermat (1645) claimed to have a proof, but the first published proof was given by Euler (1754).

PROPOSITION 39 *A positive integer n can be represented as a sum of two squares if and only if for each prime $p \equiv 3$ mod 4 that divides n, the highest power of p dividing n is even.*

Proof We observe first that, since

$$(x^2 + y^2)(u^2 + v^2) = (xu + yv)^2 + (xv - yu)^2,$$

any product of sums of two squares is again a sum of two squares.

Suppose $n = x^2 + y^2$ for some integers x,y and that n is divisible by a prime $p \equiv 3$ mod 4. Then $x^2 \equiv -y^2$ mod p. But -1 is not a square in the field $\mathbb{F}_p$, by Corollary 29. Consequently we must have $y^2 \equiv x^2 \equiv 0$ mod p. Thus p divides both x and y. Hence p^2 divides n and $(n/p)^2 = (x/p)^2 + (y/p)^2$. It follows by induction that the highest power of p which divides n is even.

Thus the condition in the statement of the proposition is necessary. Suppose now that this condition is satisfied. Then $n = qm^2$, where q is square-free and the only possible prime divisors of q are 2 and primes $p \equiv 1$ mod 4. Since $m^2 = m^2 + 0^2$ and $2 = 1^2 + 1^2$, it follows from our initial observation that n is a sum of two squares if every prime $p \equiv 1$ mod 4 is a sum of two squares. Following Gauss (1832), we will prove this with the aid of complex numbers.

A complex number $\gamma = a + bi$ is said to be a *Gaussian integer* if $a,b \in \mathbb{Z}$. The set of all Gaussian integers will be denoted by $\mathcal{G}$. Evidently $\gamma \in \mathcal{G}$ implies $\bar{\gamma} \in \mathcal{G}$, where $\bar{\gamma} = a - bi$ is the complex conjugate of γ. Moreover $\alpha,\beta \in \mathcal{G}$ implies $\alpha \pm \beta \in \mathcal{G}$ and $\alpha\beta \in \mathcal{G}$. Thus $\mathcal{G}$ is a commutative ring. In fact $\mathcal{G}$ is an integral domain, since it is a subset of the field $\mathbb{C}$. We are going to show that $\mathcal{G}$ can be given the structure of a Euclidean domain.

Define the *norm* of a complex number $\gamma = a + bi$ to be

$$N(\gamma) = \gamma\bar{\gamma} = a^2 + b^2.$$

Then $N(\gamma) \geq 0$, with equality if and only if $\gamma = 0$, and $N(\gamma_1\gamma_2) = N(\gamma_1)N(\gamma_2)$. If $\gamma \in \mathcal{G}$, then $N(\gamma)$ is an ordinary integer. Furthermore, γ is a unit in $\mathcal{G}$, i.e. γ divides 1 in $\mathcal{G}$, if and only if $N(\gamma) = 1$.

We wish to show that if $\alpha,\beta \in \mathcal{G}$ and $\alpha \neq 0$, then there exist $\kappa,\rho \in \mathcal{G}$ such that

$$\beta = \kappa\alpha + \rho, \quad N(\rho) < N(\alpha).$$

We have $\beta\alpha^{-1} = r + si$, where $r,s \in \mathbb{Q}$. Choose $a,b \in \mathbb{Z}$ so that

$$|r - a| \le 1/2, \quad |s - b| \le 1/2.$$

If $\kappa = a + bi$, then $\kappa \in \mathcal{G}$ and

$$N(\beta\alpha^{-1} - \kappa) \le 1/4 + 1/4 = 1/2 < 1.$$

Hence if $\rho = \beta - \kappa\alpha$, then $\rho \in \mathcal{G}$ and $N(\rho) < N(\alpha)$.

It follows that we can apply to $\mathcal{G}$ the whole theory of divisibility in a Euclidean domain. Now let p be a prime such that $p \equiv 1 \bmod 4$. We will show that p is a sum of two squares by constructing $\beta \in \mathcal{G}$ for which $N(\beta) = p$.

By Corollary 29, there exists an integer a such that $a^2 \equiv -1 \bmod p$. Put $\alpha = a + i$. Then $N(\alpha) = \alpha\bar{\alpha} = a^2 + 1$ is divisible by p in $\mathbb{Z}$ and hence also in $\mathcal{G}$. However, neither α nor $\bar{\alpha}$ is divisible by p in $\mathcal{G}$, since αp^{-1} and $\bar{\alpha} p^{-1}$ are not in $\mathcal{G}$. Thus p is not a prime in $\mathcal{G}$ and consequently, since $\mathcal{G}$ is a Euclidean domain, it has a factorization $p = \beta\gamma$, where neither β nor γ is a unit. Hence $N(\beta) > 1$, $N(\gamma) > 1$. Since

$$N(\beta)N(\gamma) = N(p) = p^2,$$

it follows that $N(\beta) = N(\gamma) = p$. $\quad\square$

Proposition 39 solves the problem of representing a positive integer as a sum of two squares. What if we allow more than two squares? When congruences were first introduced in §5, it was observed that a positive integer $a \equiv 7 \bmod 8$ could not be represented as a sum of three squares. It was first completely proved by Gauss (1801) that a positive integer can be represented as a sum of three squares if and only if it is not of the form $4^n a$, where $n \ge 0$ and $a \equiv 7 \bmod 8$. The proof of this result is more difficult, and will be given in Chapter VII.

It was conjectured by Bachet (1621) that *every* positive integer can be represented as a sum of four squares. Fermat claimed to have a proof, but the first published proof was given by Lagrange (1770), using earlier ideas of Euler (1751). The proof of the four-squares theorem we will give is similar to that just given for the two-squares theorem, with complex numbers replaced by quaternions.

PROPOSITION 40 *Every positive integer n can be represented as a sum of four squares.*

Proof A quaternion $\gamma = a + bi + cj + dk$ will be said to be a *Hurwitz integer* if a,b,c,d are either all integers or all halves of odd integers. The set of all Hurwitz integers will be denoted by $\mathcal{H}$. Evidently $\gamma \in \mathcal{H}$ implies $\bar{\gamma} \in \mathcal{H}$, where $\bar{\gamma} = a - bi - cj - dk$. Moreover $\alpha,\beta \in \mathcal{H}$ implies $\alpha \pm \beta \in \mathcal{H}$. We will show that $\alpha,\beta \in \mathcal{H}$ also implies $\alpha\beta \in \mathcal{H}$.

Evidently $\gamma \in \mathcal{H}$ if and only if it can be written in the form $\gamma = a_0 h + a_1 i + a_2 j + a_3 k$, where $a_0, a_1, a_2, a_3 \in \mathbb{Z}$ and $h = (1 + i + j + k)/2$. It is obvious that the product of h with i, j or k is again in $\mathcal{H}$ and it is easily verified that $h^2 = h - 1$. It follows that $\mathcal{H}$ is closed under multiplication and hence is a ring.

Define the *norm* of a quaternion $\gamma = a + bi + cj + dk$ to be

$$N(\gamma) = \gamma\bar{\gamma} = a^2 + b^2 + c^2 + d^2.$$

Then $N(\gamma) \geq 0$, with equality if and only if $\gamma = 0$. Moreover, since $\overline{\gamma_1\gamma_2} = \bar{\gamma}_2\bar{\gamma}_1$,

$$N(\gamma_1\gamma_2) = \gamma_1\gamma_2\bar{\gamma}_2\bar{\gamma}_1 = \gamma_1\bar{\gamma}_1\gamma_2\bar{\gamma}_2 = N(\gamma_1)N(\gamma_2).$$

If $\gamma \in \mathcal{H}$, then $N(\gamma) = \gamma\bar{\gamma} \in \mathcal{H}$ and hence $N(\gamma)$ is an ordinary integer. Furthermore, γ is a unit in $\mathcal{H}$, i.e. γ divides 1 in $\mathcal{H}$, if and only if $N(\gamma) = 1$.

We now show that a Euclidean algorithm may be defined on $\mathcal{H}$. Suppose $\alpha, \beta \in \mathcal{H}$ and $\alpha \neq 0$. Then

$$\beta\alpha^{-1} = r_0 + r_1 i + r_2 j + r_3 k,$$

where $r_0, r_1, r_2, r_3 \in \mathbb{Q}$. If $\kappa = a_0 h + a_1 i + a_2 j + a_3 k$, then

$$\beta\alpha^{-1} - \kappa = (r_0 - a_0/2) + (r_1 - a_0/2 - a_1)i + (r_2 - a_0/2 - a_2)j + (r_3 - a_0/2 - a_3)k.$$

We can choose $a_0 \in \mathbb{Z}$ so that $|2r_0 - a_0| \leq 1/2$ and then $a_\nu \in \mathbb{Z}$ so that $|r_\nu - a_0/2 - a_\nu| \leq 1/2$ $(\nu = 1, 2, 3)$. Then $\kappa \in \mathcal{H}$ and

$$N(\beta\alpha^{-1} - \kappa) \leq 1/16 + 3/4 = 13/16 < 1.$$

Thus if we set $\rho = \beta - \kappa\alpha$, then $\rho \in \mathcal{H}$ and

$$N(\rho) = N(\beta\alpha^{-1} - \kappa)N(\alpha) < N(\alpha).$$

By repeating this division process finitely many times we see that any $\alpha, \beta \in \mathcal{H}$ have a *greatest common right divisor* $\delta = (\alpha, \beta)_r$. Furthermore, there is a *left Bézout identity*: $\delta = \xi\alpha + \eta\beta$ for some $\xi, \eta \in \mathcal{H}$.

If a positive integer n is a sum of four squares, say $n = a^2 + b^2 + c^2 + d^2$, then $n = \gamma\bar{\gamma}$, where $\gamma = a + bi + cj + dk \in \mathcal{H}$. Since the norm of a product is the product of the norms, it follows that any product of sums of four squares is again a sum of four squares. Hence to prove the proposition we need only show that any prime p is a sum of four squares.

We show first that there exist integers a, b such that $a^2 + b^2 \equiv -1 \bmod p$. This follows from the illustration given for Proposition 34, but we will give a direct proof.

If $p = 2$, we can take $a = 1$, $b = 0$. If $p \equiv 1$ mod 4 then, by Corollary 29, there exists an integer a such that $a^2 \equiv -1$ mod p and we can take $b = 0$. Suppose now that $p \equiv 3$ mod 4. Let c be the least positive quadratic non-residue of p. Then $c \geq 2$ and $c - 1$ is a quadratic residue of p. On the other hand, -1 is a quadratic non-residue of p, by Corollary 29. Hence, by Proposition 28, $-c$ is a quadratic residue. Thus there exist integers a,b such that

$$a^2 \equiv -c, \ b^2 \equiv c - 1 \text{ mod } p,$$

and then $a^2 + b^2 \equiv -1$ mod p.

Put $\alpha = 1 + ai + bj$. Then p divides $N(\alpha) = \alpha\overline{\alpha} = 1 + a^2 + b^2$ in $\mathbb{Z}$ and hence also in $\mathcal{H}$. However, p does not divide either α or $\overline{\alpha}$ in $\mathcal{H}$, since αp^{-1} and $\overline{\alpha} p^{-1}$ are not in $\mathcal{H}$.

Let $\gamma = (p,\alpha)_r$. Then $p = \beta\gamma$ for some $\beta \in \mathcal{H}$. If β were a unit, p would be a right divisor of γ and hence also of α, which is a contradiction. Therefore $N(\beta) > 1$. Evidently $\gamma\overline{\alpha}$ is a common right divisor of $p\overline{\alpha}$ and $\alpha\overline{\alpha}$, and the Bézout representation for γ implies that $\gamma\overline{\alpha} = (p\overline{\alpha},\alpha\overline{\alpha})_r$. Since $p\overline{\alpha} = \overline{\alpha}p$ and p divides $\alpha\overline{\alpha}$, it follows that p is a right divisor of $\gamma\overline{\alpha}$. Since p does not divide $\overline{\alpha}$, γ is not a unit and hence $N(\gamma) > 1$. Since

$$N(\beta)N(\gamma) = N(p) = p^2,$$

we must have $N(\beta) = N(\gamma) = p$.

Thus if $\gamma = c_0 + c_1 i + c_2 j + c_3 k$, then $c_0^2 + c_1^2 + c_2^2 + c_3^2 = p$. If $c_0,...,c_3$ are all integers, we are finished. Otherwise $c_0,...,c_3$ are all halves of odd integers. Hence we can write $c_\nu = 2d_\nu + e_\nu$, where $d_\nu \in \mathbb{Z}$ and $e_\nu = \pm 1/2$. If we put

$$\delta = d_0 + d_1 i + d_2 j + d_3 k, \ \ \varepsilon = e_0 + e_1 i + e_2 j + e_3 k,$$

then $\gamma = 2\delta + \varepsilon$ and $N(\varepsilon) = 1$. Hence $\theta := \gamma\overline{\varepsilon} = 2\delta\overline{\varepsilon} + 1$ has all its coordinates integers and $N(\theta) = N(\gamma) = p$. $\square$

In his *Meditationes Algebraicae*, which also contains the first statement in print of Wilson's theorem, Waring (1770) stated that every positive integer is a sum of at most 4 positive integral squares, of at most 9 positive integral cubes and of at most 19 positive integral fourth powers. The statement concerning squares was proved by Lagrange in the same year, as we have seen. The statement concerning cubes was first proved by Wieferich (1909), with a gap filled by Kempner (1912), and the statement concerning fourth powers was first proved by Balasubramanian, Deshouillers and Dress (1986).

In a later edition of his book, Waring (1782) raised the same question for higher powers. *Waring's problem* was first solved by Hilbert (1909), who showed that, for each $k \in \mathbb{N}$, there exists $\gamma_k \in \mathbb{N}$ such that every positive integer is a sum of at most γ_k k-th powers. The least possible value of γ_k is traditionally denoted by $g(k)$. For example, $g(2) = 4$, since $7 = 2^2 + 3 \cdot 1^2$ is not a sum of less than 4 squares.

A lower bound for $g(k)$ was already derived by Euler (c. 1772). Let $m = \lfloor (3/2)^k \rfloor$ denote the greatest integer $\leq (3/2)^k$ and take

$$n = 2^k m - 1.$$

Since $1 \leq n < 3^k$, the only k-th powers of which n can be the sum are $0^k, 1^k$ and 2^k. Since the number of powers 2^k must be less than m, and since $n = (m - 1)2^k + (2^k - 1)1^k$, the least number of k-th powers with sum n is $m + 2^k - 2$. Hence $g(k) \geq w(k)$, where

$$w(k) = \lfloor (3/2)^k \rfloor + 2^k - 2.$$

In particular,

$$w(2) = 4, \quad w(3) = 9, \quad w(4) = 19, \quad w(5) = 37, \quad w(6) = 73.$$

By the results stated above, $g(k) = w(k)$ for $k = 2,3,4$ and this has been shown to hold also for $k = 5$ by Chen (1964) and for $k = 6$ by Pillai (1940).

Hilbert's method of proof yielded rather large upper bounds for $g(k)$. A completely new approach was developed in the 1920's by Hardy and Littlewood, using their analytic 'circle' method. They showed that, for each $k \in \mathbb{N}$, there exists $\Gamma_k \in \mathbb{N}$ such that every sufficiently large positive integer is a sum of at most Γ_k k-th powers. The least possible value of Γ_k is traditionally denoted by $G(k)$. For example, $G(2) = 4$, since no positive integer $n \equiv 7 \bmod 8$ is a sum of less than four squares. Davenport (1939) showed that $G(4) = 16$, but these are the only two values of k for which today $G(k)$ is known exactly.

It is obvious that $G(k) \leq g(k)$, and in fact $G(k) < g(k)$ for all $k > 2$. In particular, Dickson (1939) showed that 23 and 239 are the only positive integers which require the maximum 9 cubes. Hardy and Littlewood obtained the upper bound $G(k) \leq (k - 2)2^{k-1} + 5$, but this has been repeatedly improved by Hardy and Littlewood themselves, Vinogradov and others. For example, Wooley (1992) has shown that $G(k) \leq k(\log k + \log \log k + O(1))$.

By using the upper bound for $G(k)$ of Vinogradov (1935), it was shown by Dickson, Pillai and Niven (1936-1944) that $g(k) = w(k)$ for any $k > 6$, *provided*

$$(3/2)^k - \lfloor (3/2)^k \rfloor \leq 1 - \lfloor (3/2)^k \rfloor / 2^k.$$

It is possible that this inequality holds for every $k \in \mathbb{N}$. For a given k, it may be checked by direct calculation, and Kubina and Wunderlich (1990) have verified in this way that the inequality holds if $k \leq 471600000$. Furthermore, using a p-adic extension by Ridout (1957) of the theorem of Roth (1955) on the approximation of algebraic numbers by rationals, Mahler

(1957) proved that there exists $k_0 \in \mathbb{N}$ such that the inequality holds for all $k \geq k_0$. However, the proof does not provide a means of estimating k_0.

Thus we have the bizarre situation that $G(k)$ is known for only two values of k, that $g(k)$ is known for a vast number of values of k and is given by a simple formula, probably for all k, but the information about $g(k)$ is at present derived from information about $G(k)$. Is it too much to hope that an examination of the numerical data will reveal some pattern in the fractional parts of $(3/2)^k$?

7 Further remarks

There are many good introductory books on the theory of numbers, e.g. Davenport [4], LeVeque [28] and Scholz [41]. More extensive accounts are given in Hardy and Wright [15], Hua [18], Narkiewicz [33] and Niven *et al.* [34].

Historical information is provided by Dickson [5], Smith [42] and Weil [46], as well as the classics Euclid [11], Gauss [13] and Dirichlet [6]. Gauss's masterpiece is quoted here and in the text as '*D.A.*'

The reader is warned that, besides its use in §1, the word 'lattice' also has quite a different mathematical meaning, which will be encountered in Chapter VIII.

The basic theory of divisibility is discussed more thoroughly than in the usual texts by Stieltjes [43]. For Proposition 6, see Prüfer [35]. In the theory of groups, Schreier's refinement theorem and the Jordan–Hölder theorem may be viewed as generalizations of Propositions 6 and 7. These theorems are stated and proved in Chapter I, §3 of Lang [23]. The fundamental theorem of arithmetic (Proposition 7) is usually attributed to Gauss (*D.A.*, §16). However, it is really contained in Euclid's *Elements* (Book VII, Proposition 31 and Book IX, Proposition 14), except for the appropriate terminology. Perhaps this is why Euler and his contemporaries simply assumed it without proof.

Generalizations of the fundamental theorem of arithmetic to other algebraic structures are discussed in Chapter 2 of Jacobson [21]. For factorial domains, see Samuel [39].

Our discussion of the fundamental theorem did not deal with the practical problems of deciding if a given integer is prime or composite and, in the latter case, of obtaining its factorization into primes. Evidently if the integer a is composite, its least prime factor p satisfies $p^2 \leq a$. In former days one used this observation in conjunction with tables, such as [24], [25], [26]. With new methods and supercomputers, the primality of integers with hundreds of digits can now be determined without difficulty. The progress in this area may be traced through the survey articles [48], [7] and [27]. Factorization remains a more difficult problem, and this difficulty has found an important application in *public-key cryptography*; see Rivest *et al.* [37].

For Proposition 12, cf. Hillman and Hoggatt [17]. A proof that the ring of all algebraic integers is a Bézout domain is given on p. 86 of Mann [31]. The ring of all functions which are holomorphic in a given region was shown to be a Bézout domain by Wedderburn (1915); see Narasimhan [32].

For Gauss's version of Proposition 17, see *D.A.*, §42. It is natural to ask if Corollary 18 remains valid if the polynomial ring $R[t]$ is replaced by the ring $R[[t]]$ of formal power series. The ring $K[[t_1,...,t_m]]$ of all formal power series in finitely many indeterminates with coefficients from an arbitrary field K is indeed a factorial domain. However, if R is a factorial domain, the integral domain $R[[t]]$ of all formal power series in t with coefficients from R need not be factorial. For an example in which R is actually a complete local ring, see Salmon [38].

For generalizations of Eisenstein's irreducibility criterion (Proposition 19), see Gao [12]. Proposition 21 is proved in Rhai [36]. Euclidean domains are studied further in Samuel [40]. Quadratic fields $\mathbb{Q}(\sqrt{d})$ whose ring of integers $\mathcal{O}_d$ is Euclidean are discussed in Clark [3], Dubois and Steger [8] and Eggleton *et al.* [9].

Congruences are discussed in all the books on number theory cited above. In connection with Lemma 32 we mention a result of Frobenius (1895). Frobenius proved that if G is a finite group of order n and if d is a positive divisor of n, then the number of elements of G whose order divides d is a multiple of d. He conjectured that if the number is exactly d, then these elements form a (normal) subgroup of G. The conjecture can be reduced to the case where G is simple, since a counterexample of minimal order must be a noncyclic simple group. By appealing to the recent classification of all finite simple groups (see Chapter V, §7), the proof of the conjecture was completed by Iiyori and Yamaki [20].

There is a table of primitive roots on pp. 52-56 of Hua [18]. For more extensive tables, see Western and Miller [47].

It is easily seen that an even square is never a primitive root, that an odd square (including 1) is a primitive root only for the prime $p = 2$, and that -1 is a primitive root only for the primes $p = 2,3$. Artin (1927) conjectured that if the integer a is not a square or -1, then it is a primitive root for infinitely many primes p. (A quantitative form of the conjecture is considered in Chapter IX.) If the conjecture is not true, then it is almost true, since it has been shown by Heath-Brown [16] that there are at most 3 square-free positive integers a for which it fails.

A finite subgroup of the multiplicative group of a division ring need not be cyclic. For example, if $\mathbb{H}$ is the division ring of Hamilton's quaternions, $\mathbb{H}^\times$ contains the non-cyclic subgroup $\{\pm 1, \pm i, \pm j, \pm k\}$ of order 8. All possible finite subgroups of the multiplicative group of a division ring have been determined (with the aid of *class field theory*) by Amitsur [2].

For Carmichael numbers, see Alford *et al.* [1].

Galois (1830) showed that there were other finite fields besides $\mathbb{F}_p$ and indeed, as Moore (1893) later proved, he found them all. Finite fields have the following basic properties:

(i)   The number of elements in a finite field is a prime power p^n, where $n \in \mathbb{N}$ and the prime p is the characteristic of the field.

(ii)   For any prime power $q = p^n$, there is a finite field $\mathbb{F}_q$ containing exactly q elements. Moreover the field $\mathbb{F}_q$ is unique, up to isomorphism, and is the splitting field of the polynomial $t^q - t$ over $\mathbb{F}_p$.

(iii)   For any finite field $\mathbb{F}_q$, the multiplicative group $\mathbb{F}_q^\times$ of nonzero elements is cyclic.

(iv)   If $q = p^n$, the map $\sigma: a \to a^p$ is an automorphism of $\mathbb{F}_q$ and the distinct automorphisms of $\mathbb{F}_q$ are the powers σ^k ($k = 0, 1, \ldots, n - 1$).

The theorem of Chevalley and Warning (Proposition 34) extends immediately to arbitrary finite fields. Proofs and more detailed information on finite fields may be found in Lidl and Niederreiter [30] and in Joly [22].

A celebrated theorem of Wedderburn (1905) states that any finite division ring is a field, i.e. the commutative law of multiplication is a consequence of the other field axioms if the number of elements is finite. Here is a purely algebraic proof.

Assume there exists a finite division ring which is not a field and let D be one of minimum cardinality. Let C be the centre of D and $a \in D \setminus C$. The set M of all elements of D which commute with a is a field, since it is a division ring but not the whole of D. Evidently M is a maximal subfield of D which contains a. If $[D{:}C] = n$ and $[M{:}C] = m$ then, by Proposition I.32, $[D{:}M] = m$ and $n = m^2$. Thus m is independent of a.

If C has cardinality q, then D has cardinality q^n, M has cardinality q^m and the number of conjugates of a in D is $(q^n - 1)/(q^m - 1)$. Since this holds for every $a \in D \setminus C$, the partition of the multiplicative group of D into conjugacy classes shows that

$$q^n - 1 = q - 1 + r(q^n - 1)/(q^m - 1)$$

for some positive integer r. Hence $q - 1$ is divisible by

$$(q^n - 1)/(q^m - 1) = 1 + q^m + \ldots + q^{m(m-1)}.$$

Since $n > m > 1$, this is a contradiction.

For the history of the Chinese remainder theorem (not only in China), see Libbrecht [29].

We have developed the arithmetic of quaternions only as far as is needed to prove the four-squares theorem. A fuller account was given in the original (1896) paper of Hurwitz [19]. For more information about sums of squares, see Grosswald [14] and also Chapter XIII. For Waring's problem, see Waring [45], Ellison [10] and Vaughan [44].

8 Selected references

[1]   W.R. Alford, A. Granville and C. Pomerance, There are infinitely many Carmichael numbers, *Ann. Math.* **139** (1994), 703-722.

[2]   S.A. Amitsur, Finite subgroups of division rings, *Trans. Amer. Math. Soc.* **80** (1955), 361-386.

[3]   D.A. Clark, A quadratic field which is Euclidean but not norm-Euclidean, *Manuscripta Math.* **83** (1994), 327-330.

[4]   H. Davenport, *The higher arithmetic*, 7th ed., Cambridge University Press, 1999.

[5]   L.E. Dickson, *History of the theory of numbers*, 3 vols., Carnegie Institute, Washington, D.C., 1919-1923. [Reprinted, Chelsea, New York, 1992.]

[6]   P.G.L. Dirichlet, *Lectures on number theory,* with supplements by R. Dedekind, English transl. by J. Stillwell, American Mathematical Society, Providence, R.I., 1999. [German original, 1894]

[7]   J.D. Dixon, Factorization and primality tests, *Amer. Math. Monthly* **91** (1984), 333-352.

[8]   D.W. Dubois and A. Steger, A note on division algorithms in imaginary quadratic fields, *Canad. J. Math.* **10** (1958), 285-286.

[9]   R.B. Eggleton, C.B. Lacampagne and J.L. Selfridge, Euclidean quadratic fields, *Amer. Math. Monthly* **99** (1992), 829-837.

[10] W.J. Ellison, Waring's problem, *Amer. Math. Monthly* **78** (1971), 10-36.

[11] Euclid, *The thirteen books of Euclid's elements*, English translation by T.L. Heath, 2nd ed., reprinted in 3 vols., Dover, New York, 1956.

[12] S. Gao, Absolute irreducibility of polynomials via Newton polytopes, *J. Algebra* **237** (2001), 501-520.

[13] C.F. Gauss, *Disquisitiones arithmeticae*, English translation by A.A. Clarke, Yale University Press, New Haven, Conn., 1966. (This translation has some defects; see D. Shanks, *Math. Comp.* **20** (1966), 617-618.) [Latin original, 1801]

[14] E. Grosswald, *Representations of integers as sums of squares*, Springer-Verlag, New York, 1985.

[15] G.H. Hardy and E.M. Wright, *An introduction to the theory of numbers*, 5th ed., Clarendon Press, Oxford, 1979.

[16] D.R. Heath-Brown, Artin's conjecture for primitive roots, *Quart. J. Math. Oxford Ser.* (2) **37** (1986), 27-38.

[17] A.P. Hillman and V.E. Hoggatt, Exponents of primes in generalized binomial coefficients, *J. Reine Angew. Math.* **262/3** (1973), 375-380.

[18] L.K. Hua, *Introduction to number theory*, English translation by P. Shiu, Springer-Verlag, Berlin, 1982.

[19] A. Hurwitz, Über die Zahlentheorie der Quaternionen, *Mathematische Werke, Band II*, pp. 303-330, Birkhäuser, Basel, 1933.

[20] N. Iiyori and H. Yamaki, On a conjecture of Frobenius, *Bull. Amer. Math. Soc.* **25** (1991), 413-416.

[21] N. Jacobson, *Basic Algebra I*, 2nd ed., W.H. Freeman, New York, 1985.

[22] J.-R. Joly, Équations et variétés algébriques sur un corps fini, *Enseign. Math.* (2) **19** (1973), 1-117.

[23] S. Lang, *Algebra*, corrected reprint of 3rd ed., Addison-Wesley, Reading, Mass., 1994.

[24] D.H. Lehmer, *Guide to tables in the theory of numbers*, National Academy of Sciences, Washington, D.C., reprinted 1961.

[25] D.N. Lehmer, *List of prime numbers from 1 to* 10,006,721, reprinted, Hafner, New York, 1956.

[26] D.N. Lehmer, *Factor table for the first ten millions*, reprinted, Hafner, New York, 1956.

[27] A.K. Lenstra, Primality testing, *Proc. Symp. Appl. Math.* **42** (1990), 13-25.

[28] W.J. LeVeque, *Fundamentals of number theory*, reprinted Dover, Mineola, N.Y., 1996.

[29] U. Libbrecht, *Chinese mathematics in the thirteenth century*, MIT Press, Cambridge, Mass., 1973.

[30] R. Lidl and H. Niederreiter, *Finite fields*, 2nd ed., Cambridge University Press, 1997.

[31] H.B. Mann, *Introduction to algebraic number theory*, Ohio State University, Columbus, Ohio, 1955.

[32] R. Narasimhan, *Complex analysis in one variable*, Birkhäuser, Boston, Mass., 1985.

[33] W. Narkiewicz, *Number theory*, English translation by S. Kanemitsu, World Scientific, Singapore, 1983.

[34] I. Niven, H.S. Zuckerman and H.L. Montgomery, *An introduction to the theory of numbers*, 5th ed., Wiley, New York, 1991.

[35] H. Prüfer, Untersuchungen über Teilbarkeitseigenschaften, *J. Reine Angew. Math.* **168** (1932), 1-36.

[36] T.-S. Rhai, A characterization of polynomial domains over a field, *Amer. Math. Monthly* **69** (1962), 984-986.

[37] R.L. Rivest, A. Shamir and L. Adleman, A method for obtaining digital signatures and public-key cryptosystems, *Comm. ACM* **21** (1978), 120-126.

[38] P. Salmon, Sulla fattorialità delle algebre graduate e degli anelli locali, *Rend. Sem. Mat. Univ. Padova* **41** (1968), 119-138.

[39] P. Samuel, Unique factorization, *Amer. Math. Monthly* **75** (1968), 945-952.

[40] P. Samuel, About Euclidean rings, *J. Algebra* **19** (1971), 282-301.

[41] A. Scholz, *Einführung in die Zahlentheorie*, revised and edited by B. Schoeneberg, 5th ed., de Gruyter, Berlin,1973.

[42] H.J.S. Smith, Report on the theory of numbers, *Collected mathematical papers, Vol.1,* pp. 38-364, reprinted, Chelsea, New York, 1965. [Original, 1859-1865]

[43] T.J. Stieltjes, Sur la théorie des nombres, *Ann. Fac. Sci. Toulouse* **4** (1890), 1-103. [Reprinted in Tome 2, pp. 265-377 of T.J. Stieltjes, *Oeuvres complètes,* 2 vols., Noordhoff, Groningen, 1914-1918.]

[44] R.C. Vaughan, *The Hardy–Littlewood method,* 2nd ed., Cambridge Tracts in Mathematics **125**, Cambridge University Press, 1997.

[45] E. Waring, *Meditationes algebraicae,* English transl. of 1782 edition by D. Weeks, Amer. Math. Soc., Providence, R.I., 1991.

[46] A. Weil, *Number theory: an approach through history,* Birkhäuser, Boston, Mass., 1984.

[47] A.E. Western and J.C.P. Miller, *Tables of indices and primitive roots,* Royal Soc. Math. Tables, Vol. 9, Cambridge University Press, London, 1968.

[48] H.C. Williams, Primality testing on a computer, *Ars Combin.* **5** (1978), 127-185.

III
More on divisibility

In this chapter the theory of divisibility is developed further. The various sections of the chapter are to a large extent independent. We consider in turn the law of quadratic reciprocity, quadratic fields, multiplicative functions, and linear Diophantine equations.

1 The law of quadratic reciprocity

Let p be an odd prime. An integer a which is not divisible by p is said to be a *quadratic residue*, or *quadratic nonresidue*, of p according as the congruence

$$x^2 \equiv a \bmod p$$

has, or has not, a solution x. We will speak of the *quadratic nature* of $a \bmod p$, meaning whether a is a quadratic residue or nonresidue of p.

Let q be an odd prime different from p. The *law of quadratic reciprocity* connects the quadratic nature of $q \bmod p$ with the quadratic nature of $p \bmod q$. It states that if either p or q is congruent to 1 mod 4, then the quadratic nature of $q \bmod p$ is the same as the quadratic nature of $p \bmod q$, but if both p and q are congruent to 3 mod 4 then the quadratic nature of $q \bmod p$ is different from the quadratic nature of $p \bmod q$.

This remarkable result plays a key role in the arithmetic theory of quadratic forms. It was discovered empirically by Euler (1783). Legendre (1785) gave a partial proof and later (1798) introduced the convenient 'Legendre symbol'. The first complete proofs were given by Gauss (1801) in his *Disquisitiones Arithmeticae*. Indeed the result so fascinated Gauss that during the course of his lifetime he gave eight proofs, four of them resting on completely different principles: an induction argument, the theory of binary quadratic forms, properties of sums of roots of unity, and a combinatorial lemma. The proof we are now going to give is also of a combinatorial nature. Its idea originated with Zolotareff (1872), but our treatment is based on Rousseau (1994).

Let n be a positive integer and let X be the set $\{0,1,\ldots,n-1\}$. As in §7 of Chapter I, a permutation α of X is said to be *even* or *odd* according as the total number of inversions of order it induces is even or odd. If a is an integer relatively prime to n, then the map $\pi_a\colon X \to X$ defined by

$$\pi_a(x) = ax \ \mathrm{mod}\ n$$

is a permutation of X. We define the *Jacobi symbol* (a/n) to be $\mathrm{sgn}\,(\pi_a)$, i.e.

$$(a/n) \ = \ 1 \ \mathrm{or} - 1$$

according as the permutation π_a is even or odd. Thus $(a/1) = 1$, for every integer a. (The definition is sometimes extended by putting $(a/n) = 0$ if a and n are not relatively prime.)

PROPOSITION 1 *For any positive integer n and any integers a,b relatively prime to n, the Jacobi symbol has the following properties*:

(i) $(1/n) = 1$,

(ii) $(a/n) = (b/n)$ *if* $a \equiv b \ \mathrm{mod}\ n$,

(iii) $(ab/n) = (a/n)(b/n)$,

(iv) $(-1/n) = 1$ *if* $n \equiv 1$ *or* $2 \ \mathrm{mod}\ 4$ *and* $= -1$ *if* $n \equiv 3$ *or* $0 \ \mathrm{mod}\ 4$.

Proof The first two properties follow at once from the definition of the Jacobi symbol. If a and b are both relatively prime to n, then so also is their product ab. Since $\pi_{ab} = \pi_a\pi_b$, we have $\mathrm{sgn}\,(\pi_{ab}) = \mathrm{sgn}\,(\pi_a)\,\mathrm{sgn}\,(\pi_b)$, which implies (iii). We now evaluate $(-1/n)$. Since the map $\pi_{-1}\colon x \to -x \ \mathrm{mod}\ n$ fixes 0 and reverses the order of $1,\ldots,n-1$, the total number of inversions of order is $(n-2) + (n-3) + \ldots + 1 = (n-1)(n-2)/2$. It follows that $(-1/n) = (-1)^{(n-1)/2}$ or $(-1)^{(n-2)/2}$ according as n is odd or even. This proves (iv). $\square$

PROPOSITION 2 *For any relatively prime positive integers m,n*,

(i)   *if m and n are both odd, then* $(m/n)(n/m) = (-1)^{(m-1)(n-1)/4}$;

(ii)   *if m is odd and n even, then* $(m/n) = 1$ *or* $(-1)^{(m-1)/2}$ *according as* $n \equiv 2$ *or* $0 \ \mathrm{mod}\ 4$.

Proof The cyclic permutation $\tau\colon x \to x + 1 \ \mathrm{mod}\ n$ of the set $X = \{0,1,\ldots,n-1\}$ has sign $(-1)^{n-1}$, since the number of inversions of order is $n - 1$. Hence, for any integer $b \geq 0$ and any integer a relatively prime to n, the linear permutation

$$\tau^b\pi_a\colon x \to ax + b \ \mathrm{mod}\ n$$

of X has sign $(-1)^{b(n-1)}(a/n)$.

Put $Y = \{0,1,...,m-1\}$ and $P = X \times Y$. We consider two transformations μ and ν of P, defined by

$$\mu(x,y) = (mx + y \bmod n, y), \quad \nu(x,y) = (x, x + ny \bmod m).$$

For each fixed y, μ defines a permutation of the set (X,y) with sign $(-1)^{y(n-1)}(m/n)$. Since $\sum_{y=0}^{m-1} y = m(m-1)/2$, it follows that the permutation μ of P has sign

$$\text{sgn}\,(\mu) \;=\; (-1)^{m(m-1)(n-1)/2}\,(m/n)^m.$$

Similarly the permutation ν of P has sign

$$\text{sgn}\,(\nu) \;=\; (-1)^{n(m-1)(n-1)/2}\,(n/m)^n,$$

and hence $\alpha := \nu\mu^{-1}$ has sign

$$\text{sgn}\,(\alpha) \;=\; (-1)^{(m+n)(m-1)(n-1)/2}\,(m/n)^m\,(n/m)^n.$$

But α is the permutation $(mx + y \bmod n, y) \to (x, x + ny \bmod m)$ and its sign can be determined directly in the following way.

Put $Z = \{0,1,...,mn-1\}$. By Proposition II.36, for any $(x,y) \in P$ there is a unique $z \in Z$ such that

$$z \equiv x \bmod n, \quad z \equiv y \bmod m.$$

Moreover, any $z \in Z$ is obtained in this way from a unique $(x,y) \in P$. For any $z \in Z$, we will denote by $\rho(z)$ the corresponding element of P. Then the permutation α can be written in the form $\rho(mx + y) \to \rho(x + ny)$. Since ρ is a bijective map, the sign of the permutation α of P will be the same as the sign of the permutation $\beta = \rho^{-1}\alpha\rho : mx + y \to x + ny$ of Z. An inversion of order for β occurs when both $mx + y > mx' + y'$ and $x + ny < x' + ny'$, i.e. when both $m(x - x') > y' - y$ and $x - x' < n(y' - y)$. But these inequalities imply $mn(x - x') > x - x'$ and hence $x > x'$, $y' > y$. Conversely, if $x > x'$, $y' > y$, then

$$m(x - x') \geq m > y' - y, \quad n(y' - y) \geq n > x - x'.$$

Since the total number of $(x,y),(x',y') \in P$ with $x > x'$, $y < y'$ is $m(m-1)/2 \cdot n(n-1)/2$, it follows that the sign of the permutation α is $(-1)^{mn(m-1)(n-1)/4}$. Comparing this expression with the expression previously found, we obtain

$$(m/n)^m\,(n/m)^n \;=\; (-1)^{(mn+2m+2n)(m-1)(n-1)/4}.$$

This simplifies to the first statement of the proposition if m and n are both odd, and to the second statement if m is odd and n even. $\square$

COROLLARY 3 *For any odd positive integer n, $(2/n) = 1$ or -1 according as $n \equiv \pm 1$ or ± 5 mod 8.*

Proof Since the result is already known for $n = 1$, we suppose $n > 1$. Then either n or $n - 2$ is congruent to 1 mod 4 and so, by Proposition 1 and Proposition 2(i),

$$(2/n) \;=\; (-1/n)((n-2)/n) \;=\; (-1/n)(n/(n-2)) \;=\; (-1)^{(n-1)/2}\,(2/(n-2)).$$

Iterating, we obtain $(2/n) = (-1)^h$, where $h = (n-1)/2 + (n-3)/2 + \dots + 1 = (n^2 - 1)/8$. The result follows. $\square$

The value of (a/n) when n is even is completely determined by Propositions 1 and 2. The evaluation of (a/n) when n is odd reduces by these propositions and Corollary 3 to the evaluation of (m/n) for odd $m > 1$. Although Proposition 2 does not provide a formula for the Jacobi symbol in this case, it does provide a method for its rapid evaluation, as we now show.

If m and n are relatively prime odd positive integers, we can write $m = 2hn + \varepsilon_1 n_1$, where $h \in \mathbb{Z}$, $\varepsilon_1 = \pm 1$ and n_1 is an odd positive integer less than n. Then n and n_1 are also relatively prime and

$$(m/n) \;=\; (\varepsilon_1/n)(n_1/n).$$

If $n_1 = 1$, we are finished. Otherwise, using Proposition 2(i), we obtain

$$(m/n) \;=\; (-1)^{(n_1-1)(n-1)/4}\,(\varepsilon_1/n)(n/n_1) \;=\; \pm\,(n/n_1),$$

where the minus sign holds if and only if n and $\varepsilon_1 n_1$ are both congruent to 3 mod 4. The process can now be repeated with m,n replaced by n,n_1. After finitely many steps the process must terminate with $n_s = 1$.

As an example,

$$
\begin{aligned}
\left(\frac{2985}{1951}\right) &= \left(\frac{-1}{1951}\right)\left(\frac{917}{1951}\right) = -\left(\frac{1951}{917}\right)\\[2mm]
&= -\left(\frac{117}{917}\right) = -\left(\frac{917}{117}\right)\\[2mm]
&= -\left(\frac{-1}{117}\right)\left(\frac{19}{117}\right) = -\left(\frac{117}{19}\right)\\[2mm]
&= -\left(\frac{3}{19}\right) = \left(\frac{19}{3}\right) = \left(\frac{1}{3}\right) = 1.
\end{aligned}
$$

Further properties of the Jacobi symbol can be derived from those already established.

PROPOSITION 4 *If n,n' are positive integers and if a is an integer relatively prime to n such that $n' \equiv n \bmod 4a$, then $(a/n') = (a/n)$.*

Proof If $a = -1$ then, since $n' \equiv n \bmod 4$, $(a/n') = (a/n)$, by Proposition 1(iv). If $a = 2$ then, since n and n' are odd and $n' \equiv n \bmod 8$, $(a/n') = (a/n)$, by Corollary 3. Consequently, by Proposition 1(iii), it is sufficient to prove the result for odd $a > 1$.

If n is even, the result now follows from Proposition 2(ii). If n is odd, it follows from Proposition 2(i) and Proposition 1. $\square$

PROPOSITION 5 *If the integer a is relatively prime to the odd positive integers n and n', then $(a/nn') = (a/n)(a/n')$.*

Proof We have $a \equiv a' \bmod nn'$ for some $a' \in \{1,2,...,nn'\}$. Since nn' is odd, we can choose $j \in \{0,1,2,3\}$ so that $a'' = a' + jnn'$ satisfies $a'' \equiv 1 \bmod 4$. Then, by Propositions 1 and 2,

$$(a/nn') = (a''/nn') = (nn'/a'') = (n/a'')(n'/a'') = (a''/n)(a''/n') = (a/n)(a/n'). \square$$

Proposition 5 reduces the evaluation of (a/n) for odd positive n to the evaluation of (a/p), where p is an odd prime. This is where we make the connection with quadratic residues:

PROPOSITION 6 *If p is an odd prime and a an integer not divisible by p, then $(a/p) = 1$ or -1 according as a is a quadratic residue or nonresidue of p. Moreover, exactly half of the integers $1,...,p-1$ are quadratic residues of p.*

Proof If a is a quadratic residue of p, there exists an integer x such that $x^2 \equiv a \bmod p$ and hence

$$(a/p) = (x^2/p) = (x/p)(x/p) = 1.$$

Let g be a primitive root mod p. Then the integers $1,g,...,g^{p-2} \bmod p$ are just a rearrangement of the integers $1,2,...,p-1$. The permutation

$$\pi_g: x \to gx \pmod p$$

fixes 0 and cyclically permutes the remaining elements $1,g,...,g^{p-2}$. Since the number of inversions of order is $p-2$, it follows that $(g/p) = -1$. For any integer a not divisible by p there is a unique $k \in \{0,1,...,p-2\}$ such that $a \equiv g^k \bmod p$. Hence

$$(a/p) = (g^k/p) = (g/p)^k = (-1)^k.$$

Thus $(a/p) = 1$ if and only if k is even and then $a \equiv x^2 \bmod p$ with $x = g^{k/2}$.

This proves the first statement of the proposition. Since exactly half the integers in the set $\{0,1,...,p-2\}$ are even, it also proves again (cf. Proposition II.28) the second statement. $\square$

The law of quadratic reciprocity can now be established without difficulty:

THEOREM 7 *Let p and q be distinct odd primes. Then the quadratic natures of p and q are the same if $p \equiv 1$ or $q \equiv 1$ mod 4, but different if $p \equiv q \equiv 3$ mod 4.*

Proof The result follows at once from Proposition 6 since, by Proposition 2(i), $(p/q) = (q/p)$ if $p \equiv 1$ or $q \equiv 1$ mod 4, and $(p/q) = -(q/p)$ if $p \equiv q \equiv 3$ mod 4. $\square$

Legendre (1798) *defined* $(a/p) = 1$ or -1 according as a was a quadratic residue or nonresidue of p, and Jacobi (1837) extended this definition to (a/n) for any odd positive integer n relatively prime to a by setting

$$(a/n) \,=\, \Pi_p\,(a/p),$$

where p runs through the prime divisors of n, each occurring as often as its multiplicity. Propositions 5 and 6 show that these definitions of Legendre and Jacobi are equivalent to the definition adopted here. The relations $(-1/p) = (-1)^{(p-1)/2}$ and $(2/p) = (-1)^{(p^2-1)/8}$ for odd primes p are often called the *first and second supplements* to the law of quadratic reciprocity.

It should be noted that, if the congruence $x^2 \equiv a$ mod n is soluble then $(a/n) = 1$, but the converse need not hold when n is not prime. For example, if $n = 21$ and $a = 5$ then the congruence $x^2 \equiv 5$ mod 21 is insoluble, since both the congruences $x^2 \equiv 5$ mod 3 and $x^2 \equiv 5$ mod 7 are insoluble, but

$$\left(\frac{5}{21}\right) = \left(\frac{5}{3}\right)\left(\frac{5}{7}\right) = (-1)^2 = 1.$$

The Jacobi symbol finds an interesting application in the proof of the following result:

PROPOSITION 8 *If a is an integer which is not a perfect square, then there exist infinitely many primes p not dividing a for which $(a/p) = -1$.*

Proof Suppose first that $a = -1$. Since $(-1/p) = (-1)^{(p-1)/2}$, we wish to show that there are infinitely many primes $p \equiv 3$ mod 4. Clearly 7 is such a prime. Let $\{p_1,...,p_m\}$ be any finite set of such primes greater than 3. Adapting Euclid's proof of the infinity of primes (which is reproduced at the beginning of Chapter IX), we put

$$b \,=\, 4\,p_1 \cdots p_m + 3.$$

Then b is odd, but not divisible by 3 or by any of the primes $p_1,...,p_m$. Since $b \equiv 3 \bmod 4$, at least one prime divisor q of b must satisfy $q \equiv 3 \bmod 4$. Thus the set $\{3,p_1,...,p_m\}$ does not contain all primes $p \equiv 3 \bmod 4$.

Suppose next that $a = \pm 2$. Then $(a/5) = -1$. Let $\{p_1,...,p_m\}$ be any finite set of primes greater than 3 such that $(a/p_i) = -1$ $(i = 1,...,m)$ and put

$$b = 8\, p_1 \cdots p_m \pm 3,$$

where the $\pm$ sign is chosen according as $a = \pm 2$. Then b is not divisible by 3 or by any of the primes $p_1,...,p_m$. Since $b \equiv \pm 3 \bmod 8$, we have $(2/b) = -1$ and $(a/b) = -1$ in both cases. If $b = q_1 \cdots q_n$ is the representation of b as a product of primes (repetitions allowed), then

$$(a/b) = (a/q_1) \cdots (a/q_n)$$

and hence $(a/q_j) = -1$ for at least one j. Consequently the result holds also in this case.

Consider now the general case. We may assume that a is square-free, since if $a = a'b^2$, where a' is square-free, then $(a/p) = (a'/p)$ for every prime p not dividing a. Thus we can write

$$a = \varepsilon\, 2^e\, r_1 \cdots r_h,$$

where $\varepsilon = \pm 1$, $e = 0$ or 1, and $r_1,...,r_h$ are distinct odd primes. By what we have already proved, we may assume $h \geq 1$.

Let $\{p_1,...,p_m\}$ be any finite set of odd primes not containing any of the primes $r_1,...,r_h$. By Proposition 6, there exists an integer c such that $(c/r_1) = -1$. Since the moduli are relatively prime in pairs, by Corollary II.38 the simultaneous congruences

$$x \equiv 1 \bmod 8, \quad x \equiv 1 \bmod p_i \ (i = 1,...,m),$$

$$x \equiv c \bmod r_1, \quad x \equiv 1 \bmod r_j \ (j = 2,...,h),$$

have a positive solution $x = b$. Then b is not divisible by any of the odd primes $p_1,...,p_m$ or $r_1,...,r_h$. Moreover $(-1/b) = (2/b) = 1$, since $b \equiv 1 \bmod 8$. Since $(r_j/b) = (b/r_j)$ $(1 \leq j \leq h)$, it follows that

$$(a/b) = (\varepsilon/b)\, (2/b)^e\, (r_1/b) \cdots (r_h/b)$$

$$= (b/r_1)(b/r_2) \cdots (b/r_h) = (c/r_1)(1/r_2) \cdots (1/r_h) = -1.$$

As in the special case previously considered, this implies that $(a/q) = -1$ for some prime q dividing b, and the result follows. $\square$

A second proof of the law of quadratic reciprocity will now be given. Let p be an odd prime and, for any integer a not divisible by p, with Legendre *define*

$$(a/p) = 1 \text{ or} -1$$

according as a is a quadratic residue or quadratic nonresidue of p. It follows from Euler's criterion (Proposition II.28) that

$$(ab/p) = (a/p)(b/p)$$

for any integers a,b not divisible by p. Also, by Corollary II.29,

$$(-1/p) = (-1)^{(p-1)/2}.$$

Now let q be an odd prime distinct from p and let $K = \mathbb{F}_q$ be the finite field containing q elements. Since $p \neq q$, the polynomial $t^p - 1$ has no repeated factors in K and thus has p distinct roots in some field $L \supseteq K$. If ζ is any root other than 1, then the (cyclotomic) polynomial

$$f(t) = t^{p-1} + t^{p-2} + \dots + 1$$

has the roots ζ^k ($k = 1,\dots,p-1$).

Consider the *Gauss sum*

$$\tau = \sum_{x=1}^{p-1} (x/p)\zeta^x.$$

Instead of summing from 1 to $p-1$, we can just as well sum over any set of representatives of $\mathbb{F}_p^\times$:

$$\tau = \sum_{x \not\equiv 0 \bmod p} (x/p)\zeta^x.$$

Since q is odd, $(x/p)^q = (x/p)$ and hence, since L has characteristic q,

$$\tau^q = \sum_{x \not\equiv 0 \bmod p} (x/p)\zeta^{xq}.$$

If we put $y = xq$ then, since

$$(x/p) = (q^2x/p) = (qy/p) = (q/p)(y/p),$$

we obtain

$$\tau^q = \sum_{y \not\equiv 0 \bmod p} (q/p)(y/p)\zeta^y = (q/p)\tau.$$

Furthermore,

$$\tau^2 = \sum_{u,v \not\equiv 0 \bmod p} (u/p)(v/p)\zeta^u\zeta^v = \sum_{u,v \not\equiv 0 \bmod p} (uv/p)\zeta^{u+v}$$

or, putting $v = uw$,

$$\tau^2 = \sum_{w \not\equiv 0 \bmod p} (w/p) \sum_{u \not\equiv 0 \bmod p} \zeta^{u(1+w)}.$$

Since the coefficients of t^{p-1} and t^{p-2} in $f(t)$ are 1, the sum of the roots is -1 and thus

$$\sum_{u \not\equiv 0 \bmod p} \zeta^{au} = -1 \quad \text{if } a \not\equiv 0 \bmod p.$$

On the other hand, if $a \equiv 0 \bmod p$, then $\zeta^{au} = 1$ and

$$\sum_{u \not\equiv 0 \bmod p} \zeta^{au} = p - 1.$$

Hence

$$\tau^2 = (-1/p)(p-1) - \sum_{w \not\equiv 0,-1 \bmod p} (w/p) = (-1/p)p - \sum_{w \not\equiv 0 \bmod p} (w/p).$$

Since there are equally many quadratic residues and quadratic nonresidues, the last sum vanishes and we obtain finally

$$\tau^2 = (-1)^{(p-1)/2}p.$$

Thus $\tau \neq 0$ and from the previous expression for τ^q we now obtain

$$\tau^{q-1} = (q/p).$$

But

$$\tau^{q-1} = (\tau^2)^{(q-1)/2} = \{(-1)^{(p-1)/2}p\}^{(q-1)/2}$$

and $p^{(q-1)/2} = (p/q)$, by Proposition II.28 again. Hence

$$(q/p) = (-1)^{(p-1)(q-1)/4} (p/q),$$

which is the law of quadratic reciprocity.

The preceding proof is a variant of the sixth proof of Gauss (1818). Already in 1801 Gauss had shown that if p is an odd prime, then

$$\sum_{k=0}^{p-1} e^{2\pi i k^2/p} = \pm \sqrt{p} \text{ or } \pm i\sqrt{p} \quad \text{according as } p \equiv 1 \text{ or } p \equiv 3 \bmod 4.$$

After four more years of labour he managed to show that in fact the $+$ signs must be taken. From this result he obtained his fourth proof of the law of quadratic reciprocity. The sixth proof avoided this sign determination, but Gauss's result is of interest in itself. Dirichlet (1835) derived it by a powerful analytic method, which is readily generalized. Although we will make no later use of it, we now present Dirichlet's argument.

For any positive integers m,n, we define the *Gauss sum* $G(m,n)$ by

$$G(m,n) = \sum_{v=0}^{n-1} e^{2\pi i v^2 m/n}.$$

Instead of summing from 0 to $n - 1$ we can just as well sum over any complete set of representatives of the integers mod n:

$$G(m,n) = \sum_{v \bmod n} e^{2\pi i v^2 m/n}.$$

Gauss sums have a useful multiplicative property:

PROPOSITION 9 *If m,n,n' are positive integers, with n and n' relatively prime, then*

$$G(mn',n)G(mn,n') = G(m,nn').$$

Proof When v and v' run through complete sets of representatives of the integers mod n and mod n' respectively, $\mu = vn' + v'n$ runs through a complete set of representatives of the integers mod nn'. Moreover

$$\mu^2 m = (vn' + v'n)^2 m \equiv (v^2 n'^2 + v'^2 n^2)m \bmod nn'.$$

It follows that

$$G(mn',n)G(mn,n') = \sum_{v \bmod n} \sum_{v' \bmod n'} e^{2\pi i(mn'^2 v^2 + mn^2 v'^2)/nn'}$$

$$= \sum_{\mu \bmod nn'} e^{2\pi i \mu^2 m/nn'} = G(m,nn'). \quad \square$$

A deeper result is the following reciprocity formula, due to Schaar (1848):

PROPOSITION 10 *For any positive integers m,n,*

$$G(m,n) = (n/m)^{1/2} C \sum_{\mu=0}^{2m-1} e^{-\pi i \mu^2 n/2m},$$

where $C = (1 + i)/2$.

Proof Let $f: \mathbb{R} \to \mathbb{C}$ be a function which is continuously differentiable when restricted to the interval $[0,n]$ and which vanishes outside this interval. Since the sum

$$F(t) = \sum_{k=-\infty}^{\infty} f(t + k)$$

has only finitely many nonzero terms, the function F has period 1 and is continuously differentiable, except possibly for jump discontinuities when t is an integer. Therefore, by Dirichlet's convergence criterion in the theory of Fourier series,

$$\{F(+0) + F(-0)\}/2 = \lim_{N\to\infty} \sum_{h=-N}^{N} \int_0^1 e^{-2\pi i h t} F(t)\, dt.$$

But

$$\int_0^1 e^{-2\pi i h t} F(t)\, dt = \sum_{k=-\infty}^{\infty} \int_0^1 e^{-2\pi i h t} f(t + k)\, dt$$

$$= \sum_{k=-\infty}^{\infty} \int_k^{k+1} e^{-2\pi i h t} f(t)\, dt = \int_0^n e^{-2\pi i h t} f(t)\, dt.$$

Thus we obtain

$$f(0)/2 + f(1) + \dots + f(n-1) + f(n)/2 = \lim_{N\to\infty} \sum_{h=-N}^{N} \int_0^n e^{-2\pi i h t} f(t)\, dt. \qquad (*)$$

This is a simple form of *Poisson's summation formula* (which makes an appearance also in Chapters IX and X).

In particular, if we take $f(t) = e^{2\pi i t^2 m/n}$ $(0 \le t \le n)$, where m is also a positive integer, then the left side of $(*)$ is just the Gauss sum $G(m,n)$. We will now evaluate the right side of $(*)$ for this case. Put $h = 2mq + \mu$, where q and μ are integers and $0 \le \mu < 2m$. Then

$$e^{-2\pi i h t} f(t) = e^{2\pi i m(t-nq)^2/n} e^{-2\pi i \mu t}.$$

As h runs through all the integers, q does also and μ runs independently through the integers $0,\dots,2m-1$. Hence

$$\begin{aligned}
\lim_{N\to\infty} \sum_{h=-N}^{N} \int_0^n e^{-2\pi i h t} f(t)\, dt &= \sum_{\mu=0}^{2m-1} \lim_{Q\to\infty} \sum_{q=-Q}^{Q} \int_0^n e^{2\pi i m(t-nq)^2/n} e^{-2\pi i \mu t}\, dt \\
&= \sum_{\mu=0}^{2m-1} \lim_{Q\to\infty} \sum_{q=-Q}^{Q} \int_{-qn}^{-(q-1)n} e^{2\pi i t^2 m/n} e^{-2\pi i \mu t}\, dt \\
&= \sum_{\mu=0}^{2m-1} \int_{-\infty}^{\infty} e^{2\pi i t^2 m/n} e^{-2\pi i \mu t}\, dt \\
&= \sum_{\mu=0}^{2m-1} \int_{-\infty}^{\infty} e^{2\pi i m(t-\mu n/2m)^2/n} e^{-\pi i \mu^2 n/2m}\, dt \\
&= \sum_{\mu=0}^{2m-1} e^{-\pi i \mu^2 n/2m} \int_{-\infty}^{\infty} e^{2\pi i t^2 m/n}\, dt \\
&= (n/m)^{1/2} C \sum_{\mu=0}^{2m-1} e^{-\pi i \mu^2 n/2m},
\end{aligned}$$

where C is the *Fresnel integral*

$$C = \int_{-\infty}^{\infty} e^{2\pi i t^2}\, dt.$$

(This is an important example of an infinite integral which converges, although the integrand does not tend to zero.) From $(*)$ we now obtain the formula for $G(m,n)$ in the statement of the proposition. To determine the value of the constant C, take $m = 1$, $n = 3$. We obtain $i\sqrt{3} = \sqrt{3}\, C\, (1 + i)$, which simplifies to $C = (1 + i)/2$. $\square$

From Proposition 10 with $m = 1$ we obtain

$$G(1,n) = \sum_{v=0}^{n-1} e^{2\pi i v^2/n} = \begin{cases} (1+i)\sqrt{n} & \text{if } n \equiv 0 \pmod 4, \\ \sqrt{n} & \text{if } n \equiv 1 \pmod 4, \\ 0 & \text{if } n \equiv 2 \pmod 4, \\ i\sqrt{n} & \text{if } n \equiv 3 \pmod 4. \end{cases}$$

If m and n are both odd, it follows that

$$G(1,mn) = G(1,m)\,G(1,n) \ \text{if either } m \equiv 1 \text{ or } n \equiv 1 \bmod 4,$$

$$= -\,G(1,m)\,G(1,n) \ \text{if } m \equiv n \equiv 3 \bmod 4;$$

i.e.

$$G(1,mn) = (-1)^{(m-1)(n-1)/4}\,G(1,m)\,G(1,n).$$

If, in addition, m and n are relatively prime, then $G(m,n)\,G(n,m) = G(1,mn)$, by Proposition 9. Hence, if the integers m,n are odd, positive and relatively prime, then

$$G(m,n)\,G(n,m) = (-1)^{(m-1)(n-1)/4}\,G(1,m)\,G(1,n).$$

For any odd, positive relatively prime integers m,n, put

$$\rho(m,n) = G(m,n)/G(1,n).$$

Then

$$\rho(1,n) = 1,$$
$$\rho(m,n) = \rho(m',n) \ \text{if } m \equiv m' \bmod n,$$
$$\rho(m,n)\rho(n,m) = (-1)^{(m-1)(n-1)/4}.$$

We claim that $\rho(m,n)$ is just the Jacobi symbol (m/n). This is evident if $m = 1$ and, by Proposition 2(i), if $\rho(m,n) = (m/n)$, then also $\rho(n,m) = (n/m)$.

Hence if the claim is not true for all m,n, there is a pair m,n with $1 < m < n$ such that

$$\rho(m,n) \neq (m/n),$$

but $\rho(\mu,\nu) = (\mu/\nu)$ for all odd, positive relatively prime integers μ,ν with $\mu < m$. We can write $n = km + r$ for some positive integers k,r with $r < m$. Then

$$\rho(n,m) = \rho(r,m) = (r/m) = (n/m).$$

Since $\rho(m,n) \neq (m/n)$, this yields a contradiction.

Thus

$$G(m,n) = (m/n)G(1,n) \ \textit{for all odd, positive relatively prime integers } m,n.$$

In fact this relation holds also if m is negative, since

$$\overline{G(1,n)} = (-1)^{(n-1)/2}\,G(1,n) \ \text{and} \ G(-m,n) = \overline{G(m,n)}.$$

As we have already obtained an explicit formula for $G(1,n)$, we now have also an explicit evaluation of $G(m,n)$.

2 Quadratic fields

Let ζ be a complex number which is not rational, but whose square is rational. Since $\zeta \notin \mathbb{Q}$, a complex number α has at most one representation of the form $\alpha = r + s\zeta$, where $r,s \in \mathbb{Q}$. Let $\mathbb{Q}(\zeta)$ denote the set of all complex numbers α which have a representation of this form. Then $\mathbb{Q}(\zeta)$ is a *field*, since it is closed under subtraction and multiplication and since, if r and s are not both zero,

$$(r + s\zeta)^{-1} = (r - s\zeta)/(r^2 - s^2\zeta^2).$$

Evidently $\mathbb{Q}(\zeta) = \mathbb{Q}(t\zeta)$ for any nonzero rational number t. Conversely, if $\mathbb{Q}(\zeta) = \mathbb{Q}(\zeta^*)$, then $\zeta^* = t\zeta$ for some nonzero rational number t. For $\zeta^* = r + s\zeta$, where $r,s \in \mathbb{Q}$ and $s \neq 0$, and hence

$$r^2 = \zeta^{*2} - 2s\zeta\zeta^* + s^2\zeta^2.$$

Thus $\zeta\zeta^*$ is rational, and so is $\zeta\zeta^*/\zeta^2 = \zeta^*/\zeta$.

It follows that without loss of generality we may assume that $\zeta^2 = d$ is a square-free integer. Then $dt^2 \in \mathbb{Z}$ for some $t \in \mathbb{Q}$ implies $t \in \mathbb{Z}$. If $\zeta^{*2} = d^*$ is also a square-free integer, then $\mathbb{Q}(\zeta) = \mathbb{Q}(\zeta^*)$ if and only if $d = d^*$ and $\zeta^* = \pm\,\zeta$.

The *quadratic field* $\mathbb{Q}(\sqrt{d})$ is said to be *real* if $d > 0$ and *imaginary* if $d < 0$. We define the *conjugate* of an element $\alpha = r + s\sqrt{d}$ of the quadratic field $\mathbb{Q}(\sqrt{d})$ to be the element $\alpha' = r - s\sqrt{d}$. It is easily verified that

$$(\alpha + \beta)' = \alpha' + \beta', \quad (\alpha\beta)' = \alpha'\beta'.$$

Since the map $\sigma: \alpha \to \alpha'$ is also bijective, it is an *automorphism* of the field $\mathbb{Q}(\sqrt{d})$. Since $\alpha' = \alpha$ if and only if $s = 0$, the rational field $\mathbb{Q}$ is the fixed point set of σ. Since $(\alpha')' = \alpha$, the automorphism σ is an 'involution'.

We define the *norm* of an element $\alpha = r + s\sqrt{d}$ of the quadratic field $\mathbb{Q}(\sqrt{d})$ to be the rational number

$$N(\alpha) = \alpha\alpha' = r^2 - ds^2.$$

Evidently $N(\alpha) = N(\alpha')$, and $N(\alpha) = 0$ if and only if $\alpha = 0$. From the relation $(\alpha\beta)' = \alpha'\beta'$ we obtain

$$N(\alpha\beta) = N(\alpha)N(\beta).$$

An element α of the quadratic field $\mathbb{Q}(\sqrt{d})$ is said to be an *integer* of this field if it is a root of a quadratic polynomial $t^2 + at + b$ with coefficients $a,b \in \mathbb{Z}$. (Equivalently, the integers of $\mathbb{Q}(\sqrt{d})$ are the elements which are *algebraic integers*.)

It follows from Proposition II.16 that $\alpha \in \mathbb{Q}$ is an integer of the field $\mathbb{Q}(\sqrt{d})$ if and only if $\alpha \in \mathbb{Z}$. Suppose now that $\alpha = r + s\sqrt{d}$, where $r,s \in \mathbb{Q}$ and $s \neq 0$. Then α is a root of the quadratic polynomial

$$f(x) = (x - \alpha)(x - \alpha') = x^2 - 2rx + r^2 - ds^2.$$

Morover, this is the unique monic quadratic polynomial with rational coefficients which has α as a root.

Consequently, if α is an integer of $\mathbb{Q}(\sqrt{d})$, then so also is its conjugate α' and its norm $N(\alpha) = r^2 - ds^2$ is an ordinary integer.

PROPOSITION 11 *Let d be a square-free integer and define ω by*

$$\omega = \sqrt{d} \qquad \textit{if } d \equiv 2 \textit{ or } 3 \bmod 4,$$
$$\omega = (\sqrt{d} - 1)/2 \quad \textit{if } d \equiv 1 \bmod 4.$$

Then α is an integer of the quadratic field $\mathbb{Q}(\sqrt{d})$ if and only if $\alpha = a + b\omega$ for some $a,b \in \mathbb{Z}$.

Proof Suppose $\alpha = r + s\sqrt{d}$, where $r,s \in \mathbb{Q}$. As we have seen, if $s = 0$ then α is an integer of $\mathbb{Q}(\sqrt{d})$ if and only if $r \in \mathbb{Z}$. If $s \neq 0$, then α is an integer of $\mathbb{Q}(\sqrt{d})$ if and only if $a = 2r$ and $b = r^2 - ds^2$ are ordinary integers. If a is even, i.e. if $r \in \mathbb{Z}$, then $b \in \mathbb{Z}$ if and only if $ds^2 \in \mathbb{Z}$ and hence, since d is square-free, if and only if $s \in \mathbb{Z}$. If a is odd, then $a^2 \equiv 1 \bmod 4$ and hence $b \in \mathbb{Z}$ if and only if $4ds^2 \equiv 1 \bmod 4$. Since d is square-free, this implies that $2s \in \mathbb{Z}, s \notin \mathbb{Z}$. Hence $2s$ is odd and $d \equiv 1 \bmod 4$. Conversely, if $2r$ and $2s$ are odd integers and $d \equiv 1 \bmod 4$, then $r^2 - ds^2 \in \mathbb{Z}$. The result follows. $\square$

Since $\omega^2 = -\omega + (d - 1)/4$ in the case $d \equiv 1 \bmod 4$, it follows directly from Proposition 11 that the set $\mathbb{O}_d$ of all integers of the field $\mathbb{Q}(\sqrt{d})$ is closed under subtraction and multiplication and consequently is a ring. In fact $\mathbb{O}_d$ is an integral domain, since $\mathbb{O}_d \subseteq \mathbb{Q}(\sqrt{d})$.

For example, $\mathbb{O}_{-1} = \mathcal{G}$ is the ring of Gaussian integers $a + bi$, where $a,b \in \mathbb{Z}$. They form a square 'lattice' in the complex plane. Similarly $\mathbb{O}_{-3} = \mathcal{E}$ is the ring of all complex numbers $a + b\rho$, where $a,b \in \mathbb{Z}$ and $\rho = (i\sqrt{3} - 1)/2$ is a cube root of unity. These *Eisenstein integers* were studied by Eisenstein (1844). They form a hexagonal 'lattice' in the complex plane.

We have already seen in §6 of Chapter II that the ring $\mathcal{G}$ of Gaussian integers is a Euclidean domain, with $\delta(\alpha) = N(\alpha)$. It will now be shown that the ring $\mathcal{E}$ of Eisenstein integers is also a Euclidean domain, with $\delta(\alpha) = N(\alpha)$. If $\alpha,\beta \in \mathcal{E}$ and $\alpha \neq 0$, then

$$\beta\alpha^{-1} = \beta\alpha'/\alpha\alpha' = r + s\rho,$$

where $r,s \in \mathbb{Q}$. Choose $a,b \in \mathbb{Z}$ so that

$$|r - a| \leq 1/2, \quad |s - b| \leq 1/2.$$

If $\kappa = a + b\rho$, then $\kappa \in \mathscr{E}$ and

$$N(\beta\alpha^{-1} - \kappa) = \{r - a - (s - b)/2\}^2 + 3\{(s - b)/2\}^2$$
$$\leq (3/4)^2 + 3(1/4)^2 = 3/4 < 1.$$

Thus $\beta - \kappa\alpha \in \mathscr{E}$ and $N(\beta - \kappa\alpha) < N(\alpha)$.

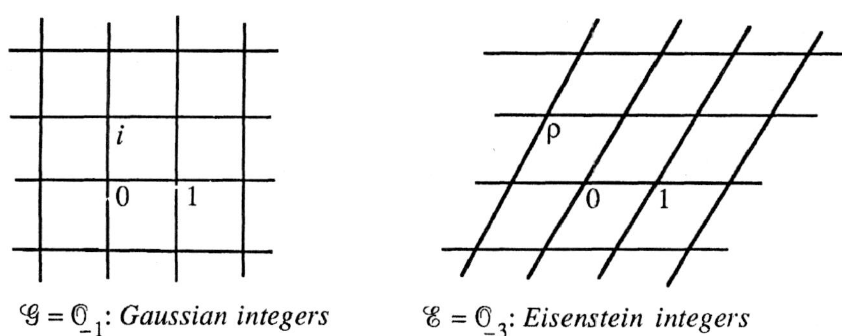

Figure 1: *Gaussian and Eisenstein integers*

Since $\mathscr{G}$ and $\mathscr{E}$ are Euclidean domains, the divisibility theory of Chapter II is valid for them. As an application, we prove

PROPOSITION 12 *The equation* $x^3 + y^3 = z^3$ *has no solutions in nonzero integers.*

Proof Assume on the contrary that such a solution exists and choose one for which $|xyz|$ is a minimum. Then $(x,y) = (x,z) = (y,z) = 1$. If 3 did not divide xyz, then x^3, y^3 and z^3 would be congruent to ± 1 mod 9, which contradicts $x^3 + y^3 = z^3$. So, without loss of generality, we may assume that $3|z$. Then $x^3 + y^3 \equiv 0$ mod 3 and, again without loss of generality, we may assume that $x \equiv 1$ mod 3, $y \equiv -1$ mod 3. This implies that

$$x^2 - xy + y^2 \equiv 3 \text{ mod } 9.$$

If $x + y$ and $x^2 - xy + y^2$ have a common prime divisor p, then p divides $3xy$, since $3xy = (x + y)^2 - (x^2 - xy + y^2)$, and this implies $p = 3$, since $(x,y) = 1$. Since

$$(x + y)(x^2 - xy + y^2) = x^3 + y^3 = z^3 \equiv 0 \text{ mod } 27,$$

it follows that

$$x + y = 9a^3, \quad x^2 - xy + y^2 = 3b^3,$$

where $a,b \in \mathbb{Z}$ and $3 \nmid b$.

We now shift operations to the Euclidean domain $\mathscr{E}$ of Eisenstein integers. We have

$$x^2 - xy + y^2 = (x + y\rho)(x + y\rho^2),$$

where $\rho = (i\sqrt{3} - 1)/2$ is a cube root of unity. Put $\lambda = 1 - \rho$, so that $(1 + \rho)\lambda^2 = 3$. Then λ is a common divisor of $x + y\rho$ and $x + y\rho^2$, since

$$x + y\rho = x + y - y\lambda, \quad x + y\rho^2 = x - 2y + y\lambda$$

and $x + y \equiv 0 \equiv x - 2y \bmod 3$. In fact λ is the greatest common divisor of $x + y\rho$ and $x + y\rho^2$ since, for all $m,n \in \mathbb{Z}$,

$$(m + n + n\rho)(x + y\rho^2) - (n + m\rho + n\rho)(x + y\rho) = (mx + ny)\lambda$$

and we can choose m,n so that $mx + ny = 1$. Since $\lambda^2 = -3\rho$ and since ρ is a unit, from $(x + y\rho)(x + y\rho^2) = 3b^3$ and the unique factorization of b in $\mathscr{E}$, we now obtain

$$x + y\rho = \varepsilon\lambda(c + d\rho)^3,$$

where $c,d \in \mathbb{Z}$ and ε is a unit. From

$$(x + y\rho)/\lambda = x - \lambda(x + y)/3 = x - 3a^3\lambda$$

and

$$(c + d\rho)^3 = c^3 - 3cd^2 + d^3 + 3cd(c - d)\rho,$$

by reducing mod 3 we get

$$\varepsilon(c^3 + d^3) \equiv 1 \bmod 3.$$

Since the units in $\mathscr{E}$ are $\pm 1, \pm\rho, \pm\rho^2$ (by the following Proposition 13), this implies $\varepsilon = \pm 1$. In fact we may suppose $\varepsilon = 1$, by changing the signs of c and d. Equating coefficients of ρ, we now get

$$a^3 = cd(c - d).$$

But $(c,d) = 1$, since $(x,y) = 1$, and hence also $(c,c - d) = (d,c - d) = 1$. It follows that $c = z_1^3$, $d = y_1^3$, $c - d = x_1^3$ for some $x_1,y_1,z_1 \in \mathbb{Z}$. Thus $x_1^3 + y_1^3 = z_1^3$ and

$$|x_1 y_1 z_1| = |a| = |z/3b| < |xyz|.$$

But this contradicts the definition of x,y,z. $\square$

The proof of Proposition 12 illustrates how problems involving ordinary integers may be better understood by viewing them as integers in a larger field of algebraic numbers.

We now return to the study of an arbitrary quadratic field $\mathbb{Q}(\sqrt{d})$, where d is a square-free integer. For convenience of writing we put $J = \mathcal{O}_d$. As in Chapter II, we say that $\varepsilon \in J$ is a *unit* if there exists $\eta \in J$ such that $\varepsilon\eta = 1$. For example, 1 and -1 are units. The set U of all units is evidently an abelian group under multiplication. Moreover, if $\varepsilon \in U$, then also $\varepsilon' \in U$.

If ε is a unit, then $N(\varepsilon) = \pm 1$, since $\varepsilon\eta = 1$ implies $N(\varepsilon)N(\eta) = 1$. Conversely, if $\varepsilon \in J$ and $N(\varepsilon) = \pm 1$, then ε is a unit, since $N(\varepsilon) = \varepsilon\varepsilon'$ and $\varepsilon' \in J$. (Note, however, that $N(\alpha) = \pm 1$ does not imply $\alpha \in J$. For example, in $\mathbb{Q}(\sqrt{-1})$, $\alpha = (3 + 4i)/5 \notin \mathcal{G}$, although $N(\alpha) = 1$.) It follows that, when $d \equiv 2$ or $3 \bmod 4$, $\alpha = a + b\sqrt{d}$ is a unit if and only if $a,b \in \mathbb{Z}$ and

$$a^2 - db^2 = \pm 1.$$

On the other hand, when $d \equiv 1 \bmod 4$, $\alpha = a + b(\sqrt{d} - 1)/2$ is a unit if and only if $a,b \in \mathbb{Z}$ and

$$(b - 2a)^2 - db^2 = \pm 4.$$

But if $b,c \in \mathbb{Z}$ and $c^2 - db^2 = \pm 4$, then $c^2 \equiv b^2 \bmod 4$ and hence $c \equiv b \bmod 2$.

Consequently, the units of J are determined by the solutions of the Diophantine equations $x^2 - dy^2 = \pm 4$ or $x^2 - dy^2 = \pm 1$, according as $d \equiv 1$ or $d \not\equiv 1 \bmod 4$. This makes it possible to determine all units, as we now show.

PROPOSITION 13 *The units of $\mathcal{O}_{-1}$ are ± 1, $\pm i$ and the units of $\mathcal{O}_{-3}$ are ± 1, $(\pm 1 \pm i\sqrt{3})/2$. For every other square-free integer $d < 0$, the only units of $\mathcal{O}_d$ are ± 1.*

For each square-free integer $d > 0$, there exists a unit $\varepsilon_0 > 1$ such that all units of $\mathcal{O}_d$ are given by $\pm \varepsilon_0^n$ ($n \in \mathbb{Z}$).

Proof Suppose first that $d < 0$. Then only the Diophantine equations with the + signs need to be considered. If $d < -4$, the only solutions of $x^2 - dy^2 = 4$ are $y = 0$, $x = \pm 2$. If $d < -4$ or if $d = -2$, the only solutions of $x^2 - dy^2 = 1$ are $y = 0$, $x = \pm 1$. In these cases the only units are ± 1. (The group U is a cyclic group of order 2, with -1 as generator.) If $d = -3$, the only solutions of $x^2 - dy^2 = 4$ are $y = 0$, $x = \pm 2$ and $y = \pm 1$, $x = \pm 1$. Hence the units are ± 1, $\pm \rho$, $\pm \rho^2$, where $\rho = (i\sqrt{3} - 1)/2$. (The group U is a cyclic group of order 6, with $-\rho$ as generator.) If $d = -1$, the only solutions of $x^2 + y^2 = 1$ are $y = 0$, $x = \pm 1$ and $y = \pm 1$, $x = 0$. Hence the units are ± 1, $\pm i$. (The group U is a cyclic group of order 4, with i as generator.)

Suppose next that $d > 0$. With the aid of continued fractions it will be shown in §4 of Chapter IV that the equation $x^2 - dy^2 = 1$ always has a solution in positive integers and, by doubling them, so also does the equation $x^2 - dy^2 = 4$. Hence there always exists a unit $\varepsilon > 1$.

For any unit $\varepsilon > 1$ we have $\varepsilon > \pm\, \varepsilon'$, since $\varepsilon' = \varepsilon^{-1}$ or $-\varepsilon^{-1}$. If $\varepsilon = a + b\omega$, where ω is defined as in Proposition 11 and $a,b \in \mathbb{Z}$, then $\varepsilon' = a - b\omega$ or $a - b - b\omega$, according as $d \not\equiv 1$ or $d \equiv 1$ mod 4. Since ω is positive, $\varepsilon > \varepsilon'$ yields $b > 0$ and $\varepsilon > -\varepsilon'$ then yields $a > 0$. Thus every unit $\varepsilon > 1$ has the form $a + b\omega$, where $a,b \in \mathbb{N}$. Consequently there is a least unit $\varepsilon_0 > 1$. Then, for any unit $\varepsilon > 1$, there is a positive integer n such that $\varepsilon_0^{\,n} \le \varepsilon < \varepsilon_0^{\,n+1}$. Since $\varepsilon\varepsilon_0^{-n}$ is a unit and $1 \le \varepsilon\varepsilon_0^{-n} < \varepsilon_0$, we must actually have $\varepsilon = \varepsilon_0^{\,n}$. (The group U is the direct product of the cyclic group of order 2 generated by -1 and the infinite cyclic group generated by ε_0.) $\square$

As an example, take $d = 2$. Then $\varepsilon_0 = 1 + \sqrt{2}$ is a unit. Since $\varepsilon_0 > 1$ and all units greater than 1 have the form $a + b\sqrt{2}$ with $a,b \in \mathbb{N}$, it follows that all units are given by $\pm\, \varepsilon_0^{\,n}$ ($n \in \mathbb{Z}$).

Having determined the units, we now consider more generally the theory of divisibility in the integral domain J. If $\alpha,\beta \in J$ and β is a proper divisor of α, then $N(\beta)$ is a proper divisor in $\mathbb{Z}$ of $N(\alpha)$ and hence $|N(\beta)| < |N(\alpha)|$. Consequently the chain condition (#) of Chapter II is satisfied. It follows that any element of J which is neither zero nor a unit is a product of finitely many irreducibles. Thus it only remains to determine the irreducibles. However, this is not such a simple matter, as the following examples indicate.

The ring $\mathscr{G}$ of Gaussian integers is a Euclidean domain. However, an ordinary prime p may or may not be irreducible in $\mathscr{G}$. For example, $2 = (1 + i)(1 - i)$ and neither factor is one of the units $\pm\, 1, \pm\, i$. On the other hand, 3 has no proper divisor $\alpha = a + bi$ which is not a unit, since $N(3) = 9$ and $N(\alpha) = a^2 + b^2 = \pm\, 3$ has no solutions in integers a,b.

Again, consider the ring $\mathcal{O}_{-5}$ of integers of the field $\mathbb{Q}(\sqrt{-5})$. An element $\alpha = a + b\sqrt{-5}$ of $\mathcal{O}_{-5}$ cannot have norm $N(\alpha) = a^2 + 5b^2$ equal to $\pm\, 2$ or $\pm\, 3$, since the square of any ordinary integer is congruent to 0,1 or 4 mod 5. It follows that, in the factorizations

$$6 = 2{\cdot}3 = (1 + \sqrt{-5})(1 - \sqrt{-5}),$$

all four factors are irreducible and the factorizations are essentially distinct, since $N(2) = 4$, $N(3) = 9$ and $N(1 \pm \sqrt{-5}) = 6$. Thus 2 is not a 'prime' in $\mathcal{O}_{-5}$ and the 'fundamental theorem of arithmetic' does not hold.

It was shown by Kummer and Dedekind in the 19th century that uniqueness of factorization could be restored by considering ideals instead of elements. Any nonzero proper ideal of $\mathcal{O}_d$ can be represented as a product of finitely many prime ideals and the representation is unique except for the order of the factors. This result will now be established.

A nonempty subset A of a commutative ring R is an *ideal* if $a,b \in A$ and $x,y \in R$ imply $ax + by \in A$. For example, R and $\{0\}$ are ideals. If $a_1,\ldots,a_m \in R$, then the set $(a_1,\ldots,a_m)$ of all elements $a_1x_1 + ... + a_mx_m$ with $x_j \in R$ ($1 \le j \le m$) is an ideal, the ideal *generated* by $a_1,...,a_m$. An ideal generated by a single element is a *principal ideal*.

If A and B are ideals in R, then the set AB of all finite sums $a_1b_1 + ... + a_nb_n$ with $a_j \in A$ and $b_j \in B$ ($1 \le j \le n$; $n \in \mathbb{N}$) is also an ideal, the *product* of A and B. For any ideals A,B,C we have

$$AB = BA, \quad (AB)C = A(BC),$$

since multiplication in R is commutative and associative.

An ideal $A \ne \{0\}$ is said to be *divisible* by an ideal B, and B is said to be a *factor* of A, if there exists an ideal C such that $A = BC$. For example, A is divisible by itself and by R, since $A = AR$. Thus R is an identity element for multiplication of ideals.

Now take $R = \mathbb{O}_d$ to be the ring of all integers of the quadratic field $\mathbb{Q}(\sqrt{d})$. We will show that in this case much more can be said.

PROPOSITION 14 *Let $A \ne \{0\}$ be an ideal in $\mathbb{O}_d$. Then there exist $\beta,\gamma \in A$ such that every $\alpha \in A$ can be uniquely represented in the form*

$$\alpha = m\beta + n\gamma \quad (m,n \in \mathbb{Z}).$$

Furthermore, if ω is defined as in Proposition 11, we may take $\beta = a$, $\gamma = b + c\omega$, where $a,b,c \in \mathbb{Z}$, $a > 0$, $c > 0$, c divides both a and b, and ac divides $\gamma\gamma'$, i.e.

$$b^2 - dc^2 \equiv 0 \ \mathrm{mod} \ ac \qquad if\ d \equiv 2\ or\ 3 \ \mathrm{mod}\ 4,$$
$$b(b-c) - (d-1)c^2/4 \equiv 0 \ \mathrm{mod} \ ac \ \ if\ d \equiv 1\ \mathrm{mod}\ 4.$$

Proof Since A is an ideal, the set J of all $z \in \mathbb{Z}$ such that $y + z\omega \in A$ for some $y \in \mathbb{Z}$ is an ideal in $\mathbb{Z}$. Moreover $J \ne \{0\}$, since $A \ne \{0\}$ and $\alpha \in A$ implies $\alpha\omega \in A$. Since $\mathbb{Z}$ is a principal ideal domain, it follows that there exists $c > 0$ such that $J = \{nc: n \in \mathbb{Z}\}$. Since $c \in J$, there exists $b \in \mathbb{Z}$ such that $\gamma := b + c\omega \in A$.

Furthermore A contains some nonzero $x \in \mathbb{Z}$, since $\alpha \in A$ implies $\alpha\alpha' \in A$. Since the set I of all $x \in \mathbb{Z} \cap A$ is an ideal in $\mathbb{Z}$, there exists $a > 0$ such that $I = \{ma: m \in \mathbb{Z}\}$. For any $\alpha = y + z\omega \in A$ we have $z = nc$ for some $n \in \mathbb{Z}$ and $\alpha - n\gamma = y - nb = ma$ for some $m \in \mathbb{Z}$. Thus $\alpha = m\beta + n\gamma$ with $\beta = a$. The representation is unique, since γ is irrational.

Since $\beta\omega \in A$, we have

$$a\omega = ra + s(b + c\omega) \ \text{ for unique } r,s \in \mathbb{Z}.$$

Thus $a = sc$ and $ra + sb = 0$, which together imply $b = -rc$. Since $\gamma\omega \in A$, we have also

$$(b + c\omega)\omega = ma + n(b + c\omega) \quad \text{for unique } m,n \in \mathbb{Z}.$$

If $d \equiv 2$ or $3 \bmod 4$, then $\omega^2 = d$. In this case $n = -r$, $cd = ma - rb$ and hence $dc^2 = mac + b^2$. If $d \equiv 1 \bmod 4$, then $\omega^2 = -\omega + (d - 1)/4$. Hence $n = -(r + 1)$, $c(d - 1)/4 = ma - rb - b$ and $(d - 1)c^2/4 = mac + b(b - c)$. $\square$

If A is an ideal in $\mathbb{O}_d$, then the set $A' = \{\alpha': \alpha \in A\}$ of all conjugates of elements of A is also an ideal in $\mathbb{O}_d$. We call A' the *conjugate* of A.

PROPOSITION 15 *If $A \neq \{0\}$ is an ideal in $\mathbb{O}_d$, then $AA' = l\mathbb{O}_d$ for some $l \in \mathbb{N}$.*

Proof Choose β,γ so that $A = \{m\beta + n\gamma: m,n \in \mathbb{Z}\}$. Then AA' consists of all integral linear combinations of $\beta\beta', \beta\gamma', \beta'\gamma$ and $\gamma\gamma'$. Furthermore $r = \beta\beta'$, $s = \beta\gamma' + \beta'\gamma$ and $t = \gamma\gamma'$ are all in $\mathbb{Z}$. If l is the greatest common divisor of r,s and t, then $l \in AA'$, by the Bézout identity, and hence $l\mathbb{O}_d \subseteq AA'$.

On the other hand, $\beta\gamma'$ and $\beta'\gamma$ are roots of the quadratic equation

$$x^2 - sx + rt = 0$$

with integer coefficients $s = \beta\gamma' + \beta'\gamma$ and $rt = \beta\beta'\gamma\gamma'$. It follows that $\beta\gamma'/l$ and $\beta'\gamma/l$ are roots of the quadratic equation

$$y^2 - (s/l)y + rt/l^2 = 0,$$

which also has integer coefficients. Since $\beta\gamma'/l$ and $\beta'\gamma/l$ are in $\mathbb{Q}(\sqrt{d})$, this means that they are in $\mathbb{O}_d$. Thus $\beta\gamma'$ and $\beta'\gamma$ are in $l\mathbb{O}_d$. Since also $\beta\beta'$ and $\gamma\gamma'$ are in $l\mathbb{O}_d$, it follows that $AA' \subseteq l\mathbb{O}_d$. $\square$

If in the proof of Proposition 15 we choose $\beta = a$ and $\gamma = b + c\omega$ as in the statement of Proposition 14, then in the statement of Proposition 15 we will have $l = ac$. Since the proof of this when $d \equiv 1 \bmod 4$ is similar, we give the proof only for $d \equiv 2$ or $3 \bmod 4$. In this case $\omega = \sqrt{d}$ and hence $r = a^2$, $s = 2ab$, $t = b^2 - dc^2$. We wish to show that ac is the greatest common divisor of r,s and t. Thus if we put

$$a = cu, \ \ b = cv, \ \ t = acw,$$

then we wish to show that $u, 2v$ and w have greatest common divisor 1. Since $uw = v^2 - d$ and d is square-free, a common divisor greater than 1 can only be 2. But if 2 were a common divisor, we would have $v^2 \equiv d \bmod 4$, which is impossible, because $d \equiv 2$ or $3 \bmod 4$.

We can now show that multiplication of ideals satisfies the cancellation law:

PROPOSITION 16 *If A,B,C are ideals in $\mathcal{O}_d$ with $A \neq \{0\}$, then $AB = AC$ implies $B = C$.*

Proof By multiplying by the conjugate A' of A we obtain $AA'B = AA'C$ and hence, by Proposition 15, $lB = lC$ for some positive integer l. But this implies $B = C$. $\square$

PROPOSITION 17 *Let A and B be nonzero ideals in $\mathcal{O}_d$. Then A is divisible by B if and only if $A \subseteq B$.*

Proof If $A = BC$ for some ideal C, then $A \subseteq B$, by the definition of the product of two ideals.

Conversely, suppose $A \subseteq B$. By Proposition 15, $BB' = l\mathcal{O}_d$ for some positive integer l. Hence $AB' \subseteq l\mathcal{O}_d$. It follows that $AB' = lC$ for some ideal C. Thus $AB' = BB'C$ and so, by Proposition 16, $A = BC$. $\square$

COROLLARY 18 *Let A and B be nonzero ideals in $\mathcal{O}_d$. If D is the set of all elements $a + b$, with $a \in A$ and $b \in B$, then D is an ideal and is a factor of both A and B. Moreover, every common factor of A and B is also a factor of D.*

Proof It follows at once from its definition that D is an ideal. Moreover D contains both A and B, since 0 is an element of any ideal. Evidently also any ideal C which contains both A and B also contains D. The result now follows from Proposition 17. $\square$

In the terminology of Chapter II, this shows that *any two nonzero ideals in $\mathcal{O}_d$ have a greatest common divisor.*

In a commutative ring R, an ideal $A \neq R,\{0\}$ is said to be *irreducible* if its only factors are A and R. It is said to be *maximal* if the only ideals containing A are A and R. It is said to be *prime* if, whenever A divides the product of two ideals, it also divides at least one of the factors.

By Proposition 17, an ideal in $\mathcal{O}_d$ is irreducible if and only if it is maximal. The existence of greatest common divisors implies, as we saw in §1 of Chapter II, that an ideal in $\mathcal{O}_d$ is irreducible if and only if it is prime. (These equivalences do not hold in arbitrary commutative rings, but they do hold for the ring of all algebraic integers in any given algebraic number field, and also for the rings associated with algebraic curves.)

PROPOSITION 19 *A nonzero ideal A in $\mathcal{O}_d$ has only finitely many factors.*

Proof Since $AA' = l\mathcal{O}_d$ for some positive integer l, any factor B of A is also a factor of $l\mathcal{O}_d$ and so contains l. Proposition 14 implies, in particular, that B is generated by two elements, say $B = (\beta_1, \beta_2)$. A fortiori, $B = (\beta_1, \beta_2, l)$ and hence, for any $\gamma_1, \gamma_2 \in \mathcal{O}_d$, also

$$B = (\beta_1 - l\gamma_1, \beta_2 - l\gamma_2, l) \, .$$

We can choose $\gamma_1 \in \mathbb{O}_d$ so that in the representation

$$\beta_1 - l\gamma_1 = a_1 + b_1\omega \quad (a_1, b_1 \in \mathbb{Z})$$

we have $0 \leq a_1, b_1 < l$. Similarly we can choose $\gamma_2 \in \mathbb{O}_d$ so that in the representation

$$\beta_2 - l\gamma_2 = a_2 + b_2\omega \quad (a_2, b_2 \in \mathbb{Z})$$

we have $0 \leq a_2, b_2 < l$. It follows that there are at most l^4 different possibilities for the ideal B.
$\square$

COROLLARY 20 *There exists no infinite sequence $\{A_n\}$ of nonzero ideals in $\mathbb{O}_d$ such that, for every n, A_{n+1} divides A_n and $A_{n+1} \neq A_n$.* $\square$

In the terminology of Chapter II, this shows that *the set of all nonzero ideals in $\mathbb{O}_d$ satisfies the chain condition* (#). Since also the conclusion of Proposition II.1 holds, the argument in §1 of Chapter II now shows that any nonzero proper ideal in $\mathbb{O}_d$ is a product of finitely many prime ideals and the representation is unique apart from the order of the factors.

It remains to determine the prime ideals. This is accomplished by the following three propositions.

PROPOSITION 21 *For each prime ideal P in $\mathbb{O}_d$ there is a unique prime number p such that P divides $p\mathbb{O}_d$. Furthermore, for any prime number p there is a prime ideal P in $\mathbb{O}_d$ such that exactly one of the following alternatives holds:*

(i) $p\mathbb{O}_d = PP'$ *and* $P \neq P'$;

(ii) $p\mathbb{O}_d = P = P'$;

(iii) $p\mathbb{O}_d = P^2$ *and* $P = P'$.

Proof If P is a prime ideal in $\mathbb{O}_d$, then $PP' = l\mathbb{O}_d$ for some positive integer l. Moreover $l > 1$, since $l \in P$. If $l = mn$, where m and n are positive integers greater than 1, then P divides either $m\mathbb{O}_d$ or $n\mathbb{O}_d$. By repeating the argument it follows that P divides $p\mathbb{O}_d$ for some prime divisor p of l. The prime number p is uniquely determined by the prime ideal P since, by the Bézout identity, if P contained distinct primes it would also contain their greatest common divisor 1.

Now let p be any prime number and let the factorisation of $p\mathbb{O}_d$ into a product of positive powers of distinct prime ideals be

$$p\mathbb{O}_d = P_1^{e_1} \cdots P_s^{e_s}.$$

If we put $Q_j = P_j'$ $(1 \le j \le s)$, then also

$$p\mathcal{O}_d = Q_1^{e_1} \cdots Q_s^{e_s}.$$

But $P_j Q_j = n_j \mathcal{O}_d$ for some integer $n_j > 1$ and hence

$$p^2 = n_1^{e_1} \cdots n_s^{e_s}.$$

Evidently the only possibilities are

(i)′ $s = 2$, $n_1 = n_2 = p$, $e_1 = e_2 = 1$;
(ii)′ $s = 1$, $n_1 = p^2$, $e_1 = 1$;
(iii)′ $s = 1$, $n_1 = p$, $e_1 = 2$.

Since the factorization is unique apart from order, this yields the three possibilities in the statement of the proposition. $\square$

Proposition 21 does not tell us which of the three possibilities holds for a given prime p. For $p \neq 2$, the next result gives an answer in terms of the Legendre symbol.

PROPOSITION 22 *Let p be an odd prime. Then, in the statement of Proposition 21, (i),(ii), or (iii) holds according as*

$$p \nmid d \text{ and } (d/p) = 1, \quad p \nmid d \text{ and } (d/p) = -1, \quad or \ p \mid d.$$

Proof Suppose first that $p \nmid d$ and that there exists $a \in \mathbb{Z}$ such that $a^2 \equiv d \bmod p$. Then $p \nmid a$ and $a^2 - d = pb$ for some $b \in \mathbb{Z}$. If $P = (p, a + \sqrt{d})$, then $P' = (p, a - \sqrt{d})$ and

$$PP' = p(p, a + \sqrt{d}, a - \sqrt{d}, b).$$

Since $(p, a + \sqrt{d}, a - \sqrt{d}, b)$ contains $2a$, which is relatively prime to p, it also contains 1. Hence $PP' = p\mathcal{O}_d$. Moreover $P \neq P'$, since $P = P'$ would imply $2a \in P$ and hence $1 \in P$. We do not need to prove that P is a prime ideal, since what we have already established is incompatible with cases (ii) and (iii) of Proposition 21.

Suppose next that $p \mid d$. Then $d = pe$ for some $e \in \mathbb{Z}$ and $p \nmid e$, since d is square-free. If $P = (p, \sqrt{d})$, then

$$P^2 = p(p, \sqrt{d}, e) = p\mathcal{O}_d,$$

since $(p, e) = 1$. Since we cannot be in cases (i) or (ii) of Proposition 21, we must be in case (iii).

Suppose conversely that either (i) or (iii) of Proposition 21 holds. Then the corresponding prime ideal P contains p. Choose $\beta = a$ and $\gamma = b + c\omega$ as in Proposition 14, so that

$$P = \{m\beta + n\gamma : m,n \in \mathbb{Z}\}.$$

In the present case we must have $a = p$, since $p \in P$ and $1 \notin P$. We must also have $c = 1$, since $PP' = p\mathbb{O}_d$ implies $ac = p$. With these values of a and c the final condition of Proposition 14 takes the form

$$b^2 \equiv d \ \mathrm{mod} \ p \qquad\qquad \text{if } d \equiv 2 \text{ or } 3 \ \mathrm{mod} \ 4,$$
$$b(b-1) \equiv (d-1)/4 \ \mathrm{mod} \ p \quad \text{if } d \equiv 1 \ \mathrm{mod} \ 4.$$

Thus in the latter case $(2b-1)^2 \equiv d \ \mathrm{mod} \ p$. In either case if $p \nmid d$, then $(d/p) = 1$.

This proves that if $p \nmid d$ and $(d/p) = -1$, then we must be in case (ii) of Proposition 21.

$\square$

PROPOSITION 23 *Let $p = 2$. Then, in the statement of Proposition 21, (i),(ii), or (iii) holds according as*

$$d \equiv 1 \ \mathrm{mod} \ 8, \ d \equiv 5 \ \mathrm{mod} \ 8, \ or \ d \equiv 2,3 \ \mathrm{mod} \ 4.$$

Proof Since the proof is similar to that of the previous proposition, we will omit some of the detail. Suppose first that $d \equiv 1 \ \mathrm{mod} \ 8$. If $P = (2,(1 - \sqrt{d})/2)$, then $P' = (2,(1 + \sqrt{d})/2)$ and

$$PP' = 2(2,(1 - \sqrt{d})/2,(2,(1 + \sqrt{d})/2,(1 - d)/8).$$

It follows that $PP' = 2\mathbb{O}_d$ and $P \neq P'$.

Suppose next that $d \equiv 2 \ \mathrm{mod} \ 4$. Then $d = 2e$, where e is odd. If $P = (2,\sqrt{d})$, then

$$P^2 = 2(2,\sqrt{d},e) = 2\mathbb{O}_d.$$

Similarly, if $d \equiv 3 \ \mathrm{mod} \ 4$ and $P = (2,1 + \sqrt{d})$, then

$$P^2 = 2(2,1 + \sqrt{d},(1 + d)/2 + \sqrt{d}) = 2\mathbb{O}_d.$$

Suppose conversely that either (i) or (iii) of Proposition 21 holds. Then the corresponding prime ideal P contains 2. Choose $\beta = a$ and $\gamma = b + c\omega$ as in Proposition 14, so that

$$P = \{m\beta + n\gamma : m,n \in \mathbb{Z}\}.$$

In the present case we must have $a = 2$, $c = 1$ and

$$b(b-1) \equiv (d-1)/4 \ \mathrm{mod} \ 2 \ \text{if } d \equiv 1 \ \mathrm{mod} \ 4.$$

Since $b(b-1)$ is even, it follows that $d \not\equiv 5 \ \mathrm{mod} \ 8$.

This proves that if $d \equiv 5 \ \mathrm{mod} \ 8$, then we must be in case (ii) of Proposition 21. $\square$

Proposition 22 uses only Legendre's definition of the Legendre symbol. What does the law of quadratic reciprocity tell us? By Proposition 4, if p and q are distinct odd primes and d an integer not divisible by p such that $q \equiv p \bmod 4d$, then $(d/p) = (d/q)$. Consequently, by Proposition 22, whether (i) or (ii) holds in Proposition 21 depends only on the residue class of $p \bmod 4d$. Thus, for given d, we need determine the behaviour of only finitely many primes p.

We mention without proof some further properties of the ring $\mathcal{O}_d$. We say that two nonzero ideals $A, \tilde{A}$ in $\mathcal{O}_d$ are *equivalent*, and we write $A \sim \tilde{A}$, if there exist nonzero principal ideals $(\alpha),(\tilde{\alpha})$ such that $(\alpha)A = (\tilde{\alpha})\tilde{A}$. It is easily verified that this is indeed an equivalence relation. Moreover, if $A \sim \tilde{A}$ and $B \sim \tilde{B}$, then $AB \sim \tilde{A}\tilde{B}$. Consequently, if we call an equivalence class of ideals an *ideal class*, we can without ambiguity define the product of two ideal classes. The set of ideal classes acquires in this way the structure of a commutative group, the ideal class containing the conjugate A' of A being the inverse of the ideal class containing A. It may be shown that this *ideal class group* is finite. The order of the group, i.e. the number of different ideal classes, is called the *class number* of the quadratic field $\mathbb{Q}(\sqrt{d})$ and is traditionally denoted by $h(d)$. The ring $\mathcal{O}_d$ is a principal ideal domain if and only if $h(d) = 1$. (Moreover, $\mathcal{O}_d$ is a factorial domain only if it is a principal ideal domain.)

The theory of quadratic fields has been extensively generalized. An *algebraic number field K* is a field containing the field $\mathbb{Q}$ of rational numbers and of finite dimension as a vector space over $\mathbb{Q}$. An *algebraic integer* is a root of a monic polynomial $x^n + a_1 x^{n-1} + ... + a_n$ with coefficients $a_1,...,a_n \in \mathbb{Z}$. The set of all algebraic integers in a given algebraic number field K is a ring $\mathcal{O}(K)$. It may be shown that, also in $\mathcal{O}(K)$, any nonzero proper ideal can be represented as a product of prime ideals and the representation is unique except for the order of the factors. One may also construct the *ideal class group* of K and show that it is finite, its order being the *class number* of K.

Some of the motivation for these generalizations came from 'Fermat's last theorem'. Fermat (c. 1640) asserted that the equation $x^n + y^n = z^n$ has no solutions in positive integers x,y,z if $n > 2$. In Proposition 12 we proved Fermat's assertion for $n = 3$. To prove the assertion in general it is sufficient to prove it when $n = 4$ and when $n = p$ is an odd prime, since if $x^{km} + y^{km} = z^{km}$, then $(x^k)^m + (y^k)^m = (z^k)^m$. Fermat himself gave a proof for $n = 4$, which is reproduced in Chapter XIII. Proofs for $n = 3,5$ and 7 were given by Euler (1760-1770), Legendre (1825) and Lamé (1839) respectively.

Kummer (1850) made a remarkable advance beyond this by proving that the assertion holds whenever $n = p$ is a 'regular' prime. Here a prime p is said to be *regular* if it does not divide the class number of the *cyclotomic field* $\mathbb{Q}(\zeta_p)$, obtained by adjoining to $\mathbb{Q}$ a p-th root of unity ζ_p. Kummer converted his result into a practical test by further proving that a prime

$p > 3$ is regular if and only if it does not divide the numerator of any of the *Bernoulli numbers* $B_2, B_4, \ldots, B_{p-3}$.

The only irregular primes less than 100 are 37, 59 and 67. Other methods for dealing with irregular primes were devised by Kummer (1857) and Vandiver (1929). By 1993 Fermat's assertion had been established in this way for all n less than four million. However, these methods did not lead to a complete proof of 'Fermat's last theorem'. As will be seen in Chapter XIII, a complete solution was first found by Wiles (1995), using quite different methods.

3   Multiplicative functions

We define a function $f: \mathbb{N} \to \mathbb{C}$ to be an *arithmetical function*. The set of all arithmetical functions can be given the structure of a commutative ring in the following way.

For any two functions $f, g: \mathbb{N} \to \mathbb{C}$, we define their *convolution* or *Dirichlet product* $f*g: \mathbb{N} \to \mathbb{C}$ by

$$f*g(n) = \sum_{d \mid n} f(d)g(n/d).$$

Dirichlet multiplication satisfies the usual commutative and associative laws:

LEMMA 24   *For any three functions $f, g, h: \mathbb{N} \to \mathbb{C}$,*

$$f*g = g*f, \quad f*(g*h) = (f*g)*h.$$

Proof   Since n/d runs through the positive divisors of n at the same time as d,

$$f*g(n) = \sum_{d \mid n} f(d)g(n/d) = \sum_{d \mid n} f(n/d)g(d) = g*f(n).$$

To prove the associative law, put $G = g*h$. Then

$$f*G(n) = \sum_{de=n} f(d)G(e) = \sum_{de=n} f(d) \sum_{d'd''=e} g(d')h(d'') = \sum_{dd'd''=n} f(d)g(d')h(d'').$$

Similarly, if we put $F = f*g$, we obtain

$$F*h(n) = \sum_{de=n} F(e)h(d) = \sum_{de=n} \sum_{d'd''=e} f(d')g(d'')h(d) = \sum_{dd'd''=n} f(d')g(d'')h(d).$$

Hence $F*h(n) = f*G(n)$. $\square$

For any two functions $f, g: \mathbb{N} \to \mathbb{C}$, we define their *sum $f + g$*: $\mathbb{N} \to \mathbb{C}$ in the natural way:

$$(f + g)(n) = f(n) + g(n).$$

It is obvious that addition is commutative and associative, and that the distributive law holds:

$$f*(g + h) = f*g + f*h.$$

The function $\delta: \mathbb{N} \to \mathbb{C}$, defined by

$$\delta(n) = 1 \text{ or } 0 \text{ according as } n = 1 \text{ or } n > 1,$$

acts as an identity element for Dirichlet multiplication:

$$\delta*f = f \text{ for every } f: \mathbb{N} \to \mathbb{C},$$

since

$$\delta*f(n) = \sum_{d \mid n} \delta(d)f(n/d) = f(n).$$

Thus the set $\mathscr{A}$ of all arithmetical functions is indeed a commutative ring.

For any function $f: \mathbb{N} \to \mathbb{C}$ which is not identically zero, put $|f| = v(f)^{-1}$, where $v(f)$ is the least positive integer n such that $f(n) \neq 0$, and put $|O| = 0$. Then

$$|f*g| = |f||g|, \quad |f + g| \le \max(|f|,|g|) \quad \text{for all } f,g \in \mathscr{A}.$$

Hence the ring $\mathscr{A}$ of all arithmetical functions is actually an integral domain. It is readily shown that the set of all $f \in \mathscr{A}$ such that $|f| < 1$ is an ideal, but not a principal ideal. (Although $\mathscr{A}$ is not a principal ideal domain, it may be shown that it is a *factorial* domain.)

The next result shows that the functions $f \in \mathscr{A}$ such that $|f| = 1$ are the *units* in the ring $\mathscr{A}$:

LEMMA 25 *For any function $f: \mathbb{N} \to \mathbb{C}$, there is a function $f^{-1}: \mathbb{N} \to \mathbb{C}$ such that $f^{-1}*f = \delta$ if and only if $f(1) \neq 0$. The inverse f^{-1} is uniquely determined and $f^{-1}(1)f(1) = 1$.*

Proof Suppose $g: \mathbb{N} \to \mathbb{C}$ has the property that $g*f = \delta$. Then $g(1)f(1) = 1$. Thus $g(1)$ is non-zero and uniquely determined. If $n > 1$, then

$$\sum_{d \mid n} g(d)f(n/d) = 0.$$

Hence

$$g(n)f(1) = -\sum_{d \mid n, d < n} g(d)f(n/d).$$

It follows by induction that $g(n)$ is uniquely determined for every $n \in \mathbb{N}$. Conversely, if g is defined inductively in this way, then $g*f = \delta$. $\quad\square$

It follows from Lemma 25 that the set of all arithmetical functions $f: \mathbb{N} \to \mathbb{C}$ such that $f(1) \neq 0$ is an abelian group under Dirichlet multiplication.

A function $f: \mathbb{N} \to \mathbb{C}$ is said to be *multiplicative* if it is not identically zero and if

$$f(mn) = f(m)f(n) \text{ for all } m,n \text{ with } (m,n) = 1.$$

It follows that $f(1) = 1$, since $f(n) \neq 0$ for some n and $f(n) = f(n)f(1)$. Any multiplicative function $f: \mathbb{N} \to \mathbb{C}$ is uniquely determined by its values at the prime powers, since if

$$m = p_1^{\alpha_1} \cdots p_s^{\alpha_s},$$

where $p_1,\ldots,p_s$ are distinct primes and $\alpha_1,\ldots,\alpha_s \in \mathbb{N}$, then

$$f(m) = f(p_1^{\alpha_1}) \cdots f(p_s^{\alpha_s}).$$

If

$$m = \prod_p p^{\alpha_p}, \quad n = \prod_p p^{\beta_p},$$

where $\alpha_p, \beta_p \geq 0$, then

$$(m,n) = \prod_p p^{\gamma_p}, \quad [m,n] = \prod_p p^{\delta_p},$$

where $\gamma_p = \min\{\alpha_p, \beta_p\}$ and $\delta_p = \max\{\alpha_p, \beta_p\}$. Since either $\gamma_p = \alpha_p$ and $\delta_p = \beta_p$, or $\gamma_p = \beta_p$ and $\delta_p = \alpha_p$, it follows that, for any multiplicative function f and all $m,n \in \mathbb{N}$,

$$f((m,n))\, f([m,n]) = \prod_p f(p^{\gamma_p})f(p^{\delta_p}) = \prod_p f(p^{\alpha_p})f(p^{\beta_p}) = f(m)f(n).$$

As we saw in §5 of Chapter II, it follows from Proposition II.4 that Euler's φ-function is multiplicative. Also, the functions $i: \mathbb{N} \to \mathbb{C}$ and $j: \mathbb{N} \to \mathbb{C}$, defined by

$$i(n) = 1, \quad j(n) = n \text{ for every } n \in \mathbb{N},$$

are obviously multiplicative. Further examples of multiplicative functions can be constructed with the aid of the next two propositions.

PROPOSITION 26 *If $f,g: \mathbb{N} \to \mathbb{C}$ are multiplicative functions, then their Dirichlet product $h = f*g$ is also multiplicative.*

Proof We have

$$h(n) = \sum_{d \mid n} f(d)g(n/d).$$

Suppose $n = n'n''$, where n' and n'' are relatively prime. Then, by Proposition II.4,

$$h(n) = \sum_{d' \mid n', d'' \mid n''} f(d'd'')g(n'n''/d'd'')$$

$$= \sum_{d' \mid n', d'' \mid n''} f(d')f(d'')g(n'/d')g(n''/d'')$$

$$= \sum_{d' \mid n'} f(d')g(n'/d') \sum_{d'' \mid n''} f(d'')g(n''/d'') = h(n')h(n''). \qquad \square$$

PROPOSITION 27 *If $f: \mathbb{N} \to \mathbb{C}$ is multiplicative, then its Dirichlet inverse $f^{-1}: \mathbb{N} \to \mathbb{C}$ is also multiplicative.*

Proof Assume on the contrary that $g: = f^{-1}$ is not multiplicative and let n',n'' be relatively prime positive integers such that $g(n'n'') \neq g(n')g(n'')$. We suppose n',n'' chosen so that the product $n = n'n''$ is minimal. Since f is multiplicative, $f(1) = 1$ and hence $g(1) = 1$. Consequently $n' > 1, n'' > 1$ and

$$0 = \Sigma_{d'|n'}\, g(d')f(n'/d') = \Sigma_{d''|n''}\, g(d'')f(n''/d'') = \Sigma_{d|n}\, g(d)f(n/d).$$

But

$$\Sigma_{d|n}\, g(d)f(n/d) = g(n)f(1) + \Sigma_{d'|n',d''|n'',d'd''<n}\, g(d'd'')f(n'n''/d'd'')$$

$$= g(n) + \Sigma_{d'|n',d''|n'',d'd''<n}\, g(d')g(d'')f(n'/d')f(n''/d'')$$

$$= g(n) - g(n')g(n'') + \Sigma_{d'|n'}\, g(d')f(n'/d') \cdot \Sigma_{d''|n''}\, g(d'')f(n''/d'')$$

$$= g(n) - g(n')g(n'').$$

Thus we have a contradiction. $\square$

It follows from Propositions 26 and 27 that under Dirichlet multiplication the multiplicative functions form a subgroup of the group of all functions $f: \mathbb{N} \to \mathbb{C}$ with $f(1) \neq 0$. The further subgroup generated by i and j contains some interesting functions. Let $\tau(n)$ denote the number of positive divisors of n, and let $\sigma(n)$ denote the sum of the positive divisors of n:

$$\tau(n) = \Sigma_{d|n}\, 1, \quad \sigma(n) = \Sigma_{d|n}\, d.$$

In other words,

$$\tau = i*i, \quad \sigma = i*j,$$

and hence, by Proposition 26, τ and σ are multiplicative functions. Thus they are uniquely determined by their values at the prime powers. But if p is prime and $\alpha \in \mathbb{N}$, the divisors of p^α are $1,p,...,p^\alpha$ and hence

$$\tau(p^\alpha) = \alpha + 1, \quad \sigma(p^\alpha) = (p^{\alpha+1} - 1)/(p - 1).$$

By Proposition II.24, Euler's φ-function satisfies $i*\varphi = j$. Thus $\varphi = i^{-1}*j$, and Propositions 26 and 27 provide a new proof that Euler's φ-function is multiplicative. Since

$$\tau*\varphi = i*i*\varphi = i*j = \sigma,$$

we also obtain the new relation

$$\sigma(n) = \sum_{d\,|\,n} \tau(n/d)\varphi(d).$$

The *Möbius function* $\mu\colon \mathbb{N} \to \mathbb{C}$ is defined to be the Dirichlet inverse i^{-1}. Thus $\mu * i = \delta$ or, in other words,

$$\sum_{d\,|\,n} \mu(d) = 1 \text{ or } 0 \text{ according as } n = 1 \text{ or } n > 1.$$

Instead of this inductive definition, we may explicitly characterize the Möbius function in the following way:

PROPOSITION 28 *For any* $n \in \mathbb{N}$,

$$\mu(n) = \begin{cases} 1 & \text{if } n = 1, \\ (-1)^s & \text{if } n \text{ is a product of } s \text{ distinct primes,} \\ 0 & \text{if } n \text{ is divisible by the square of a prime.} \end{cases}$$

Proof It is trivial that $\mu(1) = 1$. Suppose p is prime and $\alpha \in \mathbb{N}$. Since the divisors of p^{α} are $1, p, \ldots, p^{\alpha}$, we have

$$\mu(1) + \mu(p) + \ldots + \mu(p^{\alpha}) = 0.$$

Since this holds for all $\alpha \in \mathbb{N}$, it follows that $\mu(p^{\alpha}) = 0$ if $\alpha > 1$ and $\mu(p) = -\mu(1) = -1$. Since the Möbius function is multiplicative, by Proposition 27, the general formula follows. $\square$

The function defined by the statement of Proposition 28 had already appeared in work of Euler (1748), but Möbius (1832) discovered the basic property which we have adopted as a definition. From this property we can easily derive the *Möbius inversion formula*:

PROPOSITION 29 *For any function* $f\colon \mathbb{N} \to \mathbb{C}$, *if* $\hat{f}\colon \mathbb{N} \to \mathbb{C}$ *is defined by*

$$\hat{f}(n) = \sum_{d\,|\,n} f(d),$$

then

$$f(n) = \sum_{d\,|\,n} \hat{f}(d)\mu(n/d) = \sum_{d\,|\,n} \hat{f}(n/d)\mu(d).$$

Furthermore, for any function $\hat{f}\colon \mathbb{N} \to \mathbb{C}$, *there is a unique function* $f\colon \mathbb{N} \to \mathbb{C}$ *such that* $\hat{f}(n) = \sum_{d\,|\,n} f(d)$ *for every* $n \in \mathbb{N}$.

Proof Let $f\colon \mathbb{N} \to \mathbb{C}$ be given and put $\hat{f} = f * i$. Then

$$\hat{f} * \mu = f * i * \mu = f * \delta = f.$$

Conversely, let $\hat{f}\colon \mathbb{N} \to \mathbb{C}$ be given and put $f = \hat{f} * \mu$. Then $f * i = \hat{f} * \mu * i = \hat{f} * \delta = \hat{f}$. Moreover, by the first part of the proof, this is the only possible choice for f. $\square$

If we apply Proposition 29 to Euler's φ-function then, by Proposition II.24, we obtain the formula

$$\varphi(n) = n \sum_{d \mid n} \mu(d)/d.$$

In particular, if $n = p^\alpha$, where p is prime and $\alpha \in \mathbb{N}$, then

$$\varphi(p^\alpha) = \mu(1)p^\alpha + \mu(p)p^{\alpha-1} = p^\alpha(1 - 1/p).$$

Since φ is multiplicative, we recover in this way the formula

$$\varphi(n) = n \prod_{p \mid n} (1 - 1/p) \quad \text{for every } n \in \mathbb{N}.$$

The σ-function arises in the study of perfect numbers, to which the Pythagoreans attached much significance. A positive integer n is said to be *perfect* if it is the sum of its (positive) divisors other than itself, i.e. if $\sigma(n) = 2n$.

For example, 6 and 28 are perfect, since

$$6 = 1 + 2 + 3, \ 28 = 1 + 2 + 4 + 7 + 14.$$

It is an age-old conjecture that there are no odd perfect numbers. However, the even perfect numbers are characterized by the following result:

PROPOSITION 30 *An even positive integer is perfect if and only if it has the form* $2^t(2^{t+1} - 1)$, *where* $t \in \mathbb{N}$ *and* $2^{t+1} - 1$ *is prime.*

Proof Let n be any even positive integer and write $n = 2^t m$, where $t \geq 1$ and m is odd. Then, since σ is multiplicative, $\sigma(n) = d\sigma(m)$, where

$$d := \sigma(2^t) = 2^{t+1} - 1.$$

If $m = d$ and d is prime, then $\sigma(m) = 1 + d = 2^{t+1}$ and hence $\sigma(n) = 2^{t+1}m = 2n$.

On the other hand, if $\sigma(n) = 2n$, then $d\sigma(m) = 2^{t+1}m$. Since d is odd, it follows that $m = dq$ for some $q \in \mathbb{N}$. Hence

$$\sigma(m) = 2^{t+1}q = (1 + d)q = q + m.$$

Thus q is the only proper divisor of m. Hence $q = 1$ and $m = d$ is prime. $\square$

The sufficiency of the condition in Proposition 30 was proved in Euclid's *Elements* (Book IX, Proposition 36). The necessity of the condition was proved over two thousand years later by Euler. The condition is quite restrictive. In the first place, if $2^m - 1$ is prime for some $m \in \mathbb{N}$, then m must itself be prime. For, if $m = rs$, where $1 < r < m$, then with $a = 2^s$ we have

$$2^m - 1 = a^r - 1 = (a - 1)(a^{r-1} + a^{r-2} + \dots + 1).$$

A prime of the form $M_p := 2^p - 1$ is said to be a *Mersenne prime* in honour of Mersenne (1644), who gave a list of all primes $p \leq 257$ for which, he claimed, M_p was prime. However, he included two values of p for which M_p is composite and omitted three values of p for which M_p is prime. The correct list is now known to be

$$p = 2,3,5,7,13,17,19,31,61,89,107,127.$$

The first four even perfect numbers, namely 6,28,496 and 8128, which correspond to the values $p = 2,3,5$ and 7, were known to the ancient Greeks.

That M_{11} is not prime follows from $2^{11} - 1 = 2047 = 23{\times}89$. The factor 23 is not found simply by guesswork. It was already known to Fermat (1640) that if p is an odd prime, then any divisor of M_p is congruent to 1 mod $2p$. It is sufficient to establish this for prime divisors. But if q is a prime divisor of M_p, then $2^p \equiv 1$ mod q. Hence the order of 2 in $\mathbb{F}_q^{\times}$ divides p and, since it is not 1, it must be exactly p. Hence, by Fermat's little theorem (Corollary II.26), p divides $q - 1$. Thus $q \equiv 1$ mod p and actually $q \equiv 1$ mod $2p$, since q is necessarily odd.

At least 38 Mersenne primes are now known. The hunt for more uses thousands of linked personal computers and the following test, which was stated by Lucas (1878), but first completely proved by D.H. Lehmer (1930):

PROPOSITION 31 *Define the sequence* (S_n) *recurrently by*

$$S_1 = 4, \quad S_{n+1} = S_n^2 - 2 \ (n \geq 1).$$

Then, for any odd prime p, $M_p := 2^p - 1$ is prime if and only if it divides S_{p-1}.

Proof Put

$$\omega = 2 + \sqrt{3}, \quad \omega' = 2 - \sqrt{3}.$$

Since $\omega\omega' = 1$, it is easily shown by induction that

$$S_n = \omega^{2^{n-1}} + \omega'^{2^{n-1}} \quad (n \geq 1).$$

Let q be a prime and let J denote the set of all real numbers of the form $a + b\sqrt{3}$, where $a,b \in \mathbb{Z}$. Evidently J is a commutative ring. By identifying two elements $a + b\sqrt{3}$ and $\bar{a} + \bar{b}\sqrt{3}$ of J when $a \equiv \bar{a}$ and $b \equiv \bar{b}$ mod q, we obtain a finite commutative ring J_q containing q^2 elements. The set $J_q^{\times}$ of all invertible elements of J_q is a commutative group containing at most $q^2 - 1$ elements, since $0 \notin J_q^{\times}$.

Suppose first that M_p divides S_{p-1} and assume that M_p is composite. If q is the least prime divisor of M_p, then $q^2 \leq M_p$ and $q \neq 2$. By hypothesis,

$$\omega^{2^{p-2}} + \omega'^{2^{p-2}} \equiv 0 \mod q.$$

Now consider ω and ω' as elements of J_q. By multiplying by $\omega^{2^{p-2}}$, we obtain $\omega^{2^{p-1}} = -1$ and hence $\omega^{2^p} = 1$. Thus $\omega \in J_q^\times$ and the order of ω in $J_q^\times$ is exactly 2^p. Hence

$$2^p \leq q^2 - 1 \leq M_p - 1 = 2^p - 2,$$

which is a contradiction.

Suppose next that $M_p = q$ is prime. Then $q \equiv -1 \mod 8$, since $p \geq 3$. Since $(2/q) = (-1)^{(q^2-1)/8}$, it follows that 2 is a quadratic residue of q. Thus there exists an integer a such that

$$a^2 \equiv 2 \mod q.$$

Furthermore $q \equiv 1 \mod 3$, since $2^2 \equiv 1$ and hence $2^{p-1} \equiv 1 \mod 3$. Thus q is a quadratic residue of 3. Since $q \equiv -1 \mod 4$, it follows from the law of quadratic reciprocity that 3 is a quadratic nonresidue of q. Hence, by Euler's criterion (Proposition II.28),

$$3^{(q-1)/2} \equiv -1 \mod q.$$

Consider the element $\tau = a^{q-2}(1 + \sqrt{3})$ of J_q. We have

$$\tau^2 = 2^{q-2} \cdot 2\omega = \omega,$$

since $2^{q-1} \equiv 1 \mod q$. On the other hand,

$$(1 + \sqrt{3})^q = 1 + 3^{(q-1)/2}\sqrt{3} = 1 - \sqrt{3}$$

and hence

$$\tau^q = a^{q-2}(1 - \sqrt{3}).$$

Consequently,

$$\omega^{(q+1)/2} = \tau^{q+1} = a^{q-2}(1 - \sqrt{3}) \cdot a^{q-2}(1 + \sqrt{3}) = 2^{q-2}(-2) = -1.$$

Multiplying by $\omega'^{(q+1)/4}$, we obtain $\omega^{(q+1)/4} = -\omega'^{(q+1)/4}$. In other words, since $(q + 1)/4 = 2^{p-2}$,

$$S_{p-1} = \omega^{2^{p-2}} + \omega'^{2^{p-2}} \equiv 0 \mod q. \qquad \square$$

It is conjectured that there are infinitely many Mersenne primes, and hence infinitely many even perfect numbers. A heuristic argument of Gillies (1964), as modified by Wagstaff (1983), suggests that the number of primes $p \leq x$ for which M_p is prime is asymptotic to $(e^\gamma/\log 2)\log x$, where γ is Euler's constant (Chapter IX, §4) and thus $e^\gamma/\log 2 = 2.570\dots$.

We turn now from the primality of $2^m - 1$ to the primality of $2^m + 1$. It is easily seen that if $2^m + 1$ is prime for some $m \in \mathbb{N}$, then m must be a power of 2. For, if $m = rs$, where $r > 1$ is odd, then with $a = 2^s$ we have

$$2^m + 1 = a^r + 1 = (a + 1)(a^{r-1} - a^{r-2} + \dots + 1).$$

Put $F_n := 2^{2^n} + 1$. Thus, in particular,

$$F_0 = 3, \ F_1 = 5, \ F_2 = 17, \ F_3 = 257, \ F_4 = 65537.$$

Evidently $F_{n+1} - 2 = (F_n - 2)F_n$, from which it follows by induction that

$$F_n - 2 = F_0 F_1 \cdots F_{n-1} \quad (n \geq 1).$$

Since F_n is odd, this implies that $(F_m, F_n) = 1$ if $m \neq n$. As a byproduct, we have a proof that there are infinitely many primes.

It is easily verified that F_n itself is prime for $n \leq 4$. It was conjectured by Fermat that the 'Fermat numbers' F_n are all prime. However, this was disproved by Euler, who showed that 641 divides F_5. In fact

$$641 = 5 \cdot 2^7 + 1 = 5^4 + 2^4.$$

Thus $5 \cdot 2^7 \equiv -1 \bmod 641$ and hence $2^{32} \equiv -5^4 \cdot 2^{28} \equiv -(-1)^4 \equiv -1 \bmod 641$.

Fermat may have been as wrong as possible, since no F_n with $n > 4$ is known to be prime, although many have been proved to be composite. The Fermat numbers which *are* prime found an unexpected application to the construction of regular polygons by ruler and compass, the only instruments which Euclid allowed himself. It was shown by Gauss, at the age of 19, that a regular polygon of m sides can be constructed by ruler and compass if the order $\varphi(m)$ of $\mathbb{Z}_{(m)}^{\times}$ is a power of 2. It follows from the formula $\varphi(p^\alpha) = p^{\alpha-1}(p - 1)$, and the multiplicative nature of Euler's function, that $\varphi(m)$ is a power of 2 if and only if m has the form $2^k \cdot p_1 \cdots p_s$, where $k \geq 0$ and $p_1, \dots, p_s$ are distinct Fermat primes. (Wantzel (1837) showed that a regular polygon of m sides cannot be constructed by ruler and compass unless m has this form.) Gauss's result, in which he took particular pride, was a forerunner of Galois theory and is today usually established as an application of that theory.

The factor 641 of F_5 is not found simply by guesswork. Indeed, we now show that any divisor of F_n must be congruent to 1 mod 2^{n+1}. It is sufficient to establish this for prime divisors. But if p is a prime divisor of F_n, then $2^{2^n} \equiv -1 \bmod p$ and hence $2^{2^{n+1}} \equiv 1 \bmod p$. Thus the order of 2 in $\mathbb{F}_p^{\times}$ is exactly 2^{n+1}. Hence 2^{n+1} divides $p - 1$ and $p \equiv 1 \bmod 2^{n+1}$.

With a little more effort we can show that any divisor of F_n must be congruent to 1 mod 2^{n+2} if $n > 1$. For, if p is a prime divisor of F_n and $n > 1$, then $p \equiv 1$ mod 8 by what we have already proved. Hence, by Proposition II.30, 2 is a quadratic residue of p. Thus there exists an integer a such that $a^2 \equiv 2$ mod p. Since $a^{2^{n+1}} \equiv -1$ mod p and $a^{2^{n+2}} \equiv 1$ mod p, the order of a in $\mathbb{F}_p^\times$ is exactly 2^{n+2} and hence 2^{n+2} divides $p - 1$.

It follows from the preceding result that 641 is the first possible candidate for a prime divisor of F_5, since $128k + 1$ is not prime for $k = 1,3,4$ and $257 = F_3$ is relatively prime to F_5.

The hunt for Fermat primes today uses supercomputers and the following test due to Pépin (1877):

PROPOSITION 32 *If $m > 1$, then $N: = 2^m + 1$ is prime if and only if N divides $3^{(N-1)/2} + 1$.*

Proof Suppose first that N divides $a^{(N-1)/2} + 1$ for some integer a. If p is any prime divisor of N, then $a^{(N-1)/2} \equiv -1$ mod p and hence $a^{N-1} \equiv 1$ mod p. Thus, since p is necessarily odd, the order of a in $\mathbb{F}_p^\times$ divides $N - 1 = 2^m$, but does not divide $(N - 1)/2 = 2^{m-1}$. Hence the order of a must be exactly 2^m. Consequently, by Fermat's little theorem, 2^m divides $p - 1$. Thus

$$2^m \le p - 1 \le N - 1 = 2^m,$$

which implies that $N = p$ is prime.

To prove the converse we use the law of quadratic reciprocity. Suppose $N = p$ is prime. Then $p > 3$, since $m > 1$. From $2 \equiv -1$ mod 3 we obtain $p \equiv (-1)^m + 1$ mod 3. Since $3 \nmid p$, it follows that $p \equiv -1$ mod 3. Thus p is a quadratic non-residue of 3. But $p \equiv 1$ mod 4, since $m > 1$. Consequently, by the law of quadratic reciprocity, 3 is a quadratic non-residue of p. Hence, by Euler's criterion, $3^{(p-1)/2} \equiv -1$ mod p. $\square$

By means of Proposition 32 it has been shown that F_{14} is composite, even though no nontrivial factors are known!

4 Linear Diophantine equations

A *Diophantine equation* is an algebraic equation with integer coefficients of which the integer solutions are required. The name honours Diophantus of Alexandria (3rd century A.D.), who solved many problems of this type, although the six surviving books of his *Arithmetica* do not treat the linear problems with which we will be concerned.

We wish to determine integers $x_1,...,x_n$ such that

$$a_{11}x_1 + \ldots + a_{1n}x_n = c_1$$
$$a_{21}x_1 + \ldots + a_{2n}x_n = c_2$$
$$\ldots\ldots$$
$$a_{m1}x_1 + \ldots + a_{mn}x_n = c_m,$$

where the coefficients a_{jk} and the right sides c_j are given integers ($1 \le j \le m$, $1 \le k \le n$). We may write the system, in matrix notation, as

$$Ax = c.$$

The problem may also be put geometrically. A nonempty set $M \subseteq \mathbb{Z}^m$ is said to be a $\mathbb{Z}$–*module*, or simply a *module*, if $a,b \in M$ and $x,y \in \mathbb{Z}$ imply $xa + yb \in M$.

For example, if $a_1,\ldots,a_n$ is a finite subset of $\mathbb{Z}^m$, then the set M of all linear combinations $x_1a_1 + \ldots + x_na_n$ with $x_1,\ldots,x_n \in \mathbb{Z}$ is a module, the module *generated* by $a_1,\ldots,a_n$. If we take $a_1,\ldots,a_n$ to be the columns of the matrix A, then M is the set of all vectors Ax with $x \in \mathbb{Z}^n$ and the system $Ax = c$ is soluble if and only if $c \in M$.

If a module M is generated by the elements $a_1,\ldots,a_n$, then it is also generated by the elements $b_1,\ldots,b_n$, where

$$b_k = u_{1k}a_1 + \ldots + u_{nk}a_n \quad (u_{jk} \in \mathbb{Z}: 1 \le j,k \le n),$$

if the matrix $U = (u_{jk})$ is invertible. Here an $n{\times}n$ matrix U of integers is said to be *invertible* if there exists an $n{\times}n$ matrix U^{-1} of integers such that $U^{-1}U = I_n$ or, equivalently, $UU^{-1} = I_n$.

For example, if $ax + by = 1$, then the matrix

$$U = \begin{pmatrix} a & b \\ -y & x \end{pmatrix}$$

is invertible, with inverse

$$U^{-1} = \begin{pmatrix} x & -b \\ y & a \end{pmatrix}.$$

It may be shown, although we will not use it, that an $n{\times}n$ matrix U is invertible if and only if its determinant det U is a *unit*, i.e. det $U = \pm 1$. Under matrix multiplication, the set of all invertible $n{\times}n$ matrices of integers is a group, usually denoted by $GL_n(\mathbb{Z})$.

To solve the linear Diophantine system $Ax = c$ we replace it by a system $By = c$, where $B = AU$ for some invertible matrix U. The idea is to choose U so that B has such a simple form that y can be determined immediately, and then $x = Uy$.

We will use the elementary fact that interchanging two columns of a matrix A, or adding an integral multiple of one column to another column, is equivalent to postmultiplying A by a suitable invertible matrix U. In fact U is obtained by performing the same column operation on the identity matrix I_n. In the following discussion 'matrix' will mean 'matrix with entries from $\mathbb{Z}$'.

PROPOSITION 33 *If $A = (a_1 \ ... \ a_n)$ is a $1 \times n$ matrix, then there exists an invertible $n \times n$ matrix U such that*

$$AU = (d \ 0 \ ... \ 0)$$

if and only if d is a greatest common divisor of $a_1,...,a_n$.

Proof Suppose first that there exists such a matrix U. Since

$$A = (d \ 0 \ ... \ 0)U^{-1},$$

d is a common divisor of $a_1,...,a_n$. On the other hand,

$$d = a_1 b_1 + ... + a_n b_n,$$

where $b_1,...,b_n$ is the first column of U. Hence any common divisor of $a_1,...,a_n$ divides d. Thus d is a greatest common divisor of $a_1,...,a_n$.

Suppose next that $a_1,...,a_n$ have greatest common divisor d. Since there is nothing to do if $n = 1$, we assume $n > 1$ and use induction on n. Then if d' is the greatest common divisor of $a_2,...,a_n$, there exists an invertible $(n-1) \times (n-1)$ matrix V' such that

$$(a_2 \ ... \ a_n)V' = (d' \ 0 \ ... \ 0).$$

Since d is the greatest common divisor of a_1 and d', there exist integers u,v such that

$$a_1 u + d'v = d.$$

Put $V = I_1 \oplus V'$ and $W = W' \oplus I_{n-2}$, where

$$W' = \begin{pmatrix} u & -d'/d \\ v & a_1/d \end{pmatrix}.$$

Then V and W are invertible, and

$$(a_1 \ a_2 \ ... \ a_n)VW = (a_1 \ d' \ 0 \ ... \ 0)W = (d \ 0 \ ... \ 0).$$

Thus we can take $U = VW$. $\square$

COROLLARY 34 *For any given integers $a_1,...,a_n$, there exists an invertible $n \times n$ matrix U with $a_1,...,a_n$ as its first row if and only if the greatest common divisor of $a_1,...,a_n$ is 1.*

Proof An invertible matrix U has $a_1,\ldots,a_n$ as its first row if and only if

$$(a_1\, a_2 \ldots a_n) \;=\; (1\, 0 \ldots 0)U. \quad \square$$

If U is invertible, then its transpose is also invertible. It follows that there exists an invertible $n{\times}n$ matrix with $a_1,\ldots,a_n$ as its first column also if and only if the greatest common divisor of $a_1,\ldots,a_n$ is 1.

PROPOSITION 35 *For any $m{\times}n$ matrix A, there exists an invertible $n{\times}n$ matrix U such that $B = AU$ has the form*

$$B \;=\; (B_1\, O),$$

where B_1 is an $m{\times}r$ submatrix of rank r.

Proof Let A have rank r. If $r = n$, there is nothing to do. If $r < n$ and we denote the columns of A by $a_1,\ldots,a_n$, then there exist $x_1,\ldots,x_n \in \mathbb{Z}$, not all zero, such that

$$x_1 a_1 + \ldots + x_n a_n = O.$$

Moreover, we may assume that $x_1,\ldots,x_n$ have greatest common divisor 1. Then, by Corollary 34, there exists an invertible $n{\times}n$ matrix U' with $x_1,\ldots,x_n$ as its last column. Hence $A' := AU'$ has its last column zero. If $r < n - 1$, we can apply the same argument to the submatrix formed by the first $n - 1$ columns of A', and so on until we arrive at a matrix B of the required form. $\square$

The elements $b_1,\ldots,b_r$ of a module M are said to be a *basis* for M if they generate M and are linearly independent, i.e. $x_1 b_1 + \ldots + x_r b_r = O$ for some $x_1,\ldots,x_r \in \mathbb{Z}$ implies that $x_1 = \ldots = x_r = 0$. If O is the only element of M, we say also that O is a basis for M.

In geometric terms, Proposition 35 says that any finitely generated module $M \subsetneq \mathbb{Z}^m$ has a finite basis, and that a finite set of generators is a basis if and only if its elements are linearly independent over $\mathbb{Q}$. Hence any two bases have the same cardinality.

PROPOSITION 36 *For any $m{\times}n$ matrix A, the set N of all $x \in \mathbb{Z}^n$ such that $Ax = O$ is a module with a finite basis.*

Proof It is evident that N is a module. By Proposition 35, there exists an invertible $n{\times}n$ matrix U such that $AU = B = (B_1\, O)$, where B_1 is an $m{\times}r$ submatrix of rank r. Hence $By = O$ if and only if the first r coordinates of y vanish. By taking y to be the vector with k-th coordinate 1 and all other coordinates 0, for each k such that $r < k \leq n$, we see that the equation $By = O$ has $n - r$ linearly independent solutions $y^{(1)},\ldots,y^{(n-r)}$ such that all solutions are given by

$$y = z_1 y^{(1)} + \dots + z_{n-r} y^{(n-r)},$$

where $z_1,\dots,z_{n-r}$ are arbitrary integers. If we put $x^{(j)} = U y^{(j)}$, it follows that $x^{(1)},\dots,x^{(n-r)}$ are a basis for the module N. $\square$

An $m{\times}n$ matrix $B = (b_{jk})$ will be said to be in *echelon form* if the following two conditions are satisfied:

(i) $b_{jk} = 0$ for all j if $k > r$;

(ii) $b_{jk} \neq 0$ for some j if $k \le r$ and, if m_k is the least such j, then $1 \le m_1 < m_2 < \dots < m_r \le m$.

Evidently $r = \operatorname{rank} B$.

PROPOSITION 37 *For any $m{\times}n$ matrix A, there exists an invertible $n{\times}n$ matrix U such that $B = AU$ is in echelon form.*

Proof By Proposition 35, we may suppose that A has the form $(A_1\ O)$, where A_1 is an $m{\times}r$ submatrix of rank r, and by replacing A_1 by A we may suppose that A itself has rank n. We are going to show that there exists an invertible $n{\times}n$ matrix U such that, if $AU = B = (b_{jk})$, then $b_{jk} = 0$ for all $j < k$.

If $m = 1$, this follows from Proposition 33. We assume $m > 1$ and use induction on m. Then the first $m - 1$ rows of A may be assumed to have already the required triangular form. If $n \le m$, there is nothing more to do. If $n > m$, we can take $U = I_{m-1} \oplus U'$, where U' is an invertible $(n{-}m{+}1){\times}(n{-}m{+}1)$ matrix such that

$$(a_{m,m}\ a_{m,m+1}\ \dots\ a_{m,n})U' = (a'\ 0\ \dots\ 0).$$

Replacing B by A, we now suppose that for A itself we have $a_{jk} = 0$ for all $j < k$. Since A still has rank n, each column of A contains a nonzero entry. If the first nonzero entry in the k-th column appears in the m_k-th row, then $m_k \ge k$. By permuting the columns, if necessary, we may suppose in addition that $m_1 \le m_2 \le \dots \le m_n$.

Suppose $m_1 = m_2$. Let a and b be the entries in the m_1-th row of the first and second columns, and let d be their greatest common divisor. Then $d \neq 0$ and there exist integers u,v such that $au + bv = d$. If we put $U = V \oplus I_{n-2}$, where

$$V = \begin{pmatrix} u & -b/d \\ v & a/d \end{pmatrix},$$

then U is invertible. Moreover, the last $n-2$ columns of $B = AU$ are the same as in A and the first $m_1 - 1$ entries of the first two columns are still zero. However, $b_{m_1 1} = d$ and $b_{m_1 2} = 0$. By permuting the last $n-1$ columns, if necessary, we obtain a matrix A', of the same form as A, with $m_1' \le m_2' \le ... \le m_n'$, where $m_1' = m_1$ and $m_2' + ... + m_n' > m_2 + ... + m_n$.

By repeating this process finitely many times, we will obtain a matrix in echelon form. $\square$

COROLLARY 38   *If A is an $m \times n$ matrix of rank m, then there exists an invertible $n \times n$ matrix U such that $AU = B = (b_{jk})$, where*

$$b_{jj} \ne 0, \;\; b_{jk} = 0 \;\; if \; j < k \;\;\; (1 \le j \le m, 1 \le k \le n). \;\; \square$$

Before proceeding further we consider the uniqueness of the echelon form. Let $T = (t_{jk})$ be any $r \times r$ matrix which is lower triangular and invertible, i.e. $t_{jk} = 0$ if $j < k$ and the diagonal elements t_{jj} are units. It is easily seen that if $U = T \oplus I_{n-r}$, and if B is an echelon form for a matrix A with rank r, then BU is also an echelon form for A. We will show that all possible echelon forms for A are obtained in this way.

Suppose $B' = BU$ is in echelon form, for some invertible $n \times n$ matrix U, and write

$$B = (B_1 \; O),$$

where B_1 is an $m \times r$ submatrix. If

$$U = \begin{pmatrix} U_1 & U_2 \\ U_3 & U_4 \end{pmatrix},$$

then from $(B_1 \; O)U = (B_1' \; O)$ we obtain $U_2 = O$, since $B_1 U_2 = O$ and B_1 has rank r. Consequently U_1 is invertible and we can equally well take $U_3 = O$, $U_4 = I$. Let $b_1,...,b_r$ be the columns of B_1 and $b_1',...,b_r'$ the columns of B_1'. If $U_1 = (t_{jk})$, then

$$b_k' = t_{1k}b_1 + ... + t_{rk}b_r \;\;\; (1 \le k \le r).$$

Taking $k = 1$, we obtain $m_1' \ge m_1$ and so, by symmetry, $m_1' = m_1$. Since $m_k' > m_1'$ for $k > 1$, it follows that $t_{1k} = 0$ for $k > 1$. Taking $k = 2$, we now obtain in the same way $m_2' = m_2$. Proceeding in this manner, we see that U_1 is a lower triangular matrix.

We return now to the linear Diophantine equation

$$Ax = c.$$

The set of all $c \in \mathbb{Z}^m$ for which there exists a solution $x \in \mathbb{Z}^n$ is evidently a module $L \subseteq \mathbb{Z}^m$. If

U is an invertible matrix such that $B = AU$ is in echelon form, then x is a solution of the given system if and only if $y = U^{-1}x$ is a solution of the transformed system

$$By = c.$$

But the latter system is soluble if and only if c is an integral linear combination of the first r columns $b_1,...,b_r$ of B. Since $b_1,...,b_r$ are linearly independent, they form a basis for L.

To determine if a given system $Ax = c$ is soluble, we may use the customary methods of linear algebra over the field $\mathbb{Q}$ of rational numbers to test if c is linearly dependent on $b_1,...,b_r$; then express it as a linear combination of $b_1,...,b_r$, and finally check that the coefficients $y_1,...,y_r$ are all integers. The solutions of the original system are given by $x = Uy$, where y is any vector in $\mathbb{Z}^n$ whose first r coordinates are $y_1,...,y_r$.

If M_1 and M_2 are modules in $\mathbb{Z}^m$, their *intersection* $M_1 \cap M_2$ is again a module. The set of all $a \in \mathbb{Z}^m$ such that $a = a_1 + a_2$ for some $a_1 \in M_1$ and $a_2 \in M_2$ is also a module, which will be denoted by $M_1 + M_2$ and called the *sum* of M_1 and M_2. If M_1 and M_2 are finitely generated, then $M_1 + M_2$ is evidently finitely generated. We will show that $M_1 \cap M_2$ is also finitely generated.

Since $M_1 + M_2$ is a finitely generated module in $\mathbb{Z}^m$, it has a basis $a_1,...,a_n$. Since M_1 and M_2 are contained in $M_1 + M_2$, their generators $b_1,...,b_p$ and $c_1,...,c_q$ have the form

$$b_i = \sum_{k=1}^{n} u_{ki}a_k, \quad c_j = \sum_{k=1}^{n} v_{kj}a_k,$$

for some $u_{ki}, v_{kj} \in \mathbb{Z}$. Then $a \in M_1 \cap M_2$ if and only if

$$a = \sum_{i=1}^{p} y_i b_i = \sum_{j=1}^{q} z_j c_j$$

for some $y_i, z_j \in \mathbb{Z}$. Since $a_1,...,a_n$ is a basis for $M_1 + M_2$, this is equivalent to

$$\sum_{i=1}^{p} u_{ki}y_i = \sum_{j=1}^{q} v_{kj}z_j$$

or, in matrix notation, $By = Cz$. But this is equivalent to the homogeneous system $Ax = O$, where

$$A = (B\ -C), \quad x = \begin{pmatrix} y \\ z \end{pmatrix},$$

and by Proposition 36 the module of solutions of this system has a finite basis.

Suppose the modules $M_1, M_2 \subseteq \mathbb{Z}^m$ are generated by the columns of the $m \times n_1, m \times n_2$ matrices A_1, A_2. Evidently M_2 is a submodule of M_1 if and only if each column of A_2 can be expressed as a linear combination of the columns of A_1, i.e. if and only if there exists an $n_1 \times n_2$ matrix X such that

$$A_1 X = A_2.$$

We say in this case that A_1 is a *left divisor* of A_2, or that A_2 is a *right multiple* of A_1.

We may also define greatest common divisors and least common multiples for matrices. An $m \times p$ matrix D is a *greatest common left divisor* of A_1 and A_2 if it is a left divisor of both A_1 and A_2, and if every left divisor C of both A_1 and A_2 is also a left divisor of D. An $m \times q$ matrix H is a *least common right multiple* of A_1 and A_2 if it is a right multiple of both A_1 and A_2, and if every right multiple G of both A_1 and A_2 is also a right multiple of H. It will now be shown that these objects exist and have simple geometrical interpretations.

Let M_1, M_2 be the modules defined by the matrices A_1, A_2. We will show that if the sum $M_1 + M_2$ is defined by the matrix D, then D is a greatest common left divisor of A_1 and A_2. In fact D is a common left divisor of A_1 and A_2, since M_1 and M_2 are contained in $M_1 + M_2$. On the other hand, any common left divisor C of A_1 and A_2 defines a module which contains $M_1 + M_2$, since it contains both M_1 and M_2, and so C is a left divisor of D.

A similar argument shows that if the intersection $M_1 \cap M_2$ is defined by the matrix H, then H is a least common right multiple of A_1 and A_2.

The sum $M_1 + M_2$ is defined, in particular, by the block matrix $(A_1\ A_2)$. There exists an invertible $(n_1 + n_2) \times (n_1 + n_2)$ matrix U such that

$$(A_1\ A_2)U = (D'\ O),$$

where D' is an $m \times r$ submatrix of rank r. If

$$U = \begin{pmatrix} U_1 & U_2 \\ U_3 & U_4 \end{pmatrix},$$

is the corresponding partition of U, then

$$A_1 U_1 + A_2 U_3 = D'.$$

On the other hand,

$$(A_1\ A_2) = (D'\ O)U^{-1}.$$

If

$$U^{-1} = \begin{pmatrix} V_1 & V_2 \\ V_3 & V_4 \end{pmatrix}$$

is the corresponding partition of U^{-1}, then

$$A_1 = D'V_1, \quad A_2 = D'V_2.$$

Thus D' is a common left divisor of A_1 and A_2, and the previous relation implies that it is a greatest common left divisor. It follows that *any* greatest common left divisor D of A_1 and A_2 has a right 'Bézout' representation $D = A_1X_1 + A_2X_2$.

We may also define coprimeness for matrices. Two matrices A_1, A_2 of size $m \times n_1, m \times n_2$ are *left coprime* if I_m is a greatest common left divisor. If M_1, M_2 are the modules defined by A_1, A_2, this means that $M_1 + M_2 = \mathbb{Z}^m$. The definition may also be reformulated in several other ways:

PROPOSITION 39 *For any $m \times n$ matrix A, the following conditions are equivalent:*

(i) *for some, and hence every, partition $A = (A_1 \ A_2)$, the submatrices A_1 and A_2 are left coprime;*

(ii) *there exists an $n \times m$ matrix $A^\dagger$ such that $AA^\dagger = I_m$;*

(iii) *there exists an $(n - m) \times n$ matrix A^c such that*

$$\begin{pmatrix} A \\ A^c \end{pmatrix}$$

is invertible;

(iv) *there exists an invertible $n \times n$ matrix V such that $AV = (I_m \ O)$.*

Proof If $A = (A_1 \ A_2)$ for some left coprime matrices A_1, A_2, then there exist X_1, X_2 such that $A_1X_1 + A_2X_2 = I_m$ and hence (ii) holds. On the other hand, if (ii) holds then, for any partition $A = (A_1 \ A_2)$, there exist X_1, X_2 such that $A_1X_1 + A_2X_2 = I_m$ and hence A_1, A_2 are left coprime.

Thus (i) $\Leftrightarrow$ (ii). Suppose now that (ii) holds. Then A has rank m and hence there exists an invertible $n \times n$ matrix U such that $A = (D \ O)U$, where the $m \times m$ matrix D is non-singular. In fact D is invertible, since $AA^\dagger = I_m$ implies that D is a left divisor of I_m. Consequently, by changing U, we may assume $D = I_m$. If we now take $A^c = (O \ I_{n-m})U$, we see that (ii) $\Rightarrow$ (iii).

It is obvious that (iii) $\Rightarrow$ (iv) and that (iv) $\Rightarrow$ (ii). $\square$

We now consider the extension of these results to other rings besides $\mathbb{Z}$. Let R be an arbitrary ring. A nonempty set $M \subseteq R^m$ is said to be an *R-module* if $a, b \in M$ and $x, y \in R$ imply $xa + yb \in M$. The module M is *finitely generated* if it contains elements $a_1, ..., a_n$ such that every element of M has the form $x_1a_1 + ... + x_na_n$ for some $x_1, ..., x_n \in R$.

It is easily seen that if R is a *Bézout domain*, then the whole of the preceding discussion in this section remains valid if 'module' is interpreted to mean 'R-module' and 'matrix' to mean 'matrix with entries from R'. In particular, we may take $R = K[t]$ to be the ring of all polynomials in one indeterminate with coefficients from an arbitrary field K. However, both $\mathbb{Z}$ and $K[t]$ are principal ideal domains. In this case further results may be obtained.

PROPOSITION 40 *If R is a principal ideal domain and M a finitely generated R-module, then any submodule L of M is also finitely generated. Moreover, if M is generated by n elements, so also is L.*

Proof Suppose M is generated by $a_1,...,a_n$. If $n = 1$, then any $b \in L$ has the form $b = xa_1$ for some $x \in R$ and the set of all x which appear in this way is an ideal of R. Since R is a principal ideal domain, it follows that L is generated by a single element b_1, where $b_1 = x'a_1$ for some $x' \in R$.

Suppose now that $n > 1$ and that, for each $m < n$, any submodule of a module generated by m elements is also generated by m elements. Any $b \in L$ has the form

$$b = x_1a_1 + ... + x_na_n$$

for some $x_1,...,x_n \in R$ and the set of all x_1 which appear in this way is an ideal of R. Since R is a principal ideal domain, it follows that there is a fixed $b_1 \in L$ such that $b = y_1b_1 + b'$ for some $y_1 \in R$ and some b' in the module M' generated by $a_2,...,a_n$. The set of all b' which appear in this way is a submodule L' of M'. By the induction hypothesis, L' is generated by $n - 1$ elements and hence L is generated by n elements. $\square$

Just as it is useful to define vector spaces abstractly over an arbitrary field K, so it is useful to define modules abstractly over an arbitrary ring R. An abelian group M, with the group operation denoted by $+$, is said to be an *R-module* if, with any $a \in M$ and any $x \in R$, there is associated an element $xa \in M$ so that the following properties hold, for all $a,b \in M$ and all $x,y \in R$:

(i) $x(a + b) = xa + xb,$
(ii) $(x + y)a = xa + ya,$
(iii) $(xy)a = x(ya),$
(iv) $1a = a.$

The proof of Proposition 40 remains valid for modules in this abstract sense. However, a finitely generated module need not now have a basis. For, even if it is generated by a single element a, we may have $xa = O$ for some nonzero $x \in R$. Nevertheless, we are going to show that, if R is a principal ideal domain, all finitely generated R-modules can be completely characterized.

Let R be a principal ideal domain and M a finitely generated R-module, with generators $a_1,...,a_n$, say. The set N of all $x = (x_1,...,x_n) \in R^n$ such that

$$x_1 a_1 + ... + x_n a_n = O$$

is evidently a module in R^n. Hence N is finitely generated, by Proposition 40. The given module M is isomorphic to the quotient module R^n/N.

Let $f_1,...,f_m$ be a set of generators for N and let $e_1,...,e_n$ be a basis for R^n. Then

$$f_j = a_{j1}e_1 + ... + a_{jn}e_n \quad (1 \le j \le m),$$

for some $a_{jk} \in R$. The module M is completely determined by the matrix $A = (a_{jk})$. However, we can change generators and change bases.

If we put

$$f_i' = v_{i1}f_1 + ... + v_{im}f_m \quad (1 \le i \le m),$$

where $V = (v_{ij})$ is an invertible $m{\times}m$ matrix, then $f_1',...,f_m'$ is also a set of generators for N. If we put

$$e_k = u_{k1}e_1' + ... + u_{kn}e_n' \quad (1 \le k \le n),$$

where $U = (u_{k\ell})$ is an invertible $n{\times}n$ matrix, then $e_1',...,e_n'$ is also a basis for R^n. Moreover

$$f_i' = b_{i1}e_1' + ... + b_{in}e_n' \quad (1 \le i \le m),$$

where the $m{\times}n$ matrix $B = (b_{i\ell})$ is given by $B = VAU$.

The idea is to choose V and U so that B is as simple as possible. This is made precise in the next proposition, first proved by H.J.S. Smith (1861) for $R = \mathbb{Z}$. The corresponding matrix S is known as the *Smith normal form* of A.

PROPOSITION 41 *Let R be a principal ideal domain and let A be an $m{\times}n$ matrix with entries from R. If A has rank r, then there exist invertible $m{\times}m, n{\times}n$ matrices V,U with entries from R such that $S = VAU$ has the form*

$$S = \begin{pmatrix} D & O \\ O & O \end{pmatrix},$$

where $D = \mathrm{diag}\,[d_1,...,d_r]$ is a diagonal matrix such that $d_i \ne 0$ and $d_i | d_j$ for $1 \le i \le j \le r$.

Proof We show first that it is enough to obtain a matrix which satisfies all the requirements except the divisibility conditions for the d's.

If a,b are nonzero elements of R with greatest common divisor d, then there exist $u,v \in R$ such that $au + bv = d$. It is easily verified that

$$\begin{pmatrix} 1 & 1 \\ -bv/d & au/d \end{pmatrix}\begin{pmatrix} a & 0 \\ 0 & b \end{pmatrix}\begin{pmatrix} u & -b/d \\ v & a/d \end{pmatrix} = \begin{pmatrix} d & 0 \\ 0 & ab/d \end{pmatrix},$$

and the outside matrices on the left-hand side are both invertible. By applying this process finitely many times, a non-singular diagonal matrix $D' = \mathrm{diag}\,[d_1',...,d_r']$ may be transformed into a non-singular diagonal matrix $D = \mathrm{diag}\,[d_1,...,d_r]$ which satisfies $d_i|d_j$ for $1 \le i \le j \le r$.

Consider now an arbitrary matrix A. By applying Proposition 35 to the transpose of A, we may reduce the problem to the case where A has rank m and then, by Corollary 38, we may suppose further that $a_{jj} \ne 0$, $a_{jk} = 0$ for all $j < k$.

It is now sufficient to show that, for any 2×2 matrix

$$A = \begin{pmatrix} a & 0 \\ b & c \end{pmatrix},$$

with nonzero entries a,b,c, there exist invertible 2×2 matrices U,V such that VAU is a diagonal matrix. Moreover, we need only prove this when the greatest common divisor $(a,b,c) = 1$. But then there exists $q \in R$ such that $(a,b + qc) = 1$. In fact, take q to be the product of the distinct primes which divide a but not b. For any prime divisor p of a, if $p|b$, then $p \nmid c, p \nmid q$ and hence $p \nmid (b + qc)$; if $p \nmid b$, then $p|q$ and again $p \nmid (b + qc)$.

If we put $b' = b + qc$, then there exist $x,y \in R$ such that $ax + b'y = 1$, and hence $ax + by = 1 - qcy$. It is easily verified that

$$\begin{pmatrix} x & y \\ -b' & a \end{pmatrix}\begin{pmatrix} a & 0 \\ b & c \end{pmatrix}\begin{pmatrix} 1 & -cy \\ q & 1-qcy \end{pmatrix} = \begin{pmatrix} 1 & 0 \\ 0 & ac \end{pmatrix},$$

and the outside matrices on the left-hand side are both invertible. $\square$

In the important special case $R = \mathbb{Z}$, there is a more constructive proof of Proposition 41. Obviously we may suppose $A \ne O$. By interchanges of rows and columns we can arrange that a_{11} is the nonzero entry of A with minimum absolute value. If there is an entry a_{1k} $(k > 1)$ in the first row which is not divisible by a_{11}, then we can write $a_{1k} = za_{11} + a_{1k}'$, where $z,a_{1k}' \in \mathbb{Z}$ and $|a_{1k}'| < |a_{11}|$. By subtracting z times the first column from the k-th column we replace a_{1k} by a_{1k}'. Thus we obtain a new matrix A in which the minimum absolute value of the nonzero entries has been reduced.

On the other hand, if a_{11} divides a_{1k} for all $k > 1$ then, by subtracting multiples of the first column from the remaining columns, we can arrange that $a_{1k} = 0$ for all $k > 1$. If there is now an entry a_{j1} ($j > 1$) in the first column which is not divisible by a_{11} then, by subtracting a multiple of the first row from the j-th row, the minimum absolute value of the nonzero entries can again be reduced. Otherwise, by subtracting multiples of the first row from the remaining rows, we can bring A to the form

$$\begin{pmatrix} a_{11} & O \\ O & A' \end{pmatrix}.$$

Since $A \neq O$ and the minimum absolute value of the nonzero entries cannot be reduced indefinitely, we must in any event arrive at a matrix of this form after a finite number of steps. The same procedure can now be applied to the submatrix A', and so on until we obtain a matrix

$$\begin{pmatrix} D' & O \\ O & O \end{pmatrix},$$

where D' is a diagonal matrix with the same rank as A. As in the first part of the proof of Proposition 41, we can now replace D' by a diagonal matrix D which satisfies the divisibility conditions.

Clearly this constructive proof is also valid for any Euclidean domain R and, in particular, for the polynomial ring $R = K[t]$, where K is an arbitrary field.

It will now be shown that the Smith normal form of a matrix A is uniquely determined, apart from replacing each d_k by an arbitrary unit multiple. For, if S' is another Smith normal form, then $S' = V'SU'$ for some invertible $m{\times}m, n{\times}n$ matrices V',U'. Since d_1 divides all entries of S, it also divides all entries of S'. In particular, $d_1 | d_1'$. In the same way $d_1' | d_1$ and hence d_1' is a unit multiple of d_1. To show that d_k' is a unit multiple of d_k, also for $k > 1$, it is quickest to use determinants (Chapter V, §1). Since $d_1 \cdots d_k$ divides all $k{\times}k$ subdeterminants or *minors* of S, it also divides all $k{\times}k$ minors of S'. In particular, $d_1 \cdots d_k | d_1' \cdots d_k'$. Similarly, $d_1' \cdots d_k' | d_1 \cdots d_k$ and hence $d_1' \cdots d_k'$ is a unit multiple of $d_1 \cdots d_k$. The conclusion now follows by induction on k.

The products $\Delta_k = d_1 \cdots d_k$ ($1 \leq k \leq r$) are known as the *invariant factors* of the matrix A. A similar argument to that in the preceding paragraph shows that Δ_k is the greatest common divisor of all $k{\times}k$ minors of A.

Two $m{\times}n$ matrices A,B are said to be *equivalent* if there exist invertible $m{\times}m, n{\times}n$ matrices V,U such that $B = VAU$. Since equivalence is indeed an 'equivalence relation', the uniqueness of the Smith normal form implies that two $m{\times}n$ matrices A,B are equivalent if and only if they have the same rank and the same invariant factors.

We return now from matrices to modules. Let R be a principal ideal domain and M a finitely generated R-module, with generators $a_1,...,a_n$. The Smith normal form tells us that M has generators $a_1',...,a_n'$, where

$$a_k = u_{k1}a_1' + ... + u_{kn}a_n' \quad (1 \leq k \leq n)$$

for some invertible matrix $U = (u_{k\ell})$, such that $d_k a_k' = O$ $(1 \leq k \leq r)$. Moreover,

$$x_1 a_1' + ... + x_n a_n' = O$$

implies $x_k = y_k d_k$ for some $y_k \in R$ if $1 \leq k \leq r$ and $x_k = 0$ if $r < k \leq n$. In particular, $x_k a_k' = O$ for $1 \leq k \leq n$, and thus the module M is the direct sum of the submodules $M_1',...,M_n'$ generated by $a_1',...,a_n'$ respectively.

If N_k denotes the set of all $x \in R$ such that $x a_k' = O$, then N_k is the principal ideal of R generated by d_k for $1 \leq k \leq r$ and $N_k = \{0\}$ for $r < k \leq n$. The divisibility conditions on the d's imply that $N_{k+1} \subseteq N_k$ $(1 \leq k < r)$. If $N_k = R$ for some k, then a_k' contributes nothing as a generator and may be omitted.

Evidently the submodule M' generated by $a_1',...,a_r'$ consists of all $a \in M$ such that $xa = O$ for some nonzero $x \in R$, and the submodule M'' generated by $a_{r+1}',...,a_n'$ has $a_{r+1}',...,a_n'$ as a basis. Thus we have proved the *structure theorem for finitely generated modules over a principal ideal domain*:

PROPOSITION 42 *Let R be a principal ideal domain and M a finitely generated R-module. Then M is the direct sum of two submodules M' and M'', where M' consists of all $a \in M$ such that $xa = O$ for some nonzero $x \in R$ and M'' has a finite basis.*

Moreover, M' is the direct sum of s submodules $Ra_1,...,Ra_s$, such that

$$0 \subset N_s \subseteq ... \subseteq N_1 \subset R,$$

where N_k is the ideal consisting of all $x \in R$ such that $x a_k = O$ $(1 \leq k \leq s)$. $\square$

The uniquely determined submodule M' is called the *torsion submodule* of M. The *free submodule M''* is not uniquely determined, although the number of elements in a basis is uniquely determined. Of course, for a particular M one may have $M' = \{O\}$ or $M'' = \{O\}$.

Any abelian group A, with the group operation denoted by $+$, may be regarded as a $\mathbb{Z}$-module by defining na to be the sum $a + ... + a$ with n summands if $n \in \mathbb{N}$, to be O if $n = 0$, and to be $-(a + ... + a)$ with $-n$ summands if $-n \in \mathbb{N}$. The structure theorem in this case becomes the *structure theorem for finitely generated abelian groups*: any finitely generated abelian group A is the direct sum of finitely many finite or infinite cyclic subgroups. The finite cyclic subgroups have orders $d_1,...,d_s$, where $d_1 > 1$ if $s > 0$ and $d_i | d_j$ if $i \leq j$. In particular, A is the direct sum of a finite subgroup A' (of order $d_1 \cdots d_r$), its *torsion subgroup*, and a *free* subgroup A''.

The fundamental structure theorem also has an important application to linear algebra. Let V be a vector space over a field K and $T: V \to V$ a linear transformation. We can give V the structure of a $K[t]$-module by defining, for any $v \in V$ and any $f = a_0 + a_1 t + ... + a_n t^n \in K[t]$,

$$fv = a_0 v + a_1 Tv + ... + a_n T^n v.$$

If V is finite-dimensional, then for any $v \in V$ there is a nonzero polynomial f such thay $fv = O$. In this case the fundamental structure theorem says that V is the direct sum of finitely many subspaces $V_1,...,V_s$ which are invariant under T. If V_i has dimension $n_i \geq 1$, then there exists a vector $w_i \in V_i$ such that $w_i, Tw_i,...,T^{n_i-1}w_i$ are a vector space basis for V_i $(1 \leq i \leq s)$. There is a uniquely determined monic polynomial m_i of degree n_i such that $m_i(T)w_i = O$ and, finally, $m_i | m_j$ if $i \leq j$.

The Smith normal form can be used to solve systems of linear ordinary differential equations with constant coefficients. Such a system has the form

$$a_{11}(D)x_1 + ... + a_{1n}(D)x_n = c_1(t)$$
$$a_{21}(D)x_1 + ... + a_{2n}(D)x_n = c_2(t)$$
$$.....$$
$$a_{m1}(D)x_1 + ... + a_{mn}(D)x_n = c_m(t),$$

where the coefficients $a_{jk}(D)$ are polynomials in $D = d/dt$ with complex coefficients and the right sides $c_j(t)$ are, say, infinitely differentiable functions of the time t. Since $\mathbb{C}[s]$ is a Euclidean domain, we can bring the coefficient matrix $A = (a_{jk}(D))$ to Smith normal form and thus replace the given system by an equivalent system in which the variables are 'uncoupled'.

For the polynomial ring $R = K[t]$ it is possible to say more about R-modules than for an arbitrary Euclidean domain, since the absolute value

$$|f| = 2^{\partial(f)} \text{ if } f \neq O, \ |O| = 0,$$

has not only the Euclidean property, but also the properties

$$|f + g| \leq \max\{|f|,|g|\}, \ |fg| = |f||g| \text{ for any } f,g \in R.$$

For any $a \in R^m$, where $R = K[t]$, define $|a|$ to be the maximum absolute value of any of its coordinates. Then a basis for a module $M \subseteq R^m$ can be obtained in the following way. Suppose $M \neq O$ and choose a nonzero element a_1 of M for which $|a_1|$ is a minimum. If there is an element of M which is not of the form p_1a_1 with $p_1 \in R$, choose one, a_2, for which $|a_2|$ is a minimum. If there is an element of M which is not of the form $p_1a_1 + p_2a_2$ with $p_1,p_2 \in R$, choose one, a_3, for which $|a_3|$ is a minimum. And so on.

Evidently $|a_1| \leq |a_2| \leq \dots$. We will show that $a_1, a_2, \dots$ are linearly independent for as long as the the process can be continued, and thus ultimately a basis is obtained.

If this is not the case, then there exists a positive integer $k \leq m$ such that $a_1,\dots,a_k$ are linearly independent, but $a_1,\dots,a_{k+1}$ are not. Hence there exist $s_1,\dots,s_{k+1} \in R$ with $s_{k+1} \neq 0$ such that $s_1a_1 + \dots + s_{k+1}a_{k+1} = O$. For each $j \leq k$, there exist $q_j,r_j \in R$ such that

$$s_j = q_j s_{k+1} + r_j, \quad |r_j| < |s_{k+1}|.$$

Put

$$a'_{k+1} = a_{k+1} + q_1a_1 + \dots + q_ka_k, \quad b_k = r_1a_1 + \dots + r_ka_k.$$

Since a_{k+1} is not of the form $p_1a_1 + \dots + p_ka_k$, neither is a'_{k+1} and hence $|a'_{k+1}| \geq |a_{k+1}|$. Furthermore, $|b_k| \leq \max_{1 \leq j \leq k} |r_j| |a_j| < |s_{k+1}| |a_{k+1}|$. Since $b_k = -s_{k+1}a'_{k+1}$, by construction, this is a contradiction.

A basis for M which is obtained in this way will be called a *minimal basis*. It is not difficult to show that a basis $a_1,\dots,a_n$ is a minimal basis if and only if $|a_1| \leq \dots \leq |a_n|$ and the sum $|a_1| + \dots + |a_n|$ is minimal. Although a minimal basis is not uniquely determined, the values $|a_1|,\dots,|a_n|$ are uniquely determined.

5 Further remarks

For the history of the law of quadratic reciprocity, see Frei [16]. The first two proofs by Gauss of the law of quadratic reciprocity appeared in §§125-145 and §262 of [17]. A simplified account of Gauss's inductive proof has been given by Brown [7]. The proofs most commonly given use 'Gauss's lemma' and are variants of Gauss's third proof. The first proof given here, due to Rousseau [46], is of this general type, but it does not use Gauss's lemma and is based on a natural definition of the Jacobi symbol. For an extension of this definition of Zolotareff to algebraic number fields, see Cartier [9].

For Dirichlet's evaluation of Gauss sums, see [33]. A survey of Gauss sums is given in Berndt and Evans [6].

The extension of the law of quadratic reciprocity to arbitrary algebraic number fields was the subject of Hilbert's 9th Paris problem. Although such generalizations lie outside the scope of the present work, it may be worthwhile to give a brief glimpse. Let $K = \mathbb{Q}$ be the field of rational numbers and let $L = \mathbb{Q}(\sqrt{d})$ be a quadratic extension of K. If p is a prime in K, the law of quadratic reciprocity may be interpreted as describing how the ideal generated by p in L factors into prime ideals. Now let K be an arbitrary algebraic number field and let L be any finite extension of K. Quite generally, we may ask how the arithmetic of the extension L is determined by the arithmetic of K. The general reciprocity law, conjectured by Artin in 1923 and proved by him in 1927, gives an answer in the form of an isomorphism between two groups, provided the Galois group of L over K is abelian. For an introduction, see Wyman [54] and, for more detail, Tate [51]. The outstanding problem is to find a meaningful extension to the case when the Galois group is non-abelian. Some intriguing conjectures are provided by the Langlands program, for which see also Gelbart [18].

The law of quadratic reciprocity has an analogue for polynomials with coefficients from a finite field. Let $\mathbb{F}_q$ be a finite field containing q elements, where q is a power of an *odd* prime. If $g \in \mathbb{F}_q[x]$ is a monic irreducible polynomial of positive degree, then for any $f \in \mathbb{F}_q[x]$ not divisible by g we define (f/g) to be 1 if f is congruent to a square mod g, and -1 otherwise. The law of quadratic reciprocity, which was stated by Dedekind (1857) and proved by Artin (1924) in the case of prime q, says that

$$(f/g)(g/f) = (-1)^{mn(q-1)/2}$$

for any distinct monic irreducible polynomials $f,g \in \mathbb{F}_q[x]$ of positive degrees m,n. Artin also developed a theory of ideals, analogous to that for quadratic number fields, for the field obtained by adjoining to $\mathbb{F}_q[x]$ an element ω with $\omega^2 = D(x)$, where $D(x) \in \mathbb{F}_q[x]$ is square-free; see [3].

Quadratic fields are treated in the third volume of Landau [30]. There is also a useful resumé accompanying the tables in Ince [23].

A complex number is said to be *algebraic* if it is a root of a monic polynomial with rational coefficients and *transcendental* otherwise. Hence a complex number is algebraic if and only if it is an element of some algebraic number field.

For an introduction to the theory of algebraic number fields, see Samuel [47]. This vast theory may be approached in a variety of ways. For a more detailed treatment the student may choose from Hecke [22], Hasse [20], Lang [32], Narkiewicz [38] and Neukirch [39]. There are useful articles in Cassels and Fröhlich [10], and Artin [2] treats also algebraic functions.

For the early history of Fermat's last theorem, see Vandiver [52], Ribenboim [41] and Kummer [28]. Further references will be given in Chapter XIII.

Arithmetical functions are discussed in Apostol [1], McCarthy [35] and Sivaramakrishnan [48]. The term 'Dirichlet product' comes from the connection with Dirichlet series, which will be considered in Chapter IX, §6. The ring of all arithmetical functions was shown to be a factorial domain by Cashwell and Everett (1959); the result is proved in [48].

In the form $f(a \wedge b)f(a \vee b) = f(a)f(b)$, the concept of multiplicative function can be extended to any map $f: L \to \mathbb{C}$, where L is a lattice. Möbius inversion can be extended to any locally finite partially ordered set and plays a significant role in modern combinatorics; see Bender and Goldman [5], Rota [45] and Barnabei *et al.* [4].

The early history of perfect numbers and Fermat numbers is described in Dickson [13]. It has been proved that any odd perfect number, if such a thing exists, must be greater than 10^{300} and have at least 8 distinct prime factors. On the other hand, if an odd perfect number N has at most k distinct prime factors, then $N < 4^{4^k}$ and thus all such N can be found by a finite amount of computation. See te Riele [42] and Heath-Brown [21].

The proof of the Lucas–Lehmer test for Mersenne primes follows Rosen [43] and Bruce [8]. For the conjectured distribution of Mersenne primes, see Wagstaff [53]. The construction of regular polygons by ruler and compass is discussed in Hadlock [19], Jacobson [24] and Morandi [36].

Much of the material in §4 is also discussed in Macduffee [34] and Newman [40]. Corollary 34 was proved by Hermite (1849), who later (1851) also proved Corollary 38. Indeed the latter result is the essential content of *Hermite's normal form*, which will be encountered in Chapter VIII, §2.

It is clear that Corollary 34 remains valid if the underlying ring $\mathbb{Z}$ is replaced by any principal ideal domain. There have recently been some noteworthy extensions to more general rings. It may be asked, for an arbitrary commutative ring R and any $a_1,...,a_n \in R$, does there exist an invertible $n \times n$ matrix U with entries from R which has $a_1,...,a_n$ as its first row? It is obviously necessary that there exist $x_1,...,x_n \in R$ such that

$$a_1 x_1 + ... + a_n x_n = 1,$$

i.e. that the ideal generated by $a_1,...,a_n$ be the whole ring R. If $n = 2$, this necessary condition is also sufficient, by the same observation as when invertibility of matrices was first considered for $R = \mathbb{Z}$. However, if $n > 2$ there exist even factorial domains R for which the condition is not sufficient. In 1976 Quillen and Suslin independently proved the twenty-year-old conjecture that it *is* sufficient if $R = K[t_1,...,t_d]$ is the ring of polynomials in finitely many indeterminates with coefficients from an arbitrary field K.

By pursuing an analogy between projective modules in algebra and vector bundles in topology, Serre (1955) had been led to conjecture that, for $R = K[t_1,...,t_d]$, if an R-module has a finite basis and is the direct sum of two submodules, then each of these submodules has a finite basis. Seshadri (1958) proved the conjecture for $d = 2$ and in the same year Serre showed that, for arbitrary d, it would follow from the result which Quillen and Suslin subsequently proved.

For proofs of these results and for later developments, see Lam [29], Fitchas and Galligo [14], and Swan [50]. There is a short proof of the Quillen–Suslin theorem in Lang [31].

For Smith's normal form, see Smith [49] and Kaplansky [27]. It was shown by Wedderburn (1915) that Smith's normal form also holds for matrices of holomorphic functions, even though the latter do not form a principal ideal domain; see Narasimhan [37].

Finitely generated commutative groups are important, not only because more can be said about them, but also because they arise in practice. *Dirichlet's unit theorem* says that the units of an algebraic number field form a finitely generated commutative group. As will be seen in Chapter XIII, §4, *Mordell's theorem* says that the rational points of an elliptic curve also form a finitely generated commutative group.

Modules over a polynomial ring $K[s]$ play an important role in what electrical engineers call *linear systems theory*. Connected accounts are given in Kalman [26], Rosenbrock [44] and Kailath [25]. For some further mathematical developments, see Forney [15], Coppel [11], and Coppel and Cullen [12].

6 Selected references

[1] T.M. Apostol, *Introduction to analytic number theory*, Springer-Verlag, New York, 1976.

[2] E. Artin, *Algebraic numbers and algebraic functions*, Nelson, London, 1968.

[3] E. Artin, Quadratische Körper im Gebiet der höheren Kongruenzen I, II, *Collected Papers* (ed. S. Lang and J.T. Tate), pp. 1-94, reprinted, Springer-Verlag, New York, 1986.

[4] M. Barnabei, A. Brini and G.-C. Rota, The theory of Möbius functions, *Russian Math. Surveys* **41** (1986), no.3, 135-188.

[5] E.A. Bender and J.R. Goldman, On the application of Möbius inversion in combinatorial analysis, *Amer. Math. Monthly* 82 (1975), 789-803.

[6] B.C. Berndt and R.J. Evans, The determination of Gauss sums, *Bull. Amer. Math. Soc.* (*N.S.*) **5** (1981), 107-129.

[7]   E. Brown, The first proof of the quadratic reciprocity law, revisited, *Amer. Math. Monthly* **88** (1981), 257-264.

[8]   J.W. Bruce, A really trivial proof of the Lucas–Lehmer test, *Amer. Math. Monthly* **100** (1993), 370-371.

[9]   P. Cartier, Sur une généralisation des symboles de Legendre–Jacobi, *Enseign. Math.* **16** (1970), 31-48.

[10] J.W.S. Cassels and A. Fröhlich (ed.), *Algebraic number theory*, Academic Press, London, 1967.

[11] W.A. Coppel, Matrices of rational functions, *Bull. Austral. Math. Soc.* **11** (1974), 89-113.

[12] W.A. Coppel and D.J. Cullen, Strong system equivalence (II), *J. Austral. Math. Soc. B* **27** (1985), 223-237.

[13] L.E. Dickson, *History of the theory of numbers, Vol. I*, reprinted, Chelsea, New York, 1992.

[14] N. Fitchas and A. Galligo, Nullstellensatz effectif et conjecture de Serre (théorème de Quillen–Suslin) pour le calcul formel, *Math. Nachr.* **149** (1990), 231-253.

[15] G.D. Forney, Minimal bases of rational vector spaces, with applications to multivariable linear systems, *SIAM J. Control* **13** (1975), 493-520.

[16] G. Frei, The reciprocity law from Euler to Eisenstein, *The intersection of history and mathematics* (ed. C. Sasaki, M. Sugiura and J.W. Dauben), pp. 67-90, Birkhäuser, Basel, 1994.

[17] C.F. Gauss, *Disquisitiones arithmeticae*, English transl. by A.A. Clarke, Yale University Press, New Haven, 1966.

[18] S. Gelbart, An elementary introduction to the Langlands program, *Bull. Amer. Math. Soc. (N.S.)* **10** (1984), 177-219.

[19] C.R. Hadlock, *Field theory and its classical problems*, Carus Mathematical Monographs no. 19, Mathematical Association of America, Washington, D.C., 1978. [Reprinted in paperback, 2000]

[20] H. Hasse, *Number theory*, English transl. by H.G. Zimmer, Springer-Verlag, Berlin, 1980.

[21] D.R. Heath-Brown, Odd perfect numbers, *Math. Proc. Cambridge Philos. Soc.* **115** (1994), 191-196.

[22] E. Hecke, *Lectures on the theory of algebraic numbers*, English transl. by G.U. Brauer, J.R. Goldman and R. Kotzen, Springer-Verlag, New York, 1981. [German original, 1923]

[23] E.L. Ince, *Cycles of reduced ideals in quadratic fields*, Mathematical Tables Vol. IV, British Association, London, 1934.

[24] N. Jacobson, *Basic Algebra I*, 2nd ed., W.H. Freeman, New York, 1985.

[25] T. Kailath, *Linear systems*, Prentice–Hall, Englewood Cliffs, N.J., 1980.

[26] R.E. Kalman, Algebraic theory of linear systems, *Topics in mathematical system theory* (R.E. Kalman, P.L. Falb and M.A. Arbib), pp. 237-339, McGraw–Hill, New York, 1969.

[27] I. Kaplansky, Elementary divisors and modules, *Trans. Amer. Math. Soc.* **66** (1949), 464-491.

[28] E. Kummer, *Collected Papers, Vol. I* (ed. A. Weil), Springer-Verlag, Berlin, 1975.

[29] T.Y. Lam, *Serre's conjecture*, Lecture Notes in Mathematics **635**, Springer-Verlag, Berlin, 1978.

[30] E. Landau, *Vorlesungen über Zahlentheorie*, 3 vols., Hirzel, Leipzig, 1927. [Reprinted, Chelsea, New York, 1969]

[31] S. Lang, *Algebra*, corrected reprint of 3rd ed., Addison-Wesley, Reading, Mass., 1994.

[32] S. Lang, *Algebraic number theory,* 2nd ed., Springer-Verlag, New York, 1994.

[33] G. Lejeune-Dirichlet, *Werke*, Band I, pp. 237-256, reprinted Chelsea, New York, 1969.

[34] C.C. Macduffee, *The theory of matrices*, corrected reprint, Chelsea, New York, 1956.

[35] P.J. McCarthy, *Introduction to arithmetical functions*, Springer-Verlag, New York, 1986.

[36] P. Morandi, *Field and Galois theory*, Springer-Verlag, New York, 1996.

[37] R. Narasimhan, *Complex analysis in one variable*, Birkhäuser, Boston, Mass., 1985.

[38] W. Narkiewicz, *Elementary and analytic theory of algebraic numbers*, 2nd ed., Springer-Verlag, Berlin, 1990.

[39] J. Neukirch, *Algebraic number theory*, English transl. by N. Schappacher, Springer, Berlin, 1999.

[40] M. Newman, *Integral matrices*, Academic Press, New York, 1972.

[41] P. Ribenboim, *13 Lectures on Fermat's last theorem*, Springer-Verlag, New York, 1979.

[42] H.J.J. te Riele, Perfect numbers and aliquot sequences, *Computational methods in number theory* (ed. H.W. Lenstra Jr. and R. Tijdeman), Part I, pp. 141-157, Mathematical Centre Tracts **154**, Amsterdam, 1982.

[43] M.L. Rosen, A proof of the Lucas–Lehmer test, *Amer. Math. Monthly* **95** (1988), 855-856.

[44] H.H. Rosenbrock, *State-space and multivariable theory*, Nelson, London, 1970.

[45] G.-C. Rota, On the foundations of combinatorial theory I. Theory of Möbius functions, *Z. Wahrsch. Verw. Gebiete* 2 (1964), 340-368.

[46] G. Rousseau, On the Jacobi symbol, *J. Number Theory* **48** (1994), 109-111.

[47] P. Samuel, *Algebraic theory of numbers*, English transl. by A.J. Silberger, Houghton Mifflin, Boston, Mass., 1970.

[48] R. Sivaramakrishnan, *Classical theory of arithmetic functions*, M. Dekker, New York, 1989.

[49] H.J.S. Smith, *Collected mathematical papers*, *Vol. 1*, pp. 367-409, reprinted, Chelsea, New York, 1965.

[50] R.G. Swan, Gubeladze's proof of Anderson's conjecture, *Azumaya algebras, actions and modules* (ed. D. Haile and J. Osterburg), pp. 215-250, Contemporary Mathematics **124**, Amer. Math. Soc., Providence, R.I., 1992.

[51] J. Tate, Problem 9: The general reciprocity law, *Mathematical developments arising from Hilbert problems* (ed. F.E. Browder), pp. 311-322, Proc. Symp. Pure Math. **28**, Part 2, Amer. Math. Soc., Providence, Rhode Island, 1976.

[52] H.S. Vandiver, Fermat's last theorem: its history and the nature of the known results concerning it, *Amer. Math. Monthly* **53** (1946), 555-578.

[53] S.S. Wagstaff Jnr., Divisors of Mersenne numbers, *Math. Comp.* **40** (1983), 385-397.

[54] B.F. Wyman, What is a reciprocity law?, *Amer. Math. Monthly* **79** (1972), 571-586.

IV
Continued fractions and their uses

1 The continued fraction algorithm

Let $\xi = \xi_0$ be an irrational real number. Then we can write

$$\xi_0 = a_0 + \xi_1^{-1},$$

where $a_0 = \lfloor \xi_0 \rfloor$ is the greatest integer $\leq \xi_0$ and where $\xi_1 > 1$ is again an irrational number. Hence the process can be repeated indefinitely:

$$\xi_1 = a_1 + \xi_2^{-1}, \quad (a_1 = \lfloor \xi_1 \rfloor, \; \xi_2 > 1),$$
$$\xi_2 = a_2 + \xi_3^{-1}, \quad (a_2 = \lfloor \xi_2 \rfloor, \; \xi_3 > 1),$$

$$\cdots\cdots$$

By construction, $a_n \in \mathbb{Z}$ for all $n \geq 0$ and $a_n \geq 1$ if $n \geq 1$. The uniquely determined infinite sequence $[a_0,a_1,a_2,\ldots]$ is called the *continued fraction expansion* of ξ. The continued fraction expansion of ξ_n is $[a_n,a_{n+1},a_{n+2},\ldots]$.

For example, the 'golden ratio' $\tau = (1 + \sqrt{5})/2$ has the continued fraction expansion $[1,1,1,\ldots]$, since $\tau = 1 + \tau^{-1}$. Similarly, $\sqrt{2}$ has the continued fraction expansion $[1,2,2,\ldots]$, since $\sqrt{2} + 1 = 2 + 1/(\sqrt{2} + 1)$.

The relation between ξ_n and ξ_{n+1} may be written as a linear fractional transformation:

$$\xi_n = (a_n\xi_{n+1} + 1)/(1\xi_{n+1} + 0).$$

An arbitrary linear fractional transformation

$$\xi = (\alpha\xi' + \beta)/(\gamma\xi' + \delta)$$

is completely determined by its matrix

$$T = \begin{pmatrix} \alpha & \beta \\ \gamma & \delta \end{pmatrix}.$$

This description is convenient, because if we make a further linear fractional transformation

$$\xi' = (\alpha'\xi'' + \beta')/(\gamma'\xi'' + \delta')$$

with matrix

$$T' = \begin{pmatrix} \alpha' & \beta' \\ \gamma' & \delta' \end{pmatrix}$$

then, as is easily verified, the matrix

$$T'' = \begin{pmatrix} \alpha'' & \beta'' \\ \gamma'' & \delta'' \end{pmatrix}$$

of the composite transformation

$$\xi = (\alpha''\xi'' + \beta'')/(\gamma''\xi'' + \delta'')$$

is given by the matrix product $T'' = TT'$.

It follows that, if we set

$$A_k = \begin{pmatrix} a_k & 1 \\ 1 & 0 \end{pmatrix},$$

then the matrix of the linear fractional transformation which expresses ξ in terms of ξ_{n+1} is

$$T_n = A_0 \cdots A_n.$$

It is readily verified by induction that

$$T_n = \begin{pmatrix} p_n & p_{n-1} \\ q_n & q_{n-1} \end{pmatrix},$$

i.e.,

$$\xi = (p_n\xi_{n+1} + p_{n-1})/(q_n\xi_{n+1} + q_{n-1}),$$

where the elements p_n, q_n satisfy the recurrence relations

$$p_n = a_n p_{n-1} + p_{n-2}, \quad q_n = a_n q_{n-1} + q_{n-2} \quad (n \geq 0), \tag{1}$$

with the conventional starting values

$$p_{-2} = 0, \, p_{-1} = 1, \text{ resp. } q_{-2} = 1, \, q_{-1} = 0. \tag{2}$$

In particular,

$$p_0 = a_0, \, p_1 = a_1 a_0 + 1, \, q_0 = 1, \, q_1 = a_1.$$

Since $\det A_k = -1$, by taking determinants we obtain

$$p_n q_{n-1} - p_{n-1} q_n = (-1)^{n+1} \quad (n \ge 0). \tag{3}$$

By (1) also,

$$\begin{pmatrix} p_n & p_{n-1} \\ q_n & q_{n-1} \end{pmatrix} \begin{pmatrix} 1 & 1 \\ -a_n & 0 \end{pmatrix} = \begin{pmatrix} p_{n-2} & p_n \\ q_{n-2} & q_n \end{pmatrix},$$

from which, by taking determinants again, we get

$$p_n q_{n-2} - p_{n-2} q_n = (-1)^n a_n \quad (n \ge 0). \tag{4}$$

It follows from (1)-(2) that p_n and q_n are integers, and from (3) that they are coprime. Since $a_n \ge 1$ for $n \ge 1$, we have

$$1 = q_0 \le q_1 < q_2 < \dots \ .$$

Thus $q_n \ge n$ for $n \ge 1$. (In fact, since $q_n \ge q_{n-1} + q_{n-2}$ for $n \ge 1$, it is readily shown by induction that $q_n > \tau^{n-1}$ for $n > 1$, where $\tau = (1 + \sqrt{5})/2$.)

Since $q_n > 0$ for $n \ge 0$, we can rewrite (3),(4) in the forms

$$p_n/q_n - p_{n-1}/q_{n-1} = (-1)^{n+1}/q_{n-1} q_n \quad (n \ge 1), \tag{3'}$$

$$p_n/q_n - p_{n-2}/q_{n-2} = (-1)^n a_n/q_{n-2} q_n \quad (n \ge 2). \tag{4'}$$

It follows that the sequence $\{p_{2n}/q_{2n}\}$ is increasing, the sequence $\{p_{2n+1}/q_{2n+1}\}$ is decreasing, and every member of the first sequence is less than every member of the second sequence. Hence both sequences have limits and actually, since $q_n \to \infty$, the limits of the two sequences are the same.

From

$$\xi = (p_n \xi_{n+1} + p_{n-1})/(q_n \xi_{n+1} + q_{n-1})$$

we obtain

$$q_n \xi - p_n = (p_{n-1} q_n - p_n q_{n-1})/(q_n \xi_{n+1} + q_{n-1}) = (-1)^n/(q_n \xi_{n+1} + q_{n-1}).$$

Hence $\xi > p_n/q_n$ if n is even and $\xi < p_n/q_n$ if n is odd. It follows that $p_n/q_n \to \xi$ as $n \to \infty$. Consequently different irrational numbers have different continued fraction expansions.

Since ξ lies between p_n/q_n and p_{n+1}/q_{n+1}, we have

$$|p_{n+2}/q_{n+2} - p_n/q_n| < |\xi - p_n/q_n| < |p_{n+1}/q_{n+1} - p_n/q_n|.$$

By (3)' and (4)' we can rewrite this in the form

$$a_{n+2}/q_n q_{n+2} < |\xi - p_n/q_n| < 1/q_n q_{n+1} \quad (n \geq 0). \tag{5}$$

Hence

$$q_{n+2}^{-1} < |q_n \xi - p_n| < q_{n+1}^{-1},$$

which shows that $|q_n \xi - p_n|$ decreases as n increases. It follows that $|\xi - p_n/q_n|$ also decreases as n increases.

The rational number p_n/q_n is called the *n*-th *convergent* of ξ. The integers a_n are called the *partial quotients* and the real numbers ξ_n the *complete quotients* in the continued fraction expansion of ξ.

The continued fraction algorithm can be applied also when $\xi = \xi_0$ is rational, but in this case it is really the same as the Euclidean algorithm. For suppose $\xi_n = b_n/c_n$, where b_n and c_n are integers and $c_n > 0$. We can write

$$b_n = a_n c_n + c_{n+1},$$

where $a_n = \lfloor \xi_n \rfloor$ and c_{n+1} is an integer such that $0 \leq c_{n+1} < c_n$. Thus ξ_{n+1} is defined if and only if $c_{n+1} \neq 0$, and then $\xi_{n+1} = c_n/c_{n+1}$. Since the sequence of positive integers $\{c_n\}$ cannot decrease for ever, the continued fraction algorithm for a rational number ξ always terminates. At the last stage of the algorithm we have simply

$$\xi_N = a_N,$$

where $a_N > 1$ if $N > 0$. The uniquely determined finite sequence $[a_0, a_1, \ldots, a_N]$ is called the continued fraction expansion of ξ.

Convergents and *complete quotients* can be defined as before; all the properties derived for ξ irrational continue to hold for ξ rational, provided we do not go past $n = N$. The relation

$$\xi = (p_{N-1}\xi_N + p_{N-2})/(q_{N-1}\xi_N + q_{N-2})$$

now shows that

$$\xi = p_N/q_N.$$

Consequently different rational numbers have different continued fraction expansions.

Now let $a_0, a_1, a_2, \ldots$ be any infinite sequence of integers with $a_n \geq 1$ for $n \geq 1$. If we define integers p_n, q_n by the recurrence relations (1)-(2), our previous argument shows that the sequence $\{p_{2n}/q_{2n}\}$ is increasing, the sequence $\{p_{2n+1}/q_{2n+1}\}$ is decreasing, and the two sequences have a common limit ξ. If we put $\xi_0 = \xi$ and

$$\xi_{n+1} = -(q_{n-1}\xi - p_{n-1})/(q_n \xi - p_n) \quad (n \geq 0),$$

our previous argument shows also that $\xi_{n+1} > 1$ $(n \geq 0)$. Since

$$\xi_n = a_n + \xi_{n+1}^{-1},$$

it follows that $a_n = \lfloor \xi_n \rfloor$. Hence ξ is irrational and $[a_0, a_1, a_2, \ldots]$ is its continued fraction expansion.

Similarly it may be seen that, for any finite sequence of integers $a_0, a_1, \ldots, a_N$, with $a_n \geq 1$ for $1 \leq n < N$ and $a_N > 1$ if $N > 0$, there is a rational number ξ with $[a_0, a_1, \ldots, a_N]$ as its continued fraction expansion.

We will write simply $\xi = [a_0, a_1, \ldots, a_N]$ if ξ is rational and $\xi = [a_0, a_1, a_2, \ldots]$ if ξ is irrational.

We will later have need of the following result:

LEMMA 0 *Let ξ be an irrational number with complete quotients ξ_n and convergents p_n/q_n. If η is any irrational number different from ξ, and if we define η_{n+1} by*

$$\eta = (p_n \eta_{n+1} + p_{n-1})/(q_n \eta_{n+1} + q_{n-1}),$$

then $-1 < \eta_n < 0$ for all large n.

Moreover, if $\xi > 1$ and $\eta < 0$, then $-1 < \eta_n < 0$ for all $n > 0$.

Proof We have

$$\eta_{n+1} = (q_{n-1}\eta - p_{n-1})/(p_n - q_n\eta).$$

Hence

$$\begin{aligned}
\theta_{n+1} : &= q_n\eta_{n+1} + q_{n-1} \\
&= (p_n q_{n-1} - p_{n-1} q_n)/(p_n - q_n\eta) \\
&= (-1)^{n+1}/(p_n - q_n\eta) \\
&= (-1)^n/q_n(\eta - p_n/q_n).
\end{aligned}$$

Since $p_n/q_n \to \xi \neq \eta$ and $q_n \to \infty$, it follows that $\theta_n \to 0$. Since

$$\eta_{n+1} = -(q_{n-1} - \theta_{n+1})/q_n,$$

we conclude that $-1 < \eta_{n+1} < 0$ for all large n.

Suppose now that $\xi > 1$ and $\eta < 0$. It is readily verified that $\eta_n = a_n + 1/\eta_{n+1}$. But $a_n = \lfloor \xi_n \rfloor \geq 1$ for all $n \geq 0$. Consequently $\eta_n < 0$ implies $1/\eta_{n+1} < -1$ and thus $-1 < \eta_{n+1} < 0$. Since $\eta_0 < 0$, it follows by induction that $-1 < \eta_n < 0$ for all $n > 0$. $\square$

The complete quotients of a real number may be characterized in the following way:

PROPOSITION 1 *If* $\eta > 1$ *and*

$$\xi = (p\eta + p')/(q\eta + q'),$$

where p,q,p',q' *are integers such that* $pq' - p'q = \pm 1$ *and* $q > q' > 0$, *then* η *is a complete quotient of* ξ *and* p'/q', p/q *are corresponding consecutive convergents of* ξ.

Proof The relation $pq' - p'q = \pm 1$ implies that p and q are relatively prime. Since $q > 0$, p/q has a finite continued fraction expansion

$$p/q = [a_0,a_1,...,a_{n-1}] = p_{n-1}/q_{n-1}$$

and $q = q_{n-1}, p = p_{n-1}$. In fact, since $q > 1$, we have $n > 1$, $a_{n-1} \geq 2$ and $q_{n-1} > q_{n-2}$. From

$$p_{n-1}q_{n-2} - p_{n-2}q_{n-1} = (-1)^n = \varepsilon(pq' - p'q),$$

where $\varepsilon = \pm 1$, we obtain

$$p_{n-1}(q_{n-2} - \varepsilon q') = q_{n-1}(p_{n-2} - \varepsilon p').$$

Hence q_{n-1} divides $q_{n-2} - \varepsilon q'$. Since $0 < q_{n-2} < q_{n-1}$ and $0 < q' < q_{n-1}$, it follows that $q' = q_{n-2}$ if $\varepsilon = 1$ and $q' = q_{n-1} - q_{n-2}$ if $\varepsilon = -1$. Hence $p' = p_{n-2}$ if $\varepsilon = 1$ and $p' = p_{n-1} - p_{n-2}$ if $\varepsilon = -1$. Thus

$$\xi = (p_{n-1}\eta + p_{n-2})/(q_{n-1}\eta + q_{n-2}), \text{ resp. } (p_{n-1}\eta + p_{n-1} - p_{n-2})/(q_{n-1}\eta + q_{n-1} - q_{n-2}).$$

Since $\eta > 1$, its continued fraction expansion has the form $[a_n,a_{n+1},...]$, where $a_n \geq 1$. It follows that ξ has the continued fraction expansion

$$[a_0,a_1,...,a_{n-1},a_n,...], \text{ resp. } [a_0,a_1,...,a_{n-1}-1,1,a_n,...].$$

In either case p'/q' and p/q are consecutive convergents of ξ and η is the corresponding complete quotient. $\square$

A complex number ζ is said to be *equivalent* to a complex number ω if there exist integers a,b,c,d with $ad - bc = \pm 1$ such that

$$\zeta = (a\omega + b)/(c\omega + d),$$

and *properly equivalent* if actually $ad - bc = 1$. Then ω is also equivalent, resp. properly equivalent, to ζ, since

$$\omega = (d\zeta - b)/(-c\zeta + a).$$

By taking $a = d = 1$ and $b = c = 0$, we see that any complex number ζ is properly equivalent to itself. It is not difficult to verify also that if ζ is equivalent to ω and ω equivalent to χ, then ζ is equivalent to χ, and the same holds with 'equivalence' replaced by 'proper equivalence'. Thus equivalence and proper equivalence are indeed 'equivalence relations'.

For any coprime integers b,d, there exist integers a,c such that $ad - bc = 1$. Since

$$b/d \,=\, (a{\cdot}0 + b)/(c{\cdot}0 + d),$$

it follows that any rational number is properly equivalent to 0, and hence any two rational numbers are properly equivalent. The situation is more interesting for irrational numbers:

PROPOSITION 2 *Two irrational numbers ξ,η are equivalent if and only if their continued fraction expansions $[a_0,a_1,a_2,...\,]$, $[b_0,b_1,b_2,...\,]$ have the same 'tails', i.e. there exist integers $m \geq 0$ and $n \geq 0$ such that*

$$a_{m+k} \,=\, b_{n+k} \ \text{for all } k \geq 0.$$

Proof If the continued fraction expansions of ξ and η have the same tails, then some complete quotient ξ_m of ξ coincides with some complete quotient η_n of η. But ξ is equivalent to ξ_m, since $\xi = (p_{m-1}\xi_m + p_{m-2})/(q_{m-1}\xi_m + q_{m-2})$ and $p_{m-1}q_{m-2} - p_{m-2}q_{m-1} = (-1)^m$, and similarly η is equivalent to η_n. Hence ξ and η are equivalent.

Suppose on the other hand that ξ and η are equivalent. Then

$$\eta \,=\, (a\xi + b)/(c\xi + d)$$

for some integers a,b,c,d such that $ad - bc = \pm 1$. By changing the signs of all four we may suppose that $c\xi + d > 0$. From the relation

$$\xi = (p_{n-1}\xi_n + p_{n-2})/(q_{n-1}\xi_n + q_{n-2})$$

between ξ and its complete quotient ξ_n it follows that

$$\eta = (a_n\xi_n + b_n)/(c_n\xi_n + d_n),$$

where

$$a_n = ap_{n-1} + bq_{n-1}, \ \ b_n = ap_{n-2} + bq_{n-2},$$
$$c_n = cp_{n-1} + dq_{n-1}, \ \ d_n = cp_{n-2} + dq_{n-2},$$

and hence

$$a_nd_n - b_nc_n = (ad - bc)(p_{n-1}q_{n-2} - p_{n-2}q_{n-1}) = \pm 1.$$

The inequalities

$$|q_{n-1}\xi - p_{n-1}| < 1/q_n, \ \ |q_{n-2}\xi - p_{n-2}| < 1/q_{n-1}$$

imply that

$$\left|c_n - (c\xi + d)q_{n-1}\right| < |c|/q_n, \quad \left|d_n - (c\xi + d)q_{n-2}\right| < |c|/q_{n-1}.$$

Since $c\xi + d > 0$, $q_{n-1} > q_{n-2}$ and $q_n \to \infty$ as $n \to \infty$, it follows that $c_n > d_n > 0$ for sufficiently large n. Then, by Proposition 1, ξ_n is a complete quotient also of η. Thus the continued fraction expansions of ξ and η have a common tail. $\square$

2 Diophantine approximation

The subject of *Diophantine approximation* is concerned with finding integer or rational solutions for systems of inequalities. For problems in one dimension the continued fraction algorithm is a most helpful tool, as we will now see.

PROPOSITION 3 *Let p_n/q_n ($n \geq 1$) be a convergent of the real number ξ. If p,q are integers such that $0 < q \leq q_n$ and $p \neq p_n$ if $q = q_n$, then*

$$\left|q\xi - p\right| \geq \left|q_{n-1}\xi - p_{n-1}\right| > \left|q_n\xi - p_n\right|$$

and

$$\left|\xi - p/q\right| > \left|\xi - p_n/q_n\right|.$$

Proof It follows from (3) that the simultaneous linear equations

$$\lambda p_{n-1} + \mu p_n = p, \quad \lambda q_{n-1} + \mu q_n = q,$$

have integer solutions, namely

$$\lambda = (-1)^{n-1}(p_n q - q_n p), \quad \mu = (-1)^n(p_{n-1}q - q_{n-1}p).$$

The hypotheses on p,q imply that $\lambda \neq 0$. If $\mu = 0$, then

$$\left|q\xi - p\right| = \left|\lambda(q_{n-1}\xi - p_{n-1})\right| \geq \left|q_{n-1}\xi - p_{n-1}\right|.$$

Thus we now assume $\mu \neq 0$. Since $q \leq q_n$, λ and μ cannot both be positive and hence, since $q > 0$, $\lambda\mu < 0$. Then

$$q\xi - p = \lambda(q_{n-1}\xi - p_{n-1}) + \mu(q_n\xi - p_n)$$

and both terms on the right have the same sign. Hence

$$\left|q\xi - p\right| = \left|\lambda(q_{n-1}\xi - p_{n-1})\right| + \left|\mu(q_n\xi - p_n)\right| \geq \left|q_{n-1}\xi - p_{n-1}\right|.$$

This proves the first statement of the proposition. The second statement follows, since

$$|\xi - p/q| = q^{-1}|q\xi - p| > q^{-1}|q_n\xi - p_n| = (q_n/q)|\xi - p_n/q_n| \geq |\xi - p_n/q_n|. \quad \square$$

To illustrate the application of Proposition 3, consider the continued fraction expansion of $\pi = 3.14159265358...$. We easily find that it begins $[3,7,15,1,292,...]$. It follows that the first five convergents of π are

$$3/1, \ 22/7, \ 333/106, \ 355/113, \ 103993/33102.$$

Using the inequality $|\xi - p_n/q_n| < 1/q_nq_{n+1}$ and choosing $n = 3$ so that a_{n+1} is large, we obtain

$$0 < 355/113 - \pi < 0.000000267... \ .$$

The approximation $355/113$ to π was first given by the Chinese mathematician Zu Chongzhi in the 5th century A.D. Proposition 3 shows that it is a better approximation to π than any other rational number with denominator ≤ 113.

In general, a rational number p'/q', where p',q' are integers and $q' > 0$, may be said to be a *best approximation* to a real number ξ if

$$|\xi - p/q| > |\xi - p'/q'|$$

for all different rational numbers p/q whose denominator q satisfies $0 < q \leq q'$. Thus Proposition 3 says that any convergent p_n/q_n $(n \geq 1)$ of ξ is a best approximation of ξ. However, these are not the only best approximations. It may be shown that, if p_{n-2}/q_{n-2} and p_{n-1}/q_{n-1} are consecutive convergents of ξ, then any rational number of the form

$$(cp_{n-1} + p_{n-2})/(cq_{n-1} + q_{n-2}),$$

where c is an integer such that $a_n/2 < c \leq a_n$ is a best approximation of ξ. Furthermore, every best approximation of ξ has this form if, when a_n is even, one allows also $c = a_n/2$.

It follows that $355/113$ is a better approximation to π than any other rational number with denominator less than 16604, since $292/2 = 146$ and $146{\times}113 + 106 = 16604$.

The complete continued fraction expansion of π is not known. However, it was discovered by Cotes (1714) and proved by Euler (1737) that the complete continued fraction expansion of $e = 2.71828182459...$ is given by $e - 1 = [1,1,2,1,1,4,1,1,6,...]$.

The preceding results may also be applied to the construction of calendars. The solar year has a length of about 365.24219 mean solar days. The continued fraction expansion of $\lambda = (0.24219)^{-1}$ begins $[4,7,1,3,24,...]$. Hence the first five convergents of λ are

$$4/1,\ 29/7,\ 33/8,\ 128/31,\ 3105/752.$$

It follows that

$$0\ <\ 128/31 - \lambda\ <\ 0.0000428$$

and $128/31$ is a better approximation to λ than any other rational number with denominator less than 380. The Julian calendar, by adding a day every 4 years, estimated the year at 365.25 days. The Gregorian calendar, by adding 97 days every 400 years, estimates the year at 365.2425 days. Our analysis shows that, if we added instead 31 days every 128 years, we would obtain the much more precise estimate of 365.2421875 days.

Best approximations also find an application in the selection of gear ratios, and continued fractions were already used for this purpose by Huygens (1682) in constructing his planetarium (a mechanical model for the solar system).

The next proposition describes another way in which the continued fraction expansion provides good rational approximations.

PROPOSITION 4 *If p,q are coprime integers with $q > 0$ such that, for some real number ξ,*

$$|\xi - p/q| < 1/2q^2,$$

then p/q is a convergent of ξ.

Proof Let p_n/q_n be the convergents of ξ and assume that p/q is not a convergent. We show first that $q < q_N$ for some $N > 0$. This is obvious if ξ is irrational. If $\xi = p_N/q_N$ is rational, then

$$1/q_N\ \leq\ |qp_N - pq_N|/q_N\ =\ |q\xi - p|\ <\ 1/2q.$$

Hence $q < q_N$ and $N > 0$.

It follows that $q_{n-1} \leq q < q_n$ for some $n > 0$. By Proposition 3,

$$|q_{n-1}\xi - p_{n-1}|\ \leq\ |q\xi - p|\ <\ 1/2q.$$

Hence

$$
\begin{aligned}
1/qq_{n-1}\ \leq\ |qp_{n-1} - pq_{n-1}|/qq_{n-1}\ &=\ |p_{n-1}/q_{n-1} - p/q|\\
&\leq\ |p_{n-1}/q_{n-1} - \xi| + |\xi - p/q|\\
&<\ 1/2qq_{n-1} + 1/2q^2.
\end{aligned}
$$

But this implies $q < q_{n-1}$, which is a contradiction. $\square$

As an application of Proposition 4 we prove

PROPOSITION 5 *Let d be a positive integer which is not a square and m an integer such that $0 < m^2 < d$. If x,y are positive integers such that*

$$x^2 - dy^2 = m,$$

then x/y is a convergent of the irrational number $\sqrt{d}$.

Proof Suppose first that $m > 0$. Then $x/y > \sqrt{d}$ and

$$0 < x/y - \sqrt{d} = m/(xy + y^2\sqrt{d}) < \sqrt{d}/2y^2\sqrt{d} = 1/2y^2.$$

Hence x/y is a convergent of $\sqrt{d}$, by Proposition 4.

Suppose next that $m < 0$. Then $y/x > 1/\sqrt{d}$ and

$$0 < y/x - 1/\sqrt{d} = -m/d(xy + x^2/\sqrt{d}) < 1/\sqrt{d}(xy + x^2/\sqrt{d}) < 1/2x^2.$$

Hence y/x is a convergent of $1/\sqrt{d}$. But, since $1/\sqrt{d} = 0 + 1/\sqrt{d}$, the convergents of $1/\sqrt{d}$ are $0/1$ and the reciprocals of the convergents of $\sqrt{d}$. $\square$

In the next section we will show that the continued fraction expansion of $\sqrt{d}$ has a particularly simple form.

It was shown by Vahlen (1895) that at least one of any two consecutive convergents of ξ satisfies the inequality of Proposition 4. Indeed, since consecutive convergents lie on opposite sides of ξ,

$$\begin{aligned}
|p_n/q_n - \xi| + |p_{n-1}/q_{n-1} - \xi| &= |p_n/q_n - p_{n-1}/q_{n-1}| \\
&= 1/q_n q_{n-1} \leq 1/2q_n^2 + 1/2q_{n-1}^2,
\end{aligned}$$

with equality only if $q_n = q_{n-1}$. This proves the assertion, except when $n = 1$ and $q_1 = q_0 = 1$. But in this case $a_1 = 1$, $1 \leq \xi_1 < 2$ and hence

$$|\xi - p_1/q_1| = |\xi - a_0 - 1| = 1 - \xi_1^{-1} < 1/2.$$

It was shown by Borel (1903) that at least one of any three consecutive convergents of ξ satisfies the sharper inequality

$$|\xi - p/q| < 1/\sqrt{5}q^2.$$

In fact this is obtained by taking $r = 1$ in the following more general result, due to Forder (1963) and Wright (1964).

PROPOSITION 6 *Let ξ be an irrational number with the continued fraction expansion $[a_0,a_1,\ldots]$ and the convergents p_n/q_n. If, for some positive integer r,*

$$|\xi - p_n/q_n| \geq 1/(r^2 + 4)^{1/2}q_n^2 \quad \text{for } n = m - 1, m, m + 1,$$

then $a_{m+1} < r$.

Proof If we put $s = (r^2 + 4)^{1/2}/2$, then s is irrational. For otherwise $2s$ would be an integer and from $(2s + r)(2s - r) = 4$ we would obtain $2s + r = 4$, $2s - r = 1$ and hence $r = 3/2$, which is a contradiction.

By the hypotheses of the proposition,

$$1/q_{m-1}q_m \; = \; |p_{m-1}/q_{m-1} - p_m/q_m| \; = \; |\xi - p_{m-1}/q_{m-1}| + |\xi - p_m/q_m| \; \geq \; (q_{m-1}^{-2} + q_m^{-2})/2s$$

and hence

$$q_m^2 - 2sq_{m-1}q_m + q_{m-1}^2 \leq 0.$$

Furthermore, this inequality also holds when q_{m-1},q_m are replaced by q_m,q_{m+1}. Consequently q_{m-1}/q_m and q_{m+1}/q_m both satisfy the inequality $t^2 - 2st + 1 \leq 0$. Since

$$t^2 - 2st + 1 \; = \; (t - s + r/2)(t - s - r/2),$$

it follows that

$$s - r/2 \; < \; q_{m-1}/q_m \; < \; q_{m+1}/q_m \; < \; s + r/2,$$

the first and last inequalities being strict because s is irrational. Hence

$$a_{m+1} \; = \; q_{m+1}/q_m - q_{m-1}/q_m \; < \; s + r/2 - (s - r/2) \; = \; r. \quad \square$$

It follows from Proposition 6 with $r = 1$ that, for any irrational number ξ, there exist infinitely many rational numbers $p/q = p_n/q_n$ such that

$$|\xi - p/q| < 1/\sqrt{5}q^2.$$

Here the constant $\sqrt{5}$ is best possible. For take any $c > \sqrt{5}$. If there exists a rational number p/q, with $q > 0$ and $(p,q) = 1$, such that

$$|\xi - p/q| \; < \; 1/cq^2,$$

then p/q is a convergent of ξ, by Proposition 4. But for any convergent p_n/q_n we have

$$|\xi - p_n/q_n| \; = \; 1/q_n(q_n\xi_{n+1} + q_{n-1}).$$

If we take $\xi = \tau := (1 + \sqrt{5})/2$, then also $\xi_{n+1} = \tau$ and $p_n = q_{n+1}$. Hence

$$|\tau - q_{n+1}/q_n| = 1/q_n^2(\tau + q_{n-1}/q_n),$$

where $\tau + q_{n-1}/q_n \to \tau + \tau^{-1} = \sqrt{5}$, since $q_n/q_{n-1} \to \tau$. Thus, for any $c > \sqrt{5}$, there exist at most finitely many rational numbers p/q such that

$$|\tau - p/q| < 1/cq^2.$$

It follows from Proposition 6 with $r = 2$ that if

$$|\xi - p_n/q_n| \geq 1/\sqrt{8}q_n^2 \text{ for all large } n,$$

then $a_n = 1$ for all large n. The constant $\sqrt{8}$ is again best possible, since a similar argument to that just given shows that if $\sigma := 1 + \sqrt{2} = [2,2,...\,]$ then, for any $c > \sqrt{8}$, there exist at most finitely many rational numbers p/q such that

$$|\sigma - p/q| < 1/cq^2.$$

It follows from Proposition 6 with $r = 3$ that if

$$|\xi - p_n/q_n| \geq 1/\sqrt{13}q_n^2 \text{ for all large } n,$$

then $a_n \in \{1,2\}$ for all large n.

For any irrational ξ, with continued fraction expansion $[a_0,a_1,...\,]$ and convergents p_n/q_n, put

$$M(\xi) = \overline{\lim}_{n\to\infty}\, q_n^{-1}|q_n\xi - p_n|^{-1}.$$

It follows from Proposition 2 that $M(\xi) = M(\eta)$ if ξ and η are equivalent. The results just established show that $M(\xi) \geq \sqrt{5}$ for every ξ. If $M(\xi) < \sqrt{8}$, then $a_n = 1$ for all large n; hence ξ is equivalent to τ and $M(\xi) = M(\tau) = \sqrt{5}$. If $M(\xi) < \sqrt{13}$, then $a_n \in \{1,2\}$ for all large n.

An irrational number ξ is said to be *badly approximable* if $M(\xi) < \infty$. The inequalities

$$a_{n+2}/q_nq_{n+2} < |\xi - p_n/q_n| < 1/q_nq_{n+1}$$

imply

$$a_{n+1} \leq q_{n+1}/q_n < q_n^{-1}|q_n\xi - p_n|^{-1}$$

and

$$q_n^{-1}|q_n\xi - p_n|^{-1} < q_{n+2}/a_{n+2}q_n \leq q_{n+1}/q_n + 1 \leq a_{n+1} + 2.$$

Hence ξ *is badly approximable if and only if its partial quotients* a_n *are bounded.*

It is obvious that ξ is badly approximable if there exists a constant $c > 0$ such that

$$|\xi - p/q| > c/q^2$$

for every rational number p/q. Conversely, if ξ is badly approximable, then there exists such a constant $c > 0$. This is clear when p and q are coprime integers, since if p/q is *not* a convergent of ξ then, by Proposition 4,

$$|\xi - p/q| \geq 1/2q^2.$$

On the other hand, if $p = \lambda p'$, $q = \lambda q'$, where p',q' are coprime, then

$$|\xi - p/q| = |\xi - p'/q'| \geq c/q'^2 = \lambda^2 c/q^2 \geq c/q^2.$$

Some of the applications of badly approximable numbers stem from the following characterization: a real number θ is badly approximable if and only if there exists a constant $c' > 0$ such that

$$|e^{2\pi i q \theta} - 1| \geq c'/q \ \text{ for all } q \in \mathbb{N}.$$

To establish this, put $q\theta = p + \delta$, where $p \in \mathbb{Z}$ and $|\delta| \leq 1/2$. Then

$$|e^{2\pi i q \theta} - 1| = 2|\sin \pi q \theta| = 2|\sin \pi \delta|$$

and the result follows from the previous characterization, since $(\sin x)/x$ decreases from 1 to $2/\pi$ as x increases from 0 to $\pi/2$.

3 Periodic continued fractions

A complex number ζ is said to be a *quadratic irrational* if it is a root of a monic quadratic polynomial $t^2 + rt + s$ with rational coefficients r,s, but is not itself rational. Since $\zeta \notin \mathbb{Q}$, the rational numbers r,s are uniquely determined by ζ.

Equivalently, ζ is a quadratic irrational if it is a root of a quadratic polynomial

$$f(t) = At^2 + Bt + C$$

with integer coefficients A,B,C such that $B^2 - 4AC$ is not the square of an integer. The integers A,B,C are uniquely determined up to a common factor and are uniquely determined up to sign if we require that they have greatest common divisor 1. The corresponding integer $D = B^2 - 4AC$ is then uniquely determined and is called the *discriminant* of ζ. A quadratic irrational is real if and only if its discriminant is positive.

It is readily verified that if a quadratic irrational ζ is equivalent to a complex number ω, i.e. if

$$\zeta = (\alpha\omega + \beta)/(\gamma\omega + \delta),$$

where $\alpha,\beta,\gamma,\delta \in \mathbb{Z}$ and $\alpha\delta - \beta\gamma = \pm 1$, then ω is also a quadratic irrational. Moreover, if ζ is a root of the quadratic polynomial $f(t) = At^2 + Bt + C$, where A,B,C are integers with greatest common divisor 1, then ω is a root of the quadratic polynomial

$$g(t) = A't^2 + B't + C',$$

where

$$A' = \alpha^2 A + \alpha\gamma B + \gamma^2 C,$$
$$B' = 2\alpha\beta A + (\alpha\delta + \beta\gamma)B + 2\gamma\delta C,$$
$$C' = \beta^2 A + \beta\delta B + \delta^2 C,$$

and hence

$$B'^2 - 4A'C' = B^2 - 4AC = D.$$

Since

$$A = \delta^2 A' - \gamma\delta B' + \gamma^2 C',$$
$$B = -2\beta\delta A' + (\alpha\delta + \beta\gamma)B' - 2\alpha\gamma C',$$
$$C = \beta^2 A' - \alpha\beta B' + \alpha^2 C',$$

A',B',C' also have greatest common divisor 1.

If ζ is a quadratic irrational, we define the *conjugate* ζ' of ζ to be the other root of the quadratic polynomial $f(t)$ which has ζ as a root. If

$$\zeta = (\alpha\omega + \beta)/(\gamma\omega + \delta),$$

where $\alpha,\beta,\gamma,\delta \in \mathbb{Z}$ and $\alpha\delta - \beta\gamma = \pm 1$, then evidently

$$\zeta' = (\alpha\omega' + \beta)/(\gamma\omega' + \delta).$$

Suppose now that $\zeta = \xi$ is real and that the integers A,B,C are uniquely determined by requiring not only $(A,B,C) = 1$ but also $A > 0$. The real quadratic irrational ξ is said to be *reduced* if $\xi > 1$ and $-1 < \xi' < 0$. If ξ is reduced then, since $\xi > \xi'$, we must have

$$\xi = (-B + \sqrt{D})/2A, \quad \xi' = (-B - \sqrt{D})/2A.$$

Thus the inequalities $\xi > 1$ and $-1 < \xi' < 0$ imply

$$0 < \sqrt{D} + B < 2A < \sqrt{D} - B.$$

Conversely, if the coefficients A,B,C of $f(t)$ satisfy these inequalities, where $D = B^2 - 4AC > 0$, then one of the roots of $f(t)$ is reduced. For $B < 0 < A$ and so the roots ξ,ξ' of $f(t)$ have opposite signs. If ξ is the positive root, then ξ and ξ' are given by the preceding formulas and hence $\xi > 1$, $-1 < \xi' < 0$. It should be noted also that if ξ is reduced, then $B^2 < D$ and hence $C < 0$.

We return now to continued fractions. If ξ is a real quadratic irrational, then its complete quotients ξ_n are all quadratic irrationals and, conversely, if some complete quotient ξ_n is a quadratic irrational, then ξ is also a quadratic irrational.

The continued fraction expansion $[a_0,a_1,a_2,\ldots\,]$ of a real number ξ is said to be *eventually periodic* if there exist integers $m \geq 0$ and $h > 0$ such that

$$a_n = a_{n+h} \quad \text{for all } n \geq m.$$

The continued fraction expansion is then conveniently denoted by

$$[a_0,a_1,..,a_{m-1},\overline{a_m,\ldots,a_{m+h-1}}].$$

The continued fraction expansion is said to be *periodic* if it is eventually periodic with $m = 0$.

Equivalently, the continued fraction expansion of ξ is eventually periodic if $\xi_m = \xi_{m+h}$ for some $m \geq 0$ and $h > 0$, and periodic if this holds with $m = 0$. The *period* of the continued fraction expansion, in either case, is the least positive integer h with this property.

We are going to show that there is a close connection between real quadratic irrationals and eventually periodic continued fractions.

PROPOSITION 7 *A real number ξ is a reduced quadratic irrational if and only if its continued fraction expansion is periodic.*

Moreover, if $\xi = [\overline{a_0,\ldots,a_{h-1}}]$, then $-1/\xi' = [\overline{a_{h-1},\ldots,a_0}]$.

Proof Suppose first that $\xi = [\overline{a_0,\ldots,a_{h-1}}]$ has a periodic continued fraction expansion. Then $a_0 = a_h \geq 1$ and hence $\xi > 1$. Furthermore, since

$$\xi = (p_{h-1}\xi_h + p_{h-2})/(q_{h-1}\xi_h + q_{h-2})$$

and $\xi_h = \xi$, ξ is an irrational root of the quadratic polynomial

$$f(t) = q_{h-1}t^2 + (q_{h-2} - p_{h-1})t - p_{h-2}.$$

Thus ξ is a quadratic irrational. Since $f(0) = -p_{h-2} < 0$ and

$$f(-1) = q_{h-1} - q_{h-2} + p_{h-1} - p_{h-2} > 0$$

(even for $h = 1$), it follows that $-1 < \xi' < 0$. Thus ξ is reduced.

If ξ is a reduced quadratic irrational, then its complete quotients ξ_n, which are all quadratic irrationals, are also reduced, by Lemma 0 with $\eta = \xi'$. Since $\xi_n' = a_n + 1/\xi_{n+1}'$ and $-1 < \xi_n' < 0$, we have

$$a_n = \lfloor -1/\xi_{n+1}' \rfloor.$$

Thus ξ_n,ξ_n' are the roots of a uniquely determined polynomial

$$f_n(t) = A_n t^2 + B_n t + C_n,$$

where A_n,B_n,C_n are integers with greatest common divisor 1 and $A_n > 0$. Furthermore, $D = B_n^2 - 4A_n C_n$ is independent of n and positive. Since ξ_n is reduced, we have

$$\xi_n = (-B_n + \sqrt{D})/2A_n, \quad \xi_n' = (-B_n - \sqrt{D})/2A_n,$$

where

$$0 < \sqrt{D} + B_n < 2A_n < \sqrt{D} - B_n.$$

If we put $g = \lfloor \sqrt{D} \rfloor$, then $-B_n \in \{1,...,g\}$ and, for a given value of B_n, there are at most $-B_n$ possible values for A_n. Consequently the number of distinct pairs A_n,B_n does not exceed $1 + ... + g = g(g + 1)/2$. Hence we must have

$$\xi_j = \xi_k, \quad \xi_j' = \xi_k'$$

for some j,k such that $0 \le j < k \le g(g + 1)/2$. If $j = 0$, this already proves that the continued fraction expansion of ξ is periodic. If $j > 0$, then

$$a_{j-1} = \lfloor -1/\xi_j' \rfloor = \lfloor -1/\xi_k' \rfloor = a_{k-1}$$

and hence

$$\xi_{j-1} = a_{j-1} + 1/\xi_j = a_{k-1} + 1/\xi_k = \xi_{k-1}.$$

Repeating this argument j times, we obtain $\xi_0 = \xi_{k-j}$. Thus ξ has a periodic continued fraction expansion in any case.

If the period is h, so that $\xi = [\overline{a_0,...,a_{h-1}}]$, then $\xi_0' = \xi_h'$ and the relation $a_n = \lfloor -1/\xi_{n+1}' \rfloor$ implies that $-1/\xi' = [\overline{a_{h-1},...,a_0}]$. $\square$

The proof of Proposition 7 shows that the period is at most $g(g + 1)/2$ and thus is certainly less than D. By counting the pairs of integers A,B for which not only

$$0 < \sqrt{D} + B < 2A < \sqrt{D} - B,$$

but also $D \equiv B^2 \bmod 4A$, it may be shown that the period is at most $O(\sqrt{D} \log D)$. (The *Landau order symbol* used here is defined under 'Notations'.)

PROPOSITION 8 *A real number ξ is a quadratic irrational if and only if its continued fraction expansion is eventually periodic.*

Proof Suppose first that the continued fraction expansion of ξ is eventually periodic. Then some complete quotient ξ_m has a periodic continued fraction expansion and hence is a quadratic irrational, by Proposition 7. But this implies that ξ also is a quadratic irrational.

Suppose next that ξ is a quadratic irrational. We will prove that the continued fraction expansion of ξ is eventually periodic by showing that some complete quotient ξ_{n+1} is reduced. Since we certainly have $\xi_{n+1} > 1$, we need only show that $-1 < \xi_{n+1}' < 0$. But $\xi' \neq \xi$ and $\xi' = (p_n\xi_{n+1}' + p_{n-1})/(q_n\xi_{n+1}' + q_{n-1})$. Hence, by Lemma 0, $-1 < \xi_{n+1}' < 0$ for all large n. $\square$

It follows from Proposition 8 that any real quadratic irrational is badly approximable, since its partial quotients are bounded. It follows from Propositions 7 and 8 that there are only finitely many inequivalent quadratic irrationals with a given discriminant $D > 0$, since any real quadratic irrational is equivalent to a reduced one and only finitely many pairs of integers A,B satisfy the inequalities

$$0 < \sqrt{D} + B < 2A < \sqrt{D} - B.$$

Proposition 8 is due to Euler and Lagrange. It was first shown by Euler (1737) that a real number is a quadratic irrational if its continued fraction expansion is eventually periodic, and the converse was proved by Lagrange (1770). Proposition 7 was first stated and proved by Galois (1829), although it was implicit in the work of Lagrange (1773) on the reduction of binary quadratic forms. Proposition 7 provides a simple proof of the following result due to Legendre:

PROPOSITION 9 *For any real number ξ, the following two conditions are equivalent:*

(i)   *$\xi > 1, \xi$ is irrational and ξ^2 is rational;*
(ii)   *the continued fraction expansion of ξ has the form $[a_0, \overline{a_1,...,a_h}]$, where $a_h = 2a_0$ and $a_i = a_{h-i}$ for $i = 1,...,h - 1$.*

Proof Suppose first that (i) holds. Then ξ is a quadratic irrational, since it is a root of the polynomial $t^2 - \xi^2$. The continued fraction expansion of ξ cannot be periodic, by Proposition 7, since $\xi' = -\xi < -1$. However, the continued fraction expansion of ξ_1 is periodic, since $\xi_1 > 1$ and $1/\xi_1' = \xi' - a_0 < -1$. Thus $\xi_1 = [\overline{a_1,...,a_h}]$ for some $h \geq 1$. By Proposition 7 also,

$$-1/\xi_1' = [\overline{a_h,...,a_1}].$$

But

$$-1/\xi_1' = \xi + a_0 = [\overline{2a_0, a_1,...,a_h}].$$

Comparing this with the previous expression, we see that (ii) holds.

Suppose, conversely, that (ii) holds. Then ξ is irrational, $a_0 > 0$ and hence $\xi > 1$. Moreover $\xi_1 = [\,\overline{a_1,...,a_h}\,]$ is a reduced quadratic irrational and

$$- 1/\xi_1' = [\,\overline{a_h,...,a_1}\,] = [2a_0, \overline{a_1,...,a_h}\,] = a_0 + \xi.$$

Hence $\xi' = a_0 + 1/\xi_1' = -\xi$ and $\xi^2 = -\xi\xi'$ is rational. $\square$

4 Quadratic Diophantine equations

We are interested in finding all integers x,y such that

$$ax^2 + bxy + cy^2 + dx + ey + f = 0, \tag{6}$$

where $a,...,f$ are given integers. Writing (6) as a quadratic equation for x,

$$ax^2 + (by + d)x + cy^2 + ey + f = 0,$$

we see that if a solution exists for some y, then the discriminant

$$(by + d)^2 - 4a(cy^2 + ey + f)$$

must be a perfect square. Thus

$$(b^2 - 4ac)y^2 + 2(bd - 2ae)y + d^2 - 4af = z^2$$

for some integer z. If we put

$$p: = b^2 - 4ac, \quad q: = bd - 2ae, \quad r: = d^2 - 4af,$$

we have a quadratic equation for y,

$$py^2 + 2qy + r - z^2 = 0,$$

whose discriminant must also be a perfect square. Thus

$$q^2 - p(r - z^2) = w^2$$

for some integer w. Thus if (6) has a solution in integers, so also does the equation

$$w^2 - pz^2 = q^2 - pr.$$

Moreover, from all solutions in integers of the latter equation we may obtain, by retracing our steps, all solutions in integers of the original equation (6).

Thus we now restrict our attention to finding all integers x,y such that

$$x^2 - dy^2 = m, \tag{7}$$

where d and m are given integers.

The equation (7) has the remarkable property, which was known to Brahmagupta (628) and later rediscovered by Euler (1758), that if we have solutions for two values m_1, m_2 of m, then we can derive a solution for their product $m_1 m_2$. This follows from the identity

$$(x_1^2 - dy_1^2)(x_2^2 - dy_2^2) = x^2 - dy^2,$$

where

$$x = x_1 x_2 + dy_1 y_2, \quad y = x_1 y_2 + y_1 x_2.$$

(In fact, Brahmagupta's identity is just a restatement of the norm relation $N(\alpha\beta) = N(\alpha)N(\beta)$ for elements α, β of a quadratic field.) In particular, from two solutions of the equation

$$x^2 - dy^2 = 1, \tag{8}$$

a third solution can be obtained by *composition* in this way.

Composition of solutions is evidently commutative and associative. In fact the solutions of (8) form an abelian group under composition, with the trivial solution $1,0$ as identity element and the solution $x, -y$ as the inverse of the solution x,y. Also, by composing an arbitrary solution x,y of (8) with the trivial solution $-1,0$ we obtain the solution $-x, -y$.

Suppose first that $d < 0$. Evidently (7) is insoluble if $m < 0$ and $x = y = 0$ is the only solution if $m = 0$. If $m > 0$, there are at most finitely many solutions and we may find them all by testing, for each non-negative integer $y \le (-m/d)^{1/2}$, whether $m + dy^2$ is a perfect square.

Suppose now that $d > 0$. If $d = e^2$ is a perfect square, then (7) is equivalent to the finite set of simultaneous linear Diophantine equations

$$x - ey = m', \quad x + ey = m'',$$

where m', m'' are any integers such that $m'm'' = m$. Thus *we now suppose also that d is not a perfect square.* Then $\xi = \sqrt{d}$ is irrational.

If $0 < m^2 < d$ then, by Proposition 5, any positive solution x,y of (7) has the form $x = p_n$, $y = q_n$, where p_n/q_n is a convergent of ξ. In particular, all positive solutions of $x^2 - dy^2 = \pm 1$ are obtained in this way.

On the other hand, as we now show, if p_n/q_n is any convergent of ξ then

$$|p_n{}^2 - dq_n{}^2| < 2\sqrt{d}.$$

If $n = 0$, then $|p_0{}^2 - dq_0{}^2| = |a_0{}^2 - d|$, where $a_0 < \sqrt{d} < a_0 + 1$ and so $0 < d - a_0{}^2 \le 2a_0 < 2\sqrt{d}$. Now suppose $n > 0$. Then $|p_n - q_n\xi| < q_{n+1}{}^{-1}$ and hence

$$\begin{aligned}
|p_n{}^2 - dq_n{}^2| &= |p_n - q_n\xi|\,|p_n - q_n\xi + 2q_n\xi| \\
&< q_{n+1}{}^{-1}(q_{n+1}{}^{-1} + 2q_n\xi) < 2\xi.
\end{aligned}$$

An easy congruence argument shows that the equation

$$x^2 - dy^2 = -1 \tag{9}$$

has no solutions in integers if $d \equiv 3 \bmod 4$. It will now be shown that the equation (8), on the other hand, always has solutions in positive integers.

PROPOSITION 10 *Let d be a positive integer which is not a perfect square. Suppose $\xi = \sqrt{d}$ has complete quotients ξ_n, convergents p_n/q_n, and continued fraction expansion $[a_0, \overline{a_1, ..., a_h}]$ of period h.*

Then $p_n{}^2 - dq_n{}^2 = \pm 1$ if and only if $n = kh - 1$ for some positive integer k and in this case

$$p_{kh-1}{}^2 - dq_{kh-1}{}^2 = (-1)^{kh}.$$

Proof From $\xi = (p_n\xi_{n+1} + p_{n-1})/(q_n\xi_{n+1} + q_{n-1})$ we obtain

$$(p_n - q_n\xi)\xi_{n+1} = q_{n-1}\xi - p_{n-1}.$$

Multiplying by $(-1)^{n+1}(p_n + q_n\xi)$, we get

$$s_n\xi_{n+1} = \xi + r_n,$$

where

$$s_n = (-1)^{n+1}(p_n{}^2 - dq_n{}^2), \quad r_n = (-1)^n(p_{n-1}p_n - dq_{n-1}q_n).$$

Thus s_n and r_n are integers.

If $p_n{}^2 - dq_n{}^2 = \pm 1$, then actually $p_n{}^2 - dq_n{}^2 = (-1)^{n+1}$, since p_n/q_n is less than or greater than ξ according as n is even or odd. Hence $s_n = 1$ and $\xi_{n+1} = \xi + r_n$. Taking integral parts, we get $a_{n+1} = a_0 + r_n$. Consequently

$$\xi_{n+2}{}^{-1} = \xi_{n+1} - a_{n+1} = \xi - a_0 = \xi_1{}^{-1}.$$

Thus $\xi_{n+2} = \xi_1$, which implies that $n = kh - 1$ for some positive integer k.

On the other hand, if $n = kh - 1$ for some positive integer k, then $\xi_{n+2} = \xi_1$ and hence

$$\xi_{n+1} - a_{n+1} = \xi - a_0.$$

Thus $\xi_{n+1} = \xi + a_{n+1} - a_0$, which implies that $s_n = 1$, since ξ is irrational. $\square$

It follows from Proposition 10 that, if d is a positive integer which is not a perfect square, then the equation (8) always has a solution in positive integers and all such solutions are given by

$$x = p_{kh-1}, \ y = q_{kh-1} \quad (k = 1,2,...) \quad \text{if } h \text{ is even},$$
$$x = p_{2kh-1}, \ y = q_{2kh-1} \quad (k = 1,2,...) \quad \text{if } h \text{ is odd}.$$

The least solution in positive integers, obtained by taking $k = 1$, is called the *fundamental solution* of (8).

On the other hand, the equation (9) has a solution in positive integers if and only if h is odd and all such solutions are then given by

$$x = p_{kh-1}, \ y = q_{kh-1} \quad (k = 1,3,5,...).$$

The least solution in positive integers, obtained by taking $k = 1$, is called the *fundamental solution* of (9).

To illustrate these results, suppose $d = a^2 + 1$ for some $a \in \mathbb{N}$. Since $\sqrt{d} = [a,\overline{2a}]$, the equation $x^2 - dy^2 = -1$ has the fundamental solution $x = a, y = 1$ and the equation $x^2 - dy^2 = 1$ has the fundamental solution $x = 2a^2 + 1, y = 2a$. Again, suppose $d = a^2 + a$ for some $a \in \mathbb{N}$. Since $\sqrt{d} = [a,\overline{2,2a}]$, the equation $x^2 - dy^2 = -1$ is insoluble, but the equation $x^2 - dy^2 = 1$ has the fundamental solution $x = 2a + 1, y = 2$.

It is not difficult to obtain upper bounds for the fundamental solutions. Since $\xi = \sqrt{d}$ is a root of the polynomial $t^2 - d$ and since its complete quotients ξ_n are reduced for $n \geq 1$, they have the form

$$\xi_n = (-B_n + \sqrt{D})/2A_n,$$

where $D = 4d, 0 < -B_n < \sqrt{D}$ and $A_n \geq 1$. Hence $a_0 = \lfloor \xi \rfloor < \sqrt{d}$ and $a_n = \lfloor \xi_n \rfloor < 2\sqrt{d}$ for $n \geq 1$. If we put $\alpha = \lfloor 2\sqrt{d} \rfloor$, it is easily shown by induction that

$$p_n \leq (\alpha + \alpha^{-1})^{n+1}/2, \ q_n \leq (\alpha + \alpha^{-1})^n \quad (n \geq 0).$$

These inequalities may now be combined with any upper bound for the period h (cf. §3).

Under composition, the fundamental solution of (8) generates an infinite cyclic group $\mathfrak{C}$ of solutions of (8). Furthermore, by composing the fundamental solution of (9) with any element of $\mathfrak{C}$ we obtain infinitely many solutions of (9). We are going to show that, by composing also with the trivial solution $-1,0$ of (8), all integral solutions of (8) and (9) are obtained in this way.

This can be proved by means of continued fractions, but the following argument due to Nagell (1950) provides additional information.

PROPOSITION 11 *Let d be a positive integer which is not a perfect square, let m be a positive integer, and let x_0,y_0 be the fundamental solution of the equation* (8).

If the equation

$$u^2 - dv^2 = m \tag{10}$$

has an integral solution, then it has one for which $u^2 \leq m(x_0 + 1)/2$, $dv^2 \leq m(x_0 - 1)/2$.

Similarly, if the equation

$$u^2 - dv^2 = -m \tag{11}$$

has an integral solution, then it has one for which $u^2 \leq m(x_0 - 1)/2$, $dv^2 \leq m(x_0 + 1)/2$.

Proof By composing a given solution of (10) with any solution in the subgroup $\mathfrak{C}$ of solutions of (8) which is generated by the solution x_0,y_0 we obtain again a solution of (10). Let u_0,v_0 be the solution of (10) obtained in this way for which v_0 has its least non-negative value. Then $u_0^2 = m + dv_0^2$ also has its least value and by changing the sign of u_0 we may suppose $u_0 > 0$. By composing the solution u_0,v_0 of (10) with the inverse of the fundamental solution x_0,y_0 of (8) we obtain the solution

$$u = x_0 u_0 - dy_0 v_0, \quad v = x_0 v_0 - y_0 u_0$$

of (10). Since

$$u = x_0 u_0 - dy_0 v_0 = x_0 u_0 - [(x_0^2 - 1)(u_0^2 - m)]^{1/2} > 0,$$

we must have

$$x_0 u_0 - dy_0 v_0 \geq u_0.$$

Hence

$$(x_0 - 1)^2 u_0^2 \geq d^2 y_0^2 v_0^2 = (x_0^2 - 1)(u_0^2 - m).$$

Thus

$$(x_0 - 1)/(x_0 + 1) \geq 1 - m/u_0^2,$$

which implies $u_0^2 \leq m(x_0 + 1)/2$ and hence $dv_0^2 \leq m(x_0 - 1)/2$.

For the equation (11) we begin in the same way. Then from

$$(x_0 v_0)^2 = (y_0^2 + 1/d)(u_0^2 + m) > y_0^2 u_0^2$$

we obtain $v = x_0 v_0 - y_0 u_0 > 0$ and hence $x_0 v_0 - y_0 u_0 \geq v_0$. Thus

$$d(x_0 - 1)^2 v_0^2 \geq dy_0^2 u_0^2$$

and hence

$$(x_0 - 1)^2(u_0^2 + m) \geq (x_0^2 - 1)u_0^2.$$

The argument can now be completed in the same way as before. $\square$

The proof of Proposition 11 shows that if (10), or (11), has an integral solution, then we obtain all solutions by finding the *finitely many* solutions u,v which satisfy the inequalities in the statement of Proposition 11 and composing them with all solutions in $\mathfrak{C}$ of (8).

The only solutions x,y of (8) for which $x^2 \leq (x_0 + 1)/2$ are the trivial ones $x = \pm 1$, $y = 0$. Hence any solution of (8) is in $\mathfrak{C}$ or is obtained by reversing the signs of a solution in $\mathfrak{C}$.

If u,v is a positive solution of (9) such that $u^2 \leq (x_0 - 1)/2$, $dv^2 \leq (x_0 + 1)/2$, then $x = u^2 + dv^2$, $y = 2uv$ is a positive solution of (8) such that $x \leq x_0$. Hence $(x,y) = (x_0,y_0)$ is the fundamental solution of (8) and $u^2 = (x_0 - 1)/2$, $dv^2 = (x_0 + 1)/2$. Thus (u,v) is uniquely determined and is the fundamental solution of (9). Hence, if (9) has a solution, any solution is obtained by composing the fundamental solution of (9) with an element of $\mathfrak{C}$ or by reversing the signs of such a solution.

A necessary condition for the solubility in integers of the equation (9) is that d may be represented as a sum of two squares. For the period h of the continued fraction expansion $\xi = \sqrt{d} = [a_0, \overline{a_1, \ldots, a_h}]$ must be odd, say $h = 2m + 1$. It follows from Proposition 9 that

$$\xi_{m+1} = [\overline{a_m, \ldots, a_1, 2a_0, a_1, \ldots, a_m}],$$

and then from Proposition 7 that $\xi_{m+1} = -1/\xi_{m+1}'$. But, by the proof of Proposition 10,

$$s_m\xi_{m+1} = \xi + r_m,$$

where s_m and r_m are integers. Hence

$$-1 = \xi_{m+1}\xi_{m+1}' = (\xi + r_m)(-\xi + r_m)/s_m^2 = (r_m^2 - d)/s_m^2,$$

and thus $d = r_m^2 + s_m^2$. The formulas for s_m and r_m show that, if p_n/q_n are the convergents of $\sqrt{d}$, then $d = x^2 + y^2$ with

$$x = p_{m-1}p_m - dq_{m-1}q_m, \quad y = p_m^2 - dq_m^2.$$

Unfortunately, the equation (9) may be insoluble, even though d is a sum of two squares. As an example, take $d = 34 = 5^2 + 3^2$. It is easily verified that the fundamental solution of the equation $x^2 - 34y^2 = 1$ is $x_0 = 35$, $y_0 = 6$. If the equation $u^2 - 34v^2 = -1$ were soluble in integers then, by Proposition 11, it would have a solution u,v such that $34v^2 \leq 18$, which is clearly impossible.

As already observed, the equation (9) has no integral solutions if $d \equiv 3 \bmod 4$. It will now be shown that (9) does have integral solutions if $d = p$ is prime and $p \equiv 1 \bmod 4$. For let x,y be the fundamental solution of the equation (8). Since any square is congruent to 0 or 1 mod 4, we must have $y^2 \equiv 0$ and $x^2 \equiv 1$. Thus $y = 2z$ for some positive integer z and

$$(x - 1)(x + 1) = 4pz^2.$$

Since x is odd, $x - 1$ and $x + 1$ have greatest common divisor 2. It follows that there exist positive integers u,v such that

$$\text{either} \quad x - 1 = 2pu^2, x + 1 = 2v^2 \quad \text{or} \quad x - 1 = 2u^2, x + 1 = 2pv^2.$$

In the first case $v^2 - pu^2 = 1$, which contradicts the choice of x,y as the fundamental solution of (8), since $v < x$. Thus only the second case is possible and then $u^2 - pv^2 = -1$. (In fact, u,v is the fundamental solution of (9).)

This proves again that *any prime $p \equiv 1$ mod 4 may be represented as a sum of two squares,* and moreover shows that an explicit construction for this representation is provided by the continued fraction expansion of $\sqrt{p}$.

The representation of a prime $p \equiv 1$ mod 4 in the form $x^2 + y^2$ is actually unique, apart from interchanging x and y and changing their signs. For suppose

$$x^2 + y^2 \ = \ p \ = \ u^2 + v^2,$$

where x,y,u,v are all positive integers. Then

$$y^2u^2 - x^2v^2 \ = \ (p - x^2)u^2 - x^2(p - u^2) \ = \ p(u^2 - x^2).$$

Hence $yu \equiv \varepsilon xv \bmod p$, where $\varepsilon = \pm 1$. On the other hand,

$$p^2 \ = \ (x^2 + y^2)(u^2 + v^2) \ = \ (xu + \varepsilon yv)^2 + (xv - \varepsilon yu)^2.$$

Since the second term on the right is divisible by p^2, we must have $xv = \varepsilon yu$ or $xu = -\varepsilon yv$. Evidently $\varepsilon = 1$ in the first case and $\varepsilon = -1$ in the second case. Since $(x,y) = (u,v) = 1$, it follows that either $x = u$, $y = v$ or $x = v$, $y = u$.

The equation $x^2 - dy^2 = 1$, where d is a positive integer which is not a perfect square, is generally known as *Pell's equation,* following an erroneous attribution of Euler. The problem of finding its integral solutions was issued as a challenge by Fermat (1657). In the same year Brouncker and Wallis gave a method of solution which is essentially the same as the solution by continued fractions. The first proof that a solution always exists was given by Lagrange (1768).

Unknown to them all, the problem had been considered centuries earlier by Hindu mathematicians. Special cases of Pell's equation were solved by Brahmagupta (628) and a general method of solution, which was described by Bhascara II (1150), was known to Jayadeva at least a century earlier. No proofs were given, but their method is a modification of the solution by continued fractions and is often faster in practice. Bhascara found the fundamental solution of the equation $x^2 - 61y^2 = 1$, namely

$$x = 1766319049, \quad y = 226153980,$$

a remarkable achievement for the era.

5 The modular group

We recall that a complex number w is said to be *equivalent* to a complex number z if there exist integers a,b,c,d with $ad - bc = \pm 1$ such that

$$w = (az + b)/(cz + d).$$

Since we can write

$$w = (az + b)(c\bar{z} + d)/|cz + d|^2,$$

the imaginary parts are related by

$$\Im w = (ad - bc)\Im z/|cz + d|^2.$$

Consequently $\Im w$ and $\Im z$ have the same sign if $ad - bc = 1$ and opposite signs if $ad - bc = -1$. Since the map $z \to -z$ interchanges the upper and lower half-planes, we may restrict attention to z's in the *upper half-plane* $\mathfrak{H} = \{z \in \mathbb{C}: \Im z > 0\}$ and to w's which are *properly equivalent* to them, i.e. with $ad - bc = 1$.

A *modular transformation* is a map $f: \mathfrak{H} \to \mathfrak{H}$ of the form

$$f(z) = (az + b)/(cz + d),$$

where $a,b,c,d \in \mathbb{Z}$ and $ad - bc = 1$. Such a map is bijective and its inverse is again a modular transformation:

$$f^{-1}(z) = (dz - b)/(-cz + a).$$

Furthermore, if

$$g(z) = (a'z + b')/(c'z + d')$$

is another modular transformation, then the composite map $h = g \circ f$ is again a modular transformation:

$$h(z) = (a''z + b'')/(c''z + d''),$$

where

$$a'' = a'a + b'c, \quad b'' = a'b + b'd,$$
$$c'' = c'a + d'c, \quad d'' = c'b + d'd,$$

and hence

$$a''d'' - b''c'' = (a'd' - b'c')(ad - bc) = 1.$$

It follows that the set Γ of all modular transformations is a group. Moreover, composition of modular transformations corresponds to multiplication of the corresponding matrices:

$$\begin{pmatrix} a'' & b'' \\ c'' & d'' \end{pmatrix} = \begin{pmatrix} a' & b' \\ c' & d' \end{pmatrix} \begin{pmatrix} a & b \\ c & d \end{pmatrix}.$$

However, the same modular transformation is obtained if the signs of a,b,c,d are all changed (and in no other way). It follows that the *modular group* Γ is isomorphic to the factor group $SL(2,\mathbb{Z})/\{\pm I\}$ of the special linear group $SL(2,\mathbb{Z})$ of all 2×2 integer matrices with determinant 1 by its centre $\{\pm I\}$.

PROPOSITION 12 *The modular group* Γ *is generated by the transformations*

$$T(z) = z + 1, \quad S(z) = -1/z.$$

Proof It is evident that $S,T \in \Gamma$ and $S^2 = I$ is the identity transformation. Any $g \in \Gamma$ has the form

$$g(z) = (az + b)/(cz + d),$$

where $a,b,c,d \in \mathbb{Z}$ and $ad - bc = 1$. If $c = 0$, then $a = d = \pm 1$ and $g = T^m$, where $m = b/d \in \mathbb{Z}$. Similarly if $a = 0$, then $b = -c = \pm 1$ and $g = ST^m$, where $m = d/c \in \mathbb{Z}$. Suppose now that $ac \neq 0$. For any $n \in \mathbb{Z}$ we have

$$ST^{-n}g(z) = (a'z + b')/(c'z + d'),$$

where $a' = -c$, $b' = -d$, $c' = a - nc$ and $d' = b - nd$. We can choose $n = m_1$ so that for $g_1 = ST^{-m_1}g$ we have $|c'| < |a|$ and hence $|a'| + |c'| < |a| + |c|$. If $a'c' \neq 0$, the argument can be repeated with g_1 in place of g. After finitely many repetitions we must obtain

$$ST^{-m_k} \cdots ST^{-m_1}g = T^m \text{ or } ST^m.$$

Since $S^{-1} = S$ and $(T^n)^{-1} = T^{-n}$, it follows that

$$g = T^{m_1}S \cdots T^{m_k}ST^m \quad \text{or} \quad g = T^{m_1}S \cdots T^{m_k}T^m. \quad \square$$

The proof of Proposition 12 may be regarded as an analogue of the continued fraction algorithm, since

$$T^{m_1}S \cdots T^{m_k}ST^m \, z \;=\; m_1 - \cfrac{1}{m_2 - \cfrac{\ddots}{\quad - \cfrac{1}{m_k - \cfrac{1}{m+z}}}} \,.$$

Obviously Γ is also generated by S and $R := ST$. The transformation R has order 3, since

$$R(z) = -1/(z+1), \quad R^2(z) = -(z+1)/z, \quad R^3(z) = z.$$

We are going to show that all other relations between the generators S and R are consequences of the relations $S^2 = R^3 = I$, so that Γ is the *free product* of a cyclic group of order 2 and a cyclic group of order 3.

Partition the upper half-plane $\mathfrak{H}$ by putting

$$A = \{z \in \mathfrak{H}: \Re z < 0\}, \quad B = \{z \in \mathfrak{H}: \Re z \geq 0\}.$$

It is easily verified that

$$SA \subset B, \quad RB \subset A, \quad R^2B \subset A$$

(where the inclusions are strict). If $g' = SR^{\varepsilon_1}SR^{\varepsilon_2} \ldots SR^{\varepsilon_n}$ for some $n \geq 1$, where $\varepsilon_j \in \{1,2\}$, it follows that $g'B \subset B$ and $g'SA \subset B$. Similarly, if $g'' = R^{\varepsilon_1}S \ldots R^{\varepsilon_n}$, then $g''B \subset A$ and $g''SA \subset A$. By taking account of the relations $S^2 = R^3 = I$, every $g \in \Gamma$ can be written in one of the forms

$$I, S, g', g'', g'S, g''S.$$

But, by what has just been said, no element except the first is the identity transformation.

The modular group is *discrete*, since there exists a neighbourhood of the identity transformation which contains no other element of Γ.

PROPOSITION 13 *The open set*

$$F = \{z \in \mathfrak{H}: -1/2 < \Re z < 1/2, |z| > 1\}$$

(see Figure 1) *is a fundamental domain for the modular group Γ, i.e. distinct points of F are not equivalent and each point of $\mathfrak{H}$ is equivalent to some point of F or its boundary ∂F.*

Proof For any $z \in \mathbb{C}$ we write $z = x + iy$, where $x,y \in \mathbb{R}$. We show first that no two points of F are equivalent. Assume on the contrary that there exist distinct points $z,z' \in F$ with $y' \geq y$ such that

$$z' = (az + b)/(cz + d)$$

for some $a,b,c,d \in \mathbb{Z}$ with $ad - bc = 1$. If $c = 0$, then $a = d = \pm 1$, $b \neq 0$ and $z' = z + b/d$, which is impossible for $z,z' \in F$. Hence $c \neq 0$. Since

$$y' = y/|cz + d|^2,$$

we have $|cz + d| \leq 1$. Thus $|z + d/c| \leq 1/|c|$, which is impossible not only if $|c| \geq 2$ but also if $c = \pm 1$.

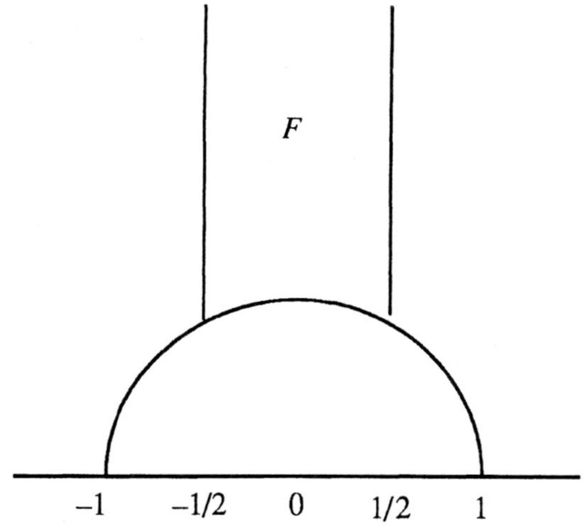

Figure 1: Fundamental domain for Γ

We show next that any $z_0 \in \mathfrak{H}$ is equivalent to a point of the *closure* $\overline{F} = F \cup \partial F$. We can choose $m_0 \in \mathbb{Z}$ so that $z_1 = z_0 + m_0$ satisfies $|x_1| \leq 1/2$. If $|z_1| \geq 1$, there is nothing more to do. Thus we now suppose $|z_1| < 1$. Put $z_2 = -1/z_1$. Then

$$y_2 = y_1/|z_1|^2 > y_1$$

and actually $y_2 \geq 2y_1$ if $y_1 \leq 1/2$, since then $|z_1|^2 \leq 1/4 + 1/4 = 1/2$. We now repeat the process, with z_2 in place of z_0, and choose $m_2 \in \mathbb{Z}$ so that $z_3 = z_2 + m_2$ satisfies $|x_3| \leq 1/2$. From $z_3 = (m_2 z_1 - 1)/z_1$ we obtain

$$|z_3|^2 = \{(m_2 x_1 - 1)^2 + (m_2 y_1)^2\}/(x_1^2 + y_1^2).$$

Assume $|z_3| < 1$. Then $m_2 \neq 0$ and also $m_2 \neq \pm 1$, since $|1 \pm x_1| \geq 1/2 \geq |x_1|$. If $|m_2| \geq 2$, then $|z_3|^2 \geq 4|y_1|^2$ and hence $y_1 < 1/2$. Thus in passing from z_1 to z_3 we obtain either $z_3 \in \overline{F}$ or $y_3 = y_2 \geq 2y_1$. Hence, after repeating the process finitely many times we must obtain a point $z_{2k+1} \in \overline{F}$. $\square$

Proposition 13 implies that the sets $\{g(\overline{F}): g \in \Gamma\}$ form a *tiling* of $\mathfrak{H}$, since

$$\mathfrak{H} = \bigcup\nolimits_{g \in \Gamma} g(\overline{F}), \quad g(F) \cap g'(F) = \varnothing \text{ if } g,g' \in \Gamma \text{ and } g \neq g'.$$

This is illustrated in Figure 2, where the domain $g(F)$ is represented simply by the group element g.

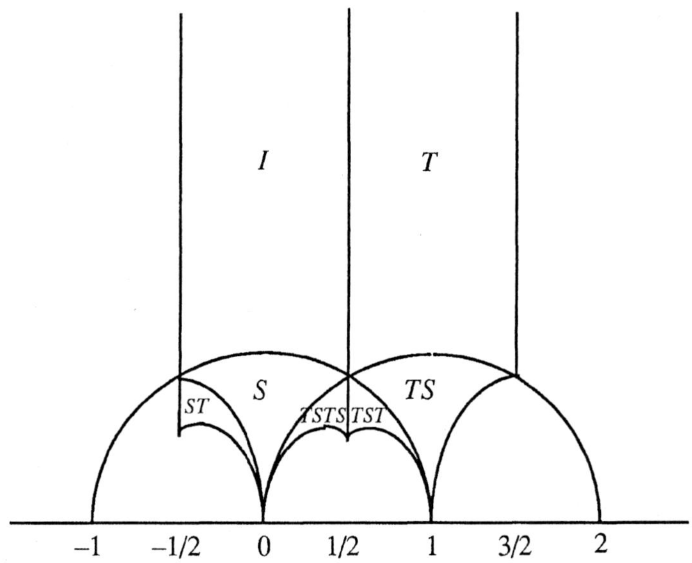

Figure 2: Tiling of $\mathfrak{H}$ by Γ

There is an interesting connection between the modular group and binary quadratic forms. The *discriminant* of a binary quadratic form

$$f = ax^2 + bxy + cy^2$$

with coefficients $a,b,c \in \mathbb{R}$ is $D := b^2 - 4ac$. The quadratic form is *indefinite* (i.e. assumes both positive and negative values) if and only if $D > 0$, and *positive definite* (i.e. assumes only positive values unless $x = y = 0$) if and only if $D < 0$, $a > 0$, which implies also $c > 0$. (If $D = 0$, the quadratic form is proportional to the square of a linear form.)

If we make a linear change of variables

$$x = \alpha x' + \beta y', \quad y = \gamma x' + \delta y',$$

where $\alpha, \beta, \gamma, \delta \in \mathbb{Z}$ and $\alpha\delta - \beta\gamma = 1$, the quadratic form f is transformed into the quadratic form

$$f' = a'x'^2 + b'x'y' + c'y'^2,$$

where

$$a' = a\alpha^2 + b\alpha\gamma + c\gamma^2,$$
$$b' = 2a\alpha\beta + b(\alpha\delta + \beta\gamma) + 2c\gamma\delta,$$
$$c' = a\beta^2 + b\beta\delta + c\delta^2,$$

and hence

$$b'^2 - 4a'c' = b^2 - 4ac = D.$$

The quadratic forms f and f' are said to be *properly equivalent*.

Thus properly equivalent forms have the same discriminant. As the name implies, proper equivalence is indeed an equivalence relation. Moreover, any form properly equivalent to an indefinite form is again indefinite, and any form properly equivalent to a positive definite form is again positive definite.

We will now show that any binary quadratic form is properly equivalent to one which is in some sense canonical. The indefinite and positive definite cases will be treated separately.

Suppose first that f is positive definite, so that $D < 0$, $a > 0$ and $c > 0$. With the quadratic form f we associate a point $\tau(f)$ of the upper half-plane $\mathfrak{H}$, namely

$$\tau(f) = (-b + i\sqrt{-D})/2a.$$

Thus $\tau(f)$ is the root with positive imaginary part of the polynomial $at^2 + bt + c$. Conversely, for any given $D < 0$ and $\tau \in \mathfrak{H}$, there is a unique positive definite quadratic form f with discriminant D such that $\tau(f) = \tau$. In fact, if $\tau = \xi + i\eta$, where $\xi, \eta \in \mathbb{R}$ and $\eta > 0$, we must take

$$a = \sqrt{(-D)}/2\eta, \quad b = -2a\xi, \quad c = (b^2 - D)/4a.$$

Let f', as above, be a form properly equivalent to f. If $t = (\alpha t' + \beta)/(\gamma t' + \delta)$, then

$$at^2 + bt + c = (a't'^2 + b't' + c')/(\gamma t' + \delta)^2.$$

It follows that if $\tau = \tau(f)$ and $\tau' = \tau(f')$, then $\tau = (\alpha \tau' + \beta)/(\gamma \tau' + \delta)$. Thus τ' is properly equivalent to τ, in the terminology introduced in Section 1.

By Proposition 13 we may choose the change of variables so that $\tau' \in \overline{F}$, i.e.

$$- 1/2 \leq \Re\tau' \leq 1/2, \ |\tau'| \geq 1.$$

It is easily verified that this is the case if and only if for f' we have

$$|b'| \leq a', \ 0 < a' \leq c'.$$

Such a quadratic form f' is said to be *reduced*. Thus every positive definite binary quadratic form is properly equivalent to a reduced form. (It is possible to ensure that every positive definite binary quadratic form is properly equivalent to a unique reduced form by slightly restricting the definition of 'reduced', but we will have no need of this.)

If the coefficients of f are integers, then so also are the coefficients of f' and τ,τ' are complex quadratic irrationals. There are only finitely many reduced forms f with integer coefficients and with a given discriminant $D < 0$. For, if f is reduced, then

$$4b^2 \ \leq \ 4a^2 \ \leq \ 4ac \ = \ b^2 - D$$

and hence $b^2 \leq -D/3$. Since $4ac = b^2 - D$, for each of the finitely many possible values of b there are only finitely many possible values for a and c.

A quadratic form $f = ax^2 + bxy + cy^2$ is said to be *primitive* if the coefficients a,b,c are integers with greatest common divisor 1. For any integer $D < 0$, let $h^\dagger(D)$ denote the number of primitive positive definite quadratic forms with discriminant D which are properly inequivalent. By what has been said, $h^\dagger(D)$ is finite.

Consider next the indefinite case:

$$f \ = \ ax^2 + bxy + cy^2$$

where $a,b,c \in \mathbb{R}$ and $D > 0$. If $a \neq 0$, we can write

$$f \ = \ a(x - \xi y)(x - \eta y),$$

where ξ,η are the distinct real roots of the polynomial $at^2 + bt + c$. It follows from Lemma 0 that, if ξ and η are irrational, then f is properly equivalent to a form f' for which $\xi' > 1$ and $-1 < \eta' < 0$. Such a quadratic form f' is said to be *reduced*. Evidently f' is reduced if and only if $-f'$ is reduced. Thus we may suppose $a' > 0$, and then f' is reduced if and only if

$$0 < \sqrt{D} + b' < 2a' < \sqrt{D} - b'.$$

If the coefficients of f are integers and the positive integer D is not a square, then $a \neq 0$ and ξ,η are conjugate real quadratic irrationals. In this case, as we already saw in Section 3, there are only finitely many reduced forms with discriminant D. For any integer $D > 0$ which is not a square, let $h^{\dagger}(D)$ denote the number of primitive quadratic forms with discriminant D which are properly inequivalent. By what has been said, $h^{\dagger}(D)$ is finite.

It should be noted that, for any quadratic form f with integer coefficients, the discriminant $D \equiv 0$ or $1 \bmod 4$. Moreover, for any $D \equiv 0$ or $1 \bmod 4$, there is a quadratic form f with integer coefficients and with discriminant D; for example,

$$f = x^2 - Dy^2/4 \qquad \text{if } D \equiv 0 \bmod 4,$$
$$f = x^2 + xy + (1-D)y^2/4 \quad \text{if } D \equiv 1 \bmod 4.$$

The preceding results for quadratic forms can also be restated in terms of quadratic fields. By making correspond to the ideal with basis $\beta = a$, $\gamma = b + c\omega$ in the quadratic field $\mathbb{Q}(\sqrt{d})$ the binary quadratic form

$$\{\beta\beta'x^2 + (\beta\gamma' + \beta'\gamma)xy + \gamma\gamma'y^2\}/ac,$$

one can establish a bijective map between 'strict' equivalence classes of ideals in $\mathbb{Q}(\sqrt{d})$ and proper equivalence classes of binary quadratic forms with discriminant D, where

$$D = 4d \quad \text{if } d \equiv 2 \text{ or } 3 \bmod 4,$$
$$D = d \quad \text{if } d \equiv 1 \bmod 4.$$

(The middle coefficient b of $f = ax^2 + bxy + cy^2$ was not required to be even in order to obtain this one-to-one correspondence.) Since any ideal class is either a strict ideal class or the union of two strict ideal classes, the finiteness of the class number $h(d)$ of the quadratic field $\mathbb{Q}(\sqrt{d})$ thus follows from the finiteness of $h^{\dagger}(D)$.

6 Non-Euclidean geometry

There is an important connection between the modular group and the non-Euclidean geometry of Bolyai (1832) and Lobachevski (1829). It was first pointed out by Beltrami (1868) that their *hyperbolic geometry* is the geometry on a manifold of constant curvature. In the model of Poincaré (1882) for two-dimensional hyperbolic geometry the underlying space is taken to be the upper half-plane $\mathfrak{H}$. A 'line' is either a semi-circle with centre on the real axis or a half-line perpendicular to the real axis. It follows that through any two distinct points there passes exactly one 'line'. However, through a given point not on a given 'line' there passes more than one 'line' having no point in common with the given 'line'.

Although Euclid's parallel axiom fails to hold, all the other axioms of Euclidean geometry are satisfied. Poincaré's model shows that if Euclidean geometry is free from contradiction, then so also is hyperbolic geometry. Before the advent of non-Euclidean geometry there had been absolute faith in Euclidean geometry. It is realized today that it is a matter for experiment to determine what kind of geometry best describes our physical world.

Poincaré's model will now be examined in more detail (with the constant curvature normalized to have the value -1). A curve γ in $\mathfrak{H}$ is specified by a continuously differentiable function $z(t) = x(t) + iy(t)$ ($a \leq t \leq b$). The (hyperbolic) *length* of γ is defined to be

$$\ell(\gamma) \;=\; \int_a^b y(t)^{-1} \, |dz/dt| \, dt \,.$$

It follows from this definition that the 'line' segment joining two points z,w of $\mathfrak{H}$ has length

$$d(z,w) \;=\; \ln \frac{|z - \overline{w}| + |z - w|}{|z - \overline{w}| - |z - w|} \,.$$

It may be shown that any other curve joining z and w has greater length. Thus the 'lines' are *geodesics*.

For any $z_0 \in \mathfrak{H}$, there is a unique geodesic through z_0 in any specified direction. Also, for any distinct real numbers ξ,η, there is a unique geodesic which intersects the real axis at ξ,η, namely the semicircle with centre at $(\xi + \eta)/2$. (By abuse of language we say 'ξ', for example, when we mean the point $(\xi,0)$.)

A linear fractional transformation

$$z' = f(z) = (az + b)/(cz + d),$$

where $a,b,c,d \in \mathbb{R}$ and $ad - bc = 1$, maps the upper half-plane $\mathfrak{H}$ onto itself and maps 'lines' onto 'lines'. Moreover, if the curve γ is mapped onto the curve γ', then $\ell(\gamma) = \ell(\gamma')$, since $\mathcal{I}f(z) = \mathcal{I}z/|cz + d|^2$ and $df/dz = 1/|cz + d|^2$. In particular,

$$d(z,w) \;=\; d(z',w').$$

Thus a linear fractional transformation of the above form is an *isometry*. It may be shown that any isometry is either a linear fractional transformation of this form or is obtained by composing such a transformation with the (orientation-reversing) transformation $x + iy \rightarrow -x + iy$. For any two 'lines' L and L', there is an isometry which maps L onto L'.

We may define *angles* to be the same as in Euclidean geometry, since any linear fractional transformation is conformal. The (hyperbolic) *area* of a domain $D \subset \mathfrak{H}$, defined by

$$\mu(D) \;=\; \iint_D y^{-2} dx dy \,,$$

is invariant under any isometry. In particular, this gives $\pi - (\alpha + \beta + \gamma)$ for the area of a 'triangle' with angles α,β,γ. Since the angles are non-negative, the area of a 'triangle' is at most π and, since the area is necessarily positive, the sum of the angles of a 'triangle' is less than π.

For example, if F is the fundamental domain of the modular group Γ, then $\overline{F}$ is a 'triangle' with angles $\pi/3,\pi/3,0$ and hence the area of $\overline{F}$ is $\pi - 2\pi/3 = \pi/3$. For any fixed $z_0 \in F$ on the imaginary axis, we may characterize F as the set of all $z \in \mathfrak{H}$ such that, for every $g \in \Gamma$ with $g \neq I$,

$$d(z,z_0) < d(z,g(z_0)) = d(g^{-1}(z),z_0).$$

By identifying two points z,z' of $\mathfrak{H}$ if $z' = g(z)$ for some $g \in \Gamma$ we obtain the *quotient space* $\mathcal{M} = \mathfrak{H}/\Gamma$. Equivalently, we may regard $\mathcal{M}$ as the closure $\overline{F}$ of the fundamental domain F with the boundary point $-1/2 + iy$ identified with the boundary point $1/2 + iy$ ($1 \le y < \infty$) and the boundary point $-e^{-i\theta}$ identified with the boundary point $e^{i\theta}$ ($0 < \theta < \pi/2$).

Since the elements of Γ are isometries of $\mathfrak{H}$, the metric on $\mathfrak{H}$ induces a metric on $\mathcal{M}$ in which the geodesics are the projections onto $\mathcal{M}$ of the geodesics in $\mathfrak{H}$. Thus if we regard $\mathcal{M}$ as $\overline{F}$ with appropriate boundary points identified, then a geodesic in $\mathcal{M}$ will be a sequence of geodesic arcs in F, each with initial point and endpoint on the boundary of F, so that the initial point of one arc is the point identified to the endpoint of the preceding arc.

Let L be a geodesic in $\mathfrak{H}$ which intersects the real axis in irrational points ξ,η such that $\xi > 1, -1 < \eta < 0$ and let

$$\xi = [a_0,a_1,a_2,\dots\,], \quad -1/\eta = [a_{-1},a_{-2},\dots\,]$$

be the continued fraction expansions of ξ and $-1/\eta$. If we choose ξ and $\eta = \xi'$ to be conjugate quadratic irrationals then, by Proposition 7, the doubly-infinite sequence

$$[\,\dots,a_{-2},a_{-1},a_0,a_1,a_2,\dots\,]$$

is periodic and it is not difficult to see that the geodesic in $\mathcal{M}$ obtained by projection from L is closed. Artin (1924) showed that there are other geodesics which behave very differently. Let the convergents of ξ be p_n/q_n and put

$$\xi = (p_{n-1}\xi_n + p_{n-2})/(q_{n-1}\xi_n + q_{n-2}), \quad \eta = (p_{n-1}\eta_n + p_{n-2})/(q_{n-1}\eta_n + q_{n-2}).$$

Then

$$\xi_n = [a_n,a_{n+1},\dots\,], \quad -1/\eta_n = [a_{n-1},a_{n-2},\dots\,],$$

and $\xi_n > 1, -1 < \eta_n < 0$. Moreover, if n is even, then ξ and η are properly equivalent to ξ_n and η_n respectively. If we choose ξ so that the sequence $a_0, a_1, a_2,...$ contains each finite sequence of positive integers (and hence contains it infinitely often), then the corresponding geodesic in $\mathcal{M}$ passes arbitrarily close to every point of $\mathcal{M}$ and to every direction at that point.

Some much studied subgroups of the modular group are the *congruence subgroups* $\Gamma(n)$, consisting of all linear fractional transformations $z \to (az + b)/(cz + d)$ in Γ congruent to the identity transformation, i.e.

$$a \equiv d \equiv \pm 1, \quad b \equiv c \equiv 0 \mod n.$$

We may in the same way investigate the geodesics in the *quotient space* $\mathfrak{H}/\Gamma(n)$. In the case $n = 3$ it has been shown by Lehner and Sheingorn (1984) that there is an interesting connection with the *Markov spectrum*.

In Section 2 we defined, for any irrational number ξ with convergents p_n/q_n,

$$M(\xi) = \overline{\lim}_{n \to \infty} q_n^{-1}|q_n\xi - p_n|^{-1},$$

and we noted that $M(\xi) = M(\eta)$ if ξ and η are equivalent. It is not difficult to show that there are uncountably many inequivalent ξ for which $M(\xi) = 3$. However, it was shown by Markov (1879/80) that there is a sequence of real quadratic irrationals $\xi^{(k)}$ such that $M(\xi) < 3$ if and only if ξ is equivalent to $\xi^{(k)}$ for some k. If $\mu_k = M(\xi^{(k)})$, then $\mu_1 < \mu_2 < \mu_3 < ...$ and $\mu_k \to 3$ as $k \to \infty$. Although μ_k is irrational, μ_k^2 is rational. The first few values are

$$\mu_1 = 5^{1/2} = 2.236... , \quad \mu_2 = 8^{1/2} = 2.828... ,$$
$$\mu_3 = (221)^{1/2}/5 = 2.973... , \quad \mu_4 = (1517)^{1/2}/13 = 2.996... .$$

As we already showed in Section 2, we can take $\xi^{(1)} = (1 + \sqrt{5})/2$ and $\xi^{(2)} = 1 + \sqrt{2}$.

Lehner and Sheingorn showed that the simple closed geodesics in $\mathfrak{H}/\Gamma(3)$ are just the projections of the geodesics in $\mathfrak{H}$ whose endpoints ξ,η on the real axis are conjugate quadratic irrationals equivalent to $\xi^{(k)}$ for some k.

There is a recursive procedure for calculating the quantities μ_k and $\xi^{(k)}$. A *Markov triple* is a triple (u,v,w) of positive integers such that

$$u^2 + v^2 + w^2 = 3uvw.$$

If (u,v,w) is a Markov triple, then so also are $(3uw - v,u,w)$ and $(3uv - w,u,v)$. They are distinct from the original triple if $u = \max(u,v,w)$, since then $u < 3uw - v$ and $u < 3uv - w$. They are also distinct from one another if $w < v$. Starting from the trivial triple $(1,1,1)$, all Markov triples can be obtained by repeated applications of this process. The successive values of $u = \max(u,v,w)$ are $1,2,5,13,29,\ldots$. The numbers μ_k and $\xi^{(k)}$ are the corresponding successive values of $(9 - 4/u^2)^{1/2}$ and $(9 - 4/u^2)^{1/2}/2 + 1/2 + v/uw$.

It was conjectured by Frobenius (1913) that a Markov triple is uniquely determined by its greatest element. This has been verified whenever the greatest element does not exceed 10^{140}. It has also been proved when the greatest element is a prime (and in some other cases) by Baragar (1996), using the theory of quadratic fields.

7 Complements

There is an important analogue of the continued fraction algorithm for infinite series. Let K be an arbitrary field and let F denote the set of all formal Laurent series

$$f = \sum_{n \in \mathbb{Z}} \alpha_n t^n$$

with coefficients $\alpha_n \in K$ such that $\alpha_n \neq 0$ for at most finitely many $n > 0$. If

$$g = \sum_{n \in \mathbb{Z}} \beta_n t^n$$

is also an element of F, and if we define addition and multiplication by

$$f + g = \sum_{n \in \mathbb{Z}} (\alpha_n + \beta_n) t^n, \quad fg = \sum_{n \in \mathbb{Z}} \gamma_n t^n,$$

where $\gamma_n = \sum_{j+k=n} \alpha_j \beta_k$, then F acquires the structure of a commutative ring. In fact, F is a field. For, if $f = \sum_{n \leq \nu} \alpha_n t^n$, where $\alpha_\nu \neq 0$, we obtain $g = \sum_{n \leq -\nu} \beta_n t^n$ such that $fg = 1$ by solving successively the equations

$$
\begin{aligned}
\alpha_\nu \beta_{-\nu} &= 1 \\
\alpha_\nu \beta_{-\nu-1} + \alpha_{\nu-1} \beta_{-\nu} &= 0 \\
\alpha_\nu \beta_{-\nu-2} + \alpha_{\nu-1} \beta_{-\nu-1} + \alpha_{\nu-2} \beta_{-\nu} &= 0
\end{aligned}
$$

$$\ldots\ldots$$

Define the absolute value of an element $f = \sum_{n \in \mathbb{Z}} \alpha_n t^n$ of F by putting

$$|O| = 0, \quad |f| = 2^{\nu(f)} \ \text{if} \ f \neq O,$$

where $\nu(f)$ is the greatest integer n such that $\alpha_n \neq 0$. It is easily verified that

$$|fg| = |f|\,|g|, \quad |f + g| \leq \max(|f|,|g|),$$

and $|f + g| = \max(|f|,|g|)$ if $|f| \neq |g|$.

For any $f = \sum_{n \in \mathbb{Z}} \alpha_n t^n \in F$, let

$$\lfloor f \rfloor = \sum_{n \geq 0} \alpha_n t^n, \quad \{f\} = \sum_{n < 0} \alpha_n t^n$$

denote respectively its polynomial and strictly proper parts. Then $|\{f\}| < 1$, and $|\lfloor f \rfloor| = |f|$ if $|f| \geq 1$, i.e. if $\lfloor f \rfloor \neq O$.

If $f_0 := f$ is not the formal Laurent series of a rational function, we can write

$$f_0 = a_0 + 1/f_1,$$

where $a_0 = \lfloor f_0 \rfloor$ and $|f_1| > 1$. In the same way,

$$f_1 = a_1 + 1/f_2,$$

where $a_1 = \lfloor f_1 \rfloor$ and $|f_2| > 1$. Continuing in this way, we obtain the *continued fraction expansion* $[a_0, a_1, a_2, \ldots]$ of f. In the same way as for real numbers, if we define polynomials p_n, q_n by the recurrence relations

$$p_n = a_n p_{n-1} + p_{n-2}, \quad q_n = a_n q_{n-1} + q_{n-2} \quad (n \geq 0),$$

with $p_{-2} = q_{-1} = 0$, $p_{-1} = q_{-2} = 1$, then

$$p_n q_{n-1} - p_{n-1} q_n = (-1)^{n+1} \quad (n \geq 0),$$

$$f = (p_n f_{n+1} + p_{n-1})/(q_n f_{n+1} + q_{n-1}) \quad (n \geq 0),$$

and so on. In addition, however, we now have

$$|a_n| = |f_n| > 1 \quad (n \geq 1),$$

from which we obtain by induction

$$|p_n| = |a_n|\,|p_{n-1}| > |p_{n-1}|, \quad |q_n| = |a_n|\,|q_{n-1}| > |q_{n-1}| \quad (n \geq 1).$$

Hence

$$|p_n| = |a_0 a_1 \ldots a_n|, \quad |q_n| = |a_1 \ldots a_n| \quad (n \geq 1).$$

From the relation $q_n f - p_n = (-1)^n/(q_n f_{n+1} + q_{n-1})$ we further obtain

$$|q_n f - p_n| = |q_{n+1}|^{-1},$$

since

$$|q_n f_{n+1} + q_{n-1}| = |q_n f_{n+1}| = |q_n|\,|a_{n+1}| = |q_{n+1}|.$$

In particular, $|q_n f - p_n| < 1$ and hence

$$p_n = \lfloor q_n f \rfloor, \quad |\{q_n f\}| = |q_{n+1}|^{-1} \quad (n \geq 1).$$

Thus p_n is readily determined from q_n. Furthermore,

$$|f - p_n/q_n| = |q_n|^{-1} |q_{n+1}|^{-1} \to 0 \text{ as } n \to \infty.$$

The rational function p_n/q_n is called the *n-th convergent* of f. The polynomials a_n are called the *partial quotients*, and the Laurent series f_n the *complete quotients*, in the continued fraction expansion of f.

The continued fraction algorithm can also be applied when f is the formal Laurent expansion of a rational function, but in this case the process terminates after a finite number of steps. If $a_0, a_1, a_2, \ldots$ is any finite or infinite sequence of polynomials with $|a_n| > 1$ for $n \geq 1$, there is a unique formal Laurent series f with $[a_0, a_1, a_2, \ldots]$ as its continued fraction expansion.

For formal Laurent series there are sharper Diophantine properties than for real numbers:

PROPOSITION 14 *Let f be a formal Laurent series with convergents p_n/q_n and let p,q be polynomials with $q \neq 0$.*

(i) *If $|q| < |q_{n+1}|$ and $p/q \neq p_n/q_n$, then*

$$|qf - p| \geq |q_{n-1}f - p_{n-1}| = |q_n|^{-1}.$$

(ii) *If $|qf - p| < |q|^{-1}$, then p/q is a convergent of f.*

Proof (i) Assume on the contrary that $|qf - p| < |q_n|^{-1}$. Since

$$q_n(qf - p) - q(q_n f - p_n) = qp_n - pq_n \neq 0$$

and $|q_n| \, |qf - p| < 1$, we must have

$$|q| \, |q_{n+1}|^{-1} = |q| \, |q_n f - p_n| = |qp_n - pq_n| \geq 1,$$

which is contrary to hypothesis.

(ii) Assume that p/q is not a convergent of f. If $f = p_N/q_N$ is a rational function then $|q| < |q_N|$, since

$$1 \leq |qp_N - pq_N| = |qf - p| \, |q_N| < |q|^{-1} \, |q_N|.$$

Thus, whether or not f is rational, we can choose n so that $|q_n| \leq |q| < |q_{n+1}|$. Hence, by (i),

$$|qf - p| \geq |q_n|^{-1} \geq |q|^{-1},$$

which is a contradiction. $\square$

It was shown by Abel (1826) that, for any complex polynomial $D(t)$ which is not a square, the 'Pell' equation $X^2 - D(t)Y^2 = 1$ has a solution in polynomials $X(t),Y(t)$ of positive degree if and only if $\sqrt{D(t)}$ may be represented as a periodic continued fraction: $\sqrt{D(t)} = [a_0,\overline{a_1,...,a_h}\,]$, where $a_h = 2a_0$ and $a_i = a_{h-i}$ ($i = 1,...,h - 1$) are polynomials of positive degree. By differentiation one obtains

$$XX'/Y = Y'D + (1/2)YD'.$$

It follows that Y divides X', since X and Y are relatively prime, and

$$(X + Y\sqrt{D})' = (X + Y\sqrt{D})X'/Y\sqrt{D}.$$

Thus the 'abelian' integral

$$\int X'(t)dt/Y(t)\sqrt{D(t)}$$

is actually the elementary function $\log\{X(t) + Y(t)\sqrt{D(t)}\}$.

Some remarkable results have recently been obtained on the approximation of algebraic numbers by rational numbers, which deserve to be mentioned here, even though the proofs are beyond our scope.

A complex number ζ is said to be an *algebraic number*, or simply *algebraic, of degree d* if it is a root of a polynomial of degree d with rational coefficients which is irreducible over the rational field $\mathbb{Q}$. Thus an algebraic number of degree 2 is just a quadratic irrational.

For any irrational number ξ, there exist infinitely many rational numbers p/q such that

$$|\xi - p/q| < 1/q^2,$$

since the inequality is satisfied by any convergent of ξ. It was shown by Roth (1955) that if ξ is a real algebraic number of degree $d \geq 2$ then, for any given $\varepsilon > 0$, there exist only finitely many rational numbers p/q with $q > 0$ such that

$$|\xi - p/q| < 1/q^{2+\varepsilon}.$$

The proof does not provide a bound for the magnitude of the rational numbers which satisfy the inequality, but it does provide a bound for their number. Roth's result was the culmination of a line of research that was begun by Thue (1909), and further developed by Siegel (1921) and Dyson (1947).

A sharpening of Roth's result has been *conjectured* by Lang (1965): if ξ is a real algebraic number of degree $d \geq 2$ then, for any given $\varepsilon > 0$, there exist only finitely many rational numbers p/q with $q > 1$ such that

$$|\xi - p/q| < 1/q^2(\log q)^{1+\varepsilon}.$$

An even stronger sharpening has been conjectured by P.M. Wong (1989) in which $(\log q)^{1+\varepsilon}$ is replaced by $(\log q)(\log \log q)^{1+\varepsilon}$ with $q > 2$.

For real algebraic numbers of degree 2 we already know more than this. For, if ξ is a real quadratic irrational, its partial quotients are bounded and so there exists a constant $c = c(\xi) > 0$ such that $|\xi - p/q| > c/q^2$ for every rational number p/q. It is a long-standing conjecture that this is false for any real algebraic number ξ of degree $d > 2$.

It is not difficult to show that Roth's theorem may be restated in the following homogeneous form: if

$$L_1(u,v) = \alpha u + \beta v, \quad L_2(u,v) = \gamma u + \delta v,$$

are linearly independent linear forms with algebraic coefficients $\alpha,\beta,\gamma,\delta$ then, for any given $\varepsilon > 0$, there exist at most finitely many integers x,y, not both zero, such that

$$|L_1(x,y)L_2(x,y)| < \max(|x|,|y|)^{-\varepsilon}.$$

The *subspace theorem* of W. Schmidt (1972) generalizes Roth's theorem in this form to higher dimensions. In the stronger form given it by Vojta (1989) it says: if $L_1(u),...,L_n(u)$ are linearly independent linear forms in n variables $u = (u_1,...,u_n)$ with (real or complex) algebraic coefficients, then there exist finitely many proper linear subspaces $V_1,...,V_h$ of $\mathbb{Q}^n$ such that every nonzero $x = (x_1,...,x_n) \in \mathbb{Z}^n$ for which

$$|L_1(x)...L_n(x)| < \|x\|^{-\varepsilon},$$

where $\|x\| = \max(|x_1|,...,|x_n|)$, is contained in some subspace V_i, except for finitely many points whose number may depend on ε. A new proof of Schmidt's subspace theorem has been given by Faltings and Wüstholz (1994). The subspace theorem has also been given a more quantitative form by Schmidt (1989) and Evertse (1996). These results have immediate applications to the simultaneous approximation of several algebraic numbers.

Vojta (1987) has developed a remarkable analogy between the approximation of algebraic numbers by rationals and the theory of Nevanlinna (1925) on the value distribution of meromorphic functions, in which Roth's theorem corresponds to Nevanlinna's second main theorem. Although the analogy is largely formal, it is suggestive in both directions. It has already led to new proofs for the theorems of Roth and Schmidt, and to a proof of the Mordell conjecture (discussed below) which is quite different from the original proof by Faltings.

Roth's theorem has an interesting application to Diophantine equations. Let

$$f(z) = a_0 z^n + a_1 z^{n-1} + ... + a_n$$

be a polynomial of degree $n \geq 3$ with integer coefficients whose roots are distinct and not rational. Let

$$f(u,v) \;=\; a_0 u^n + a_1 u^{n-1} v + \ldots + a_n v^n$$

be the corresponding homogeneous polynomial and let $g(u,v)$ be a polynomial of degree $m \geq 0$ with integer coefficients. We will deduce from Roth's theorem that the equation

$$f(x,y) \;=\; g(x,y)$$

has at most finitely many solutions in integers if $m \leq n - 3$. This was already proved by Thue for $m = 0$.

Assume on the contrary that there exist infinitely many solutions in integers. Without loss of generality we may assume that there exist infinitely many integer solutions x,y for which $|x| \leq |y|$. Then there exists a constant $c_1 > 0$ such that

$$|g(x,y)| \leq c_1 \, |y|^m.$$

Over the complex field $\mathbb{C}$ the homogeneous polynomial $f(u,v)$ has a factorization

$$f(u,v) \;=\; a_0 \, \prod_{j=1}^{n} (u - \zeta_j v),$$

where $\zeta_1,\ldots,\zeta_n$ are distinct algebraic numbers which are not rational. For at least one j we must have, for infinitely many x,y,

$$|a_0| \, |x - \zeta_j y|^n \;\leq\; c_1 \, |y|^m$$

and hence

$$|x - \zeta_j y| \;\leq\; c_2 \, |y|^{m/n},$$

where $c_2 = (c_1/|a_0|)^{1/n}$. If $k \neq j$, then

$$
\begin{aligned}
|x - \zeta_k y| &\geq |(\zeta_j - \zeta_k)y| - |x - \zeta_j y| \\
&\geq c_3 \, |y| - c_2 \, |y|^{m/n} \;\geq\; c_4 \, |y|,
\end{aligned}
$$

where c_3, c_4 are positive constants. It follows that

$$|a_0| \, |x - \zeta_j y| \, c_4^{\,n-1} \, |y|^{n-1} \;\leq\; |f(x,y)| \;=\; |g(x,y)| \;\leq\; c_1 \, |y|^m$$

and hence

$$|\zeta_j - x/y| \;\leq\; c_5 / |y|^{n-m},$$

where the positive constant c_5 depends only on the coefficients of f and g. Evidently this implies that ζ_j is real. Since ζ_j is not rational and $m \leq n - 3$, we now obtain a contradiction to Roth's theorem.

It is actually possible to characterize all polynomial Diophantine equations with infinitely many solutions. Let $F(x,y)$ be a polynomial with rational coefficients which is irreducible over $\mathbb{C}$. It was shown by Siegel (1929), by combining his own results on the approximation of algebraic numbers with results of Mordell and Weil concerning the rational points on elliptic curves and Jacobian varieties, that if the equation

$$F(x,y) = 0 \qquad\qquad (*)$$

has infinitely many integer solutions, then there exist polynomials or Laurent polynomials $\phi(t),\psi(t)$ (not both constant) with coefficients from either the rational field $\mathbb{Q}$ or a real quadratic field $\mathbb{Q}(\sqrt{d})$, where $d > 0$ is a square-free integer, such that $F(\phi(t),\psi(t))$ is identically zero. If $\phi(t),\psi(t)$ are Laurent polynomials with coefficients from $\mathbb{Q}(\sqrt{d})$, they may be chosen to be invariant when t is replaced by t^{-1} and the coefficients are replaced by their conjugates in $\mathbb{Q}(\sqrt{d})$.

This implies, in particular, that the algebraic curve defined by (*) may be transformed by a birational transformation with rational coefficients into either a linear equation $ax + by + c = 0$ or a Pellian equation $x^2 - dy^2 - m = 0$. It is not significant that the birational transformation has rational, rather than integral, coefficients since, by combining a result of Mahler (1934) with the *Mordell conjecture*, it may be seen that the same conclusions hold if the equation (*) has infinitely many solutions in rational numbers whose denominators involve only finitely many primes.

The conjecture of Mordell (1922) says that the equation (*) has at most finitely many *rational* solutions if the algebraic curve defined by (*) has genus $g > 1$. (The concept of *genus* will not be formally defined here, but we mention that the genus of an irreducible plane algebraic curve may be calculated by a procedure due to M. Noether.) The conjecture has now been proved by Faltings (1983), as will be mentioned in Chapter XIII. As mentioned also at the end of Chapter XIII, if the algebraic curve defined by (*) has genus 1, then explicit bounds may be obtained for the number of integral points. It was already shown by Hilbert and Hurwitz (1890) that the algebraic curve defined by (*) has genus 0 if and only if it is birationally equivalent over $\mathbb{Q}$ either to a line or to a conic. There then exist rational functions $\phi(t),\psi(t)$ (not both constant) with coefficients either from $\mathbb{Q}$ or from a quadratic extension of $\mathbb{Q}$ such that $F(\phi(t),\psi(t))$ is identically zero. The coefficients may be taken from $\mathbb{Q}$ if the curve has at least one non-singular rational point.

Thus in retrospect, and quite unfairly, Siegel's remarkable result may be seen as simply picking out those curves of genus 0 which have infinitely many integral points, a problem which had already been treated by Maillet (1919).

In this connection it may be mentioned that the formula for Pythagorean triples given in §5 of Chapter II may be derived from the parametrization of the unit circle $x^2 + y^2 = 1$ by the rational functions

$$x(t) = (1 - t^2)/(1 + t^2), \quad y(t) = 2t/(1 + t^2).$$

8 Further remarks

More extensive accounts of the theory of continued fractions are given in the books of Rockett and Szusz [45] and Perron [41]. Many historical references are given in Brezinski [12]. The first systematic account of the subject, which it is still a delight to read, was given in 1774 by Lagrange [32] in his additions to the French translation of Euler's *Algebra*.

The continued fraction algorithm is such a useful tool that there have been many attempts to generalize it to higher dimensions. Jacobi, in a paper published posthumously (1868), defined a continued fraction algorithm in $\mathbb{R}^2$. Perron (1907) extended his definition to $\mathbb{R}^n$ and proved that convergence holds in the following weak sense: for a given nonzero $x \in \mathbb{R}^n$, the Jacobi–Perron algorithm constructs recursively a sequence of bases $\mathcal{B}^k = \{b_1{}^k,...,b_n{}^k\}$ of $\mathbb{Z}^n$ such that, for each $j \in \{1,...,n\}$, the angle between the line Ob_j^k and the line Ox tends to zero as $k \to \infty$. More recently, other algorithms have been proposed for which convergence holds in the strong sense that, for each $j \in \{1,...,n\}$, the distance of b_j^k from the line Ox tends to zero as $k \to \infty$. See Brentjes [11], Ferguson [22], Just [28] and Lagarias [31].

Proposition 2 was first proved by Serret [51]. Proposition 3 was proved by Lagrange. The complete characterization of best approximations is proved in the book of Perron.

Lambert (1766) proved that π was irrational by using a continued fraction expansion for $\tan x$. For the continued fraction expansion of π, see Choong *et al.* [15]. Badly approximable numbers are thoroughly surveyed by Shallit [52].

The theory of Diophantine approximation is treated more comprehensively in the books of Koksma [30], Cassels [13] and Schmidt [47].

The estimate $O(\sqrt{D} \log D)$ for the period of the continued fraction expansion of a quadratic irrational with discriminant D is proved by elementary means in the book of Rockett and Szusz. Further references are given in Podsypanin [42].

The ancient Hindu method of solving Pell's equation is discussed in Selenius [49]. Tables for solving the Diophantine equation $x^2 - dy^2 = m$, where $m^2 < d$, are given in Patz [39]. Pell's equation plays a role in the negative solution of Hilbert's tenth problem, which asks for an algorithm to determine whether an arbitrary polynomial Diophantine equation is solvable in integers. See Davis *et al.* [18] and Jones and Matijasevic [26].

The continued fraction construction for the representation of a prime $p \equiv 1 \bmod 4$ as a sum of two squares is due to Legendre. Some other constructions are given in Chapter V of Davenport [17] and in Wagon [61]. A construction for the representation of any positive integer as a sum of four squares is given by Rousseau [46].

The modular group is the basic example of a *Fuchsian group*, i.e. a discrete subgroup of the group $PSL_2(\mathbb{R})$ of all linear fractional transformations $z \to (az + b)/(cz + d)$, where $a,b,c,d \in \mathbb{R}$ and $ad - bc = 1$. Fuchsian groups are studied from different points of view in the books of Katok [29], Beardon [7], Lehner [36], and Vinberg and Shvartsman [58].

The significance of Fuchsian groups stems in part from the uniformization theorem, which characterizes Riemann surfaces. A *Riemann surface* is a 1-dimensional complex manifold. Two Riemann surfaces are *conformally equivalent* if there is a bijective holomorphic map from one to the other. The *uniformization theorem*, first proved by Koebe and Poincaré independently in 1907, says that any Riemann surface is conformally equivalent to exactly one of the following:

(i) the complex plane $\mathbb{C}$,

(ii) the Riemann sphere $\mathbb{C} \cup \{\infty\}$,

(iii) the cylinder $\mathbb{C}/G$, where G is the cyclic group generated by the translation $z \to z + 1$,

(iv) a torus $\mathbb{C}/G$, where G is the abelian group generated by the translations $z \to z + 1$ and $z \to z + \tau$ for some $\tau \in \mathfrak{H}$ (the upper half-plane),

(v) a quotient space $\mathfrak{H}/G$, where G is a Fuchsian group which acts *freely* on $\mathfrak{H}$, i.e. if $z \in \mathfrak{H}$, $g \in G$ and $g \neq I$, then $g(z) \neq z$.

(It should be noted that, since the modular group does not act freely on $\mathfrak{H}$, the corresponding 'Riemann surface' is *ramified*.) For more information on the uniformization theorem, see Abikoff [1], Bers [9], Farkas and Kra [21], Jost [27], Beardon and Stephenson [8], and He and Schramm [24].

For the equivalence between quadratic fields and binary quadratic forms, see Zagier [63]. The class number $h(d)$ of the quadratic field $\mathbb{Q}(\sqrt{d})$ has been deeply investigated, originally by exploiting this equivalence. Dirichlet (1839) obtained an analytic formula for $h(d)$ with the aid of his theorem on primes in an arithmetic progression (which will be proved in Chapter X). A clearly motivated proof of Dirichlet's formula is given in Hasse [23], and there are some interesting observations on the formula in Stark [56].

It was conjectured by Gauss (1801), in the language of quadratic forms, that $h(d) \to \infty$ as $d \to -\infty$. This was first proved by Heilbronn (1934). Siegel (1935) showed that actually

$$\log h(d)/\log |d| \to 1/2 \quad \text{as} \quad d \to -\infty.$$

Generalizations of these results to arbitrary algebraic number fields are given in books on algebraic number theory, e.g. Narkiewicz [38].

Siegel (1943) has given a natural generalization of the modular group to higher dimensions. Instead of the upper half-plane $\mathfrak{H}$, we consider the space $\mathfrak{H}_n$ of all complex $n{\times}n$ matrices $Z = X + iY$, where X,Y are real symmetric matrices and Y is positive definite. If the real $2n{\times}2n$ matrix

$$M = \begin{pmatrix} A & B \\ C & D \end{pmatrix}$$

is *symplectic*, i.e. if $M^t J M = J$, where

$$J = \begin{pmatrix} O & I \\ -I & O \end{pmatrix},$$

then the linear fractional transformation $Z \to (AZ + B)(CZ + D)^{-1}$, maps $\mathfrak{H}_n$ onto itself. Siegel's modular group Γ_n is the group of all such transformations. The generalized upper half-plane $\mathfrak{H}_n$ is itself just a special case of the vast theory of symmetric Riemannian spaces initiated by E. Cartan (1926/7). See Siegel [54] and Helgason [25].

The development of non-Euclidean geometry is traced in Bonola [10]. (This edition also contains translations of works by Bolyai and Lobachevski.) The basic properties of Poincaré's model, here only stated, are proved in the books of Katok [29] and Beardon [7].

For the connection between continued fractions and geodesics, see Artin [5] and Sheingorn [53]. For the Markov spectrum see not only the books of Cassels [13] and Rockett and Szusz [45], but also Cusick and Flahive [16] and Baragar [6].

The theory of continued fractions for formal Laurent series is developed further in de Mathan [37]. The corresponding theory of Diophantine approximation is surveyed in Lasjaunias [35]. The polynomial Pell equation is discussed by Schmidt [48]. For formal Laurent series there is a multidimensional generalization which is quite different from those for real numbers; see Antoulas [4].

Roth's theorem and Schmidt's subspace theorem are proved in Schmidt [47]. See also Faltings and Wüstholz [20] and Evertse [19]. Nevanlinna's theory of the value distribution of meromorphic functions is treated in the recent book of Cherry and Ye [14]. For Vojta's work see, for example, [59] and [60]. It should be noted, though, that this area is still in a state of flux, besides using techniques beyond our scope. For an overview, see Lang [34].

Siegel's theorem on Diophantine equations with infinitely many solutions is proved with the aid of non-standard analysis by Robinson and Roquette [44]; the proof is reproduced in Stepanov [57]. The theorem is discussed from the standpoint of *Diophantine geometry* in Serre [50]. Any algebraic curve over $\mathbb{Q}$ of genus zero which has a nonsingular rational point can be parametrized by rational functions *effectively*; see Poulakis [43].

It is worth noting that if $F(x,y)$ is a polynomial with rational coefficients which is irreducible over $\mathbb{Q}$, but not over $\mathbb{C}$, then the curve $F(x,y) = 0$ has at most finitely many rational points. For any rational point is a common root of at least two distinct complex-irreducible factors of F and any two such factors have at most finitely many common complex roots.

In conclusion we mention some further applications of continued fractions. A procedure, due to Vincent (1836), for separating the roots of a polynomial with integer coefficients has acquired some practical value with the advent of modern computers. See Alesina and Galuzzi [3].

Continued fractions play a role in the small divisor problems of classical mechanics. As an example, suppose the function f is holomorphic in some neighbourhood of the origin and $f(z) = \lambda z + O(z^2)$, where $\lambda = e^{2\pi i\theta}$ for some irrational θ. It is readily shown that there exists a formal power series h which linearizes f, i.e. $f(h(z)) = h(\lambda z)$. Brjuno (1971) proved that this formal power series converges in a neighbourhood of the origin if $\sum_{n\geq 0} (\log q_{n+1})/q_n < \infty$, where q_n is the denominator of the n-th convergent of θ. It was shown by Yoccoz (1995) that this condition is also necessary. In fact, if $\sum_{n\geq 0} (\log q_{n+1})/q_n = \infty$, the conclusion fails even for $f(z) = \lambda z(1 - z)$. See Yoccoz [62] and Pérez-Marco [40].

Our discussion of continued fractions has neglected their analytic theory. The outstanding work of Stieltjes (1894) on the *problem of moments*, which was extended by Hamburger (1920) and R. Nevanlinna (1922) from the half-line to the whole line, not only gave birth to the Stieltjes integral but also contributed to the development of functional analysis. For modern accounts, see Akhiezer [2], Landau [33] and Simon [55].

9   Selected references

[1]   W. Abikoff, The uniformization theorem, *Amer. Math. Monthly* **88** (1981), 574-592.

[2]   N.I. Akhiezer, *The classical moment problem*, Hafner, New York, 1965.

[3] A. Alesina and M. Galuzzi, A new proof of Vincent's theorem, *Enseign. Math.* **44** (1998), 219-256.

[4] A.C. Antoulas, On recursiveness and related topics in linear systems, *IEEE Trans. Automat. Control* **31** (1986), 1121-1135.

[5] E. Artin, Ein mechanisches System mit quasiergodischen Bahnen, *Abh. Math. Sem. Univ. Hamburg* **3** (1924), 170-175. [*Collected Papers*, pp. 499-504, Addison-Wesley, Reading, Mass., 1965.]

[6] A. Baragar, On the unicity conjecture for Markoff numbers, *Canad. Math. Bull.* **39** (1996), 3-9.

[7] A.F. Beardon, *The geometry of discrete groups*, Springer-Verlag, New York, 1983.

[8] A.F. Beardon and K. Stephenson, The uniformization theorem for circle packings, *Indiana Univ. Math. J.* **39** (1990), 1383-1425.

[9] L. Bers, On Hilbert's 22nd problem, *Mathematical developments arising from Hilbert problems* (ed. F.E. Browder), pp. 559-609, Proc. Symp. Pure Math. **28**, Part 2, Amer. Math. Soc., Providence, R.I., 1976.

[10] R. Bonola, *Non-Euclidean geometry*, English transl. by H.S. Carslaw, reprinted Dover, New York, 1955.

[11] A.J. Brentjes, *Multi-dimensional continued fraction algorithms*, Mathematics Centre Tracts **145**, Amsterdam, 1981.

[12] C. Brezinski, *History of continued fractions and Padé approximants*, Springer-Verlag, Berlin, 1991.

[13] J.W.S. Cassels, *An introduction to Diophantine approximation*, Cambridge University Press, 1957.

[14] W. Cherry and Z. Ye, *Nevanlinna's theory of value distribution*, Springer-Verlag, New York, 2000.

[15] K.Y. Choong, D.E. Daykin and C.R. Rathbone, Rational approximations to π, *Math. Comp.* **25** (1971), 387-392.

[16] T.W. Cusick and M.E. Flahive, *The Markoff and Lagrange spectra*, Mathematical Surveys and Monographs **30**, Amer. Math. Soc., Providence, R.I., 1989.

[17] H. Davenport, *The higher arithmetic*, 7th ed., Cambridge University Press, 1999.

[18] M. Davis, Y. Matijasevic and J. Robinson, Hilbert's tenth problem. Diophantine equations: positive aspects of a negative solution, *Mathematical developments arising from Hilbert problems* (ed. F.E. Browder), pp. 323-378, Proc. Symp. Pure Math. **28**, Part 2, Amer. Math. Soc., Providence, R.I., 1976.

[19] J.H. Evertse, An improvement of the quantitative subspace theorem, *Compositio Math.* **101** (1996), 225-311.

[20] G. Faltings and G. Wüstholz, Diophantine approximation on projective spaces, *Invent. Math.* **116** (1994), 109-138.

[21] H.M. Farkas and I. Kra, *Riemann surfaces*, Springer-Verlag, New York, 1980.

[22] H. Ferguson, A short proof of the existence of vector Euclidean algorithms, *Proc. Amer. Math. Soc.* **97** (1986), 8-10.

[23] H. Hasse, *Vorlesungen über Zahlentheorie*, Zweite Auflage, Springer-Verlag, Berlin, 1964.

[24] Z.-H. He and O. Schramm, On the convergence of circle packings to the Riemann map, *Invent. Math.* **125** (1996), 285-305.

[25] S. Helgason, *Differential geometry, Lie groups, and symmetric spaces*, Academic Press, New York, 1978. [Corrected reprint, Amer. Math. Soc., Providence, R.I., 2001]

[26] J.P. Jones and Y.V. Matijasevic, Proof of recursive insolvability of Hilbert's tenth problem, *Amer. Math. Monthly* **98** (1991) 689-709.

[27] J. Jost, *Compact Riemann surfaces*, transl. by R.R. Simha, Springer-Verlag, Berlin, 1997.

[28] B. Just, Generalizing the continued fraction algorithm to arbitrary dimensions, *SIAM J. Comput.* **21** (1992), 909-926.

[29] S. Katok, *Fuchsian groups*, University of Chicago Press, 1992.

[30] J.F. Koksma, *Diophantische Approximationen*, Springer-Verlag, Berlin, 1936.

[31] J.C. Lagarias, Geodesic multidimensional continued fractions, *Proc. London Math. Soc.* (3) **69** (1994), 464-488.

[32] J.L. Lagrange, *Oeuvres*, t. VII, pp. 5-180, reprinted Olms Verlag, Hildesheim, 1973.

[33] H.J. Landau, The classical moment problem: Hilbertian proofs, *J. Funct. Anal.* **38** (1980), 255-272.

[34] S. Lang, *Number Theory III: Diophantine geometry*, Encyclopaedia of Mathematical Sciences vol. 60, Springer-Verlag, Berlin, 1991.

[35] A. Lasjaunias, A survey of Diophantine approximation in fields of power series, *Monatsh. Math.* **130** (2000), 211-229.

[36] J. Lehner, *Discontinuous groups and automorphic functions*, Mathematical Surveys VIII, Amer. Math. Soc., Providence, R.I., 1964.

[37] B. de Mathan, Approximations diophantiennes dans un corps local, *Bull. Soc. Math. France Suppl. Mém.* **21** (1970), Chapitre IV.

[38] W. Narkiewicz, *Elementary and analytic theory of algebraic numbers*, 2nd ed., Springer-Verlag, Berlin, 1990.

[39] W. Patz, *Tafel der regelmässigen Kettenbrüche und ihrer vollständigen Quotienten für die Quadratwurzeln aus den natürlichen Zahlen von 1-10000*, Akademie-Verlag, Berlin, 1955.

[40] R. Pérez-Marco, Fixed points and circle maps, *Acta Math.* **179** (1997), 243-294.

[41] O. Perron, *Die Lehre von den Kettenbrüchen*, Dritte Auflage, Teubner, Stuttgart, Band I, 1954; Band II, 1957. (Band II treats the analytic theory of continued fractions.)

[42] E.V. Podsypanin, Length of the period of a quadratic irrational, *J. Soviet Math.* **18** (1982), 919-923.

[43] D. Poulakis, Bounds for the minimal solution of genus zero Diophantine equations, *Acta Arith.* **86** (1998), 51-90.

[44] A. Robinson and P. Roquette, On the finiteness theorem of Siegel and Mahler concerning Diophantine equations, *J. Number Theory* **7** (1975), 121-176.

[45] A.M Rockett and P. Szusz, *Continued fractions*, World Scientific, River Edge, N.J., 1992.

[46] G. Rousseau, On a construction for the representation of a positive integer as the sum of four squares, *Enseign. Math.* (2) **33** (1987), 301-306.

[47] W.M. Schmidt, *Diophantine approximation*, Lecture Notes in Mathematics **785**, Springer-Verlag, Berlin, 1980.

[48] W.M. Schmidt, On continued fractions and diophantine approximation in power series fields, *Acta Arith.* **95** (2000), 139-166.

[49] C.-O. Selenius, Rationale of the chakravala process of Jayadeva and Bhaskara II, *Historia Math.* **2** (1975), 167-184.

[50] J.-P. Serre, *Lectures on the Mordell–Weil theorem*, English transl. by M. Brown from notes by M. Waldschmidt, Vieweg & Sohn, Braunschweig, 1989.

[51] J.A. Serret, Developpements sur une classe d'équations, *J. Math. Pures Appl.* **15** (1850), 152-168.

[52] J. Shallit, Real numbers with bounded partial quotients, *Enseign. Math.* **38** (1992), 151-187.

[53] M. Sheingorn, Continued fractions and congruence subgroup geodesics, *Number theory with an emphasis on the Markoff spectrum* (ed. A.D. Pollington and W. Moran), pp. 239-254, Lecture Notes in Pure and Applied Mathematics **147**, Dekker, New York, 1993.

[54] C.L. Siegel, Symplectic geometry, *Amer. J. Math.* **65** (1943), 1-86. [*Gesammelte Abhandlungen, Band II,* pp. 274-359, Springer-Verlag, Berlin, 1966.]

[55] B. Simon, The classical moment problem as a self-adjoint finite difference operator, *Adv. in Math.* **137** (1998), 82-203.

[56] H.M. Stark, Dirichlet's class-number formula revisited, *A tribute to Emil Grosswald: Number theory and related analysis* (ed. M. Knopp and M. Sheingorn), pp. 571-577, Contemporary Mathematics **143**, Amer. Math. Soc., Providence, R.I., 1993.

[57] S.A. Stepanov, *Arithmetic of algebraic curves*, English transl. by I. Aleksanova, Consultants Bureau, New York, 1994.

[58] E.B. Vinberg and O.V. Shvartsman, *Discrete groups of motions of spaces of constant curvature*, Geometry II, pp. 139-248, Encyclopaedia of Mathematical Sciences Vol. 29, Springer-Verlag, Berlin, 1993.

[59] P. Vojta, *Diophantine approximations and value distribution theory*, Lecture Notes in Mathematics **1239**, Springer-Verlag, Berlin, 1987.

[60] P. Vojta, A generalization of theorems of Faltings and Thue–Siegel–Roth–Wirsing, *J. Amer. Math. Soc.* **5** (1992), 763-804.

[61] S. Wagon, The Euclidean algorithm strikes again, *Amer. Math. Monthly* **97** (1990), 125-129.

[62] J.-C. Yoccoz, Théorème de Siegel, nombres de Bruno et polynômes quadratiques, *Astérisque* **231** (1995), 3-88.

[63] D.B. Zagier, *Zetafunktionen und quadratische Körper*, Springer-Verlag, Berlin, 1981.

V
Hadamard's determinant problem

It was shown by Hadamard (1893) that, if all elements of an $n \times n$ matrix of complex numbers have absolute value at most μ, then the determinant of the matrix has absolute value at most $\mu^n n^{n/2}$. For each positive integer n there exist complex $n \times n$ matrices for which this upper bound is attained. For example, the upper bound is attained for $\mu = 1$ by the matrix (ω^{jk}) $(1 \leq j,k \leq n)$, where ω is a primitive n-th root of unity. This matrix is real for $n = 1,2$. However, Hadamard also showed that if the upper bound is attained for a real $n \times n$ matrix, where $n > 2$, then n is divisible by 4.

Without loss of generality one may suppose $\mu = 1$. A real $n \times n$ matrix for which the upper bound $n^{n/2}$ is attained in this case is today called a *Hadamard matrix*. It is still an open question whether an $n \times n$ Hadamard matrix exists for every positive integer n divisible by 4.

Hadamard's inequality played an important role in the theory of linear integral equations created by Fredholm (1900), and partly for this reason many proofs and generalizations were soon given. Fredholm's approach to linear integral equations has been superseded, but Hadamard's inequality has found connections with several other branches of mathematics, such as number theory, combinatorics and group theory. Hadamard matrices have been used to enhance the precision of spectrometers, to design agricultural experiments and to correct errors in messages transmitted by spacecraft.

The moral is that a good mathematical problem will in time find applications. Although the case where n is divisible by 4 has a richer theory, we will also treat other cases of Hadamard's determinant problem, since progress with them might lead to progress also for Hadamard matrices.

1 What is a determinant?

The system of two simultaneous linear equations

$$\alpha_{11}\xi_1 + \alpha_{12}\xi_2 = \beta_1$$
$$\alpha_{21}\xi_1 + \alpha_{22}\xi_2 = \beta_2$$

has, if $\delta_2 = \alpha_{11}\alpha_{22} - \alpha_{12}\alpha_{21}$ is nonzero, the unique solution

$$\xi_1 = (\beta_1\alpha_{22} - \beta_2\alpha_{12})/\delta_2, \quad \xi_2 = -(\beta_1\alpha_{21} - \beta_2\alpha_{11})/\delta_2.$$

If $\delta_2 = 0$, then either there is no solution or there is more than one solution.

Similarly the system of three simultaneous linear equations

$$\alpha_{11}\xi_1 + \alpha_{12}\xi_2 + \alpha_{13}\xi_3 = \beta_1$$
$$\alpha_{21}\xi_1 + \alpha_{22}\xi_2 + \alpha_{23}\xi_3 = \beta_2$$
$$\alpha_{31}\xi_1 + \alpha_{32}\xi_2 + \alpha_{33}\xi_3 = \beta_3$$

has a unique solution if and only if $\delta_3 \neq 0$, where

$$\delta_3 = \alpha_{11}\alpha_{22}\alpha_{33} + \alpha_{12}\alpha_{23}\alpha_{31} + \alpha_{13}\alpha_{21}\alpha_{32}$$
$$- \alpha_{11}\alpha_{23}\alpha_{32} - \alpha_{12}\alpha_{21}\alpha_{33} - \alpha_{13}\alpha_{22}\alpha_{31}.$$

These considerations may be extended to any finite number of simultaneous linear equations. The system

$$\alpha_{11}\xi_1 + \alpha_{12}\xi_2 + \ldots + \alpha_{1n}\xi_n = \beta_1$$
$$\alpha_{21}\xi_1 + \alpha_{22}\xi_2 + \ldots + \alpha_{2n}\xi_n = \beta_2$$
$$\ldots\ldots$$
$$\alpha_{n1}\xi_1 + \alpha_{n2}\xi_2 + \ldots + \alpha_{nn}\xi_n = \beta_n$$

has a unique solution if and only if $\delta_n \neq 0$, where

$$\delta_n = \sum \pm \alpha_{1k_1}\alpha_{2k_2} \cdots \alpha_{nk_n},$$

the sum being taken over all $n!$ permutations $k_1, k_2, \ldots, k_n$ of $1, 2, \ldots, n$ and the sign chosen being $+$ or $-$ according as the permutation is even or odd, as defined in Chapter I, §7.

It has been tacitly assumed that the given quantities α_{jk}, β_j $(j,k = 1,\ldots,n)$ are real numbers, in which case the solution ξ_k $(k = 1,\ldots,n)$ also consists of real numbers. However, everything that has been said remains valid if the given quantities are elements of an arbitrary field F, in which case the solution also consists of elements of F. Since δ_n is an element of F which is uniquely determined by the matrix

$$A = \begin{bmatrix} \alpha_{11} & \ldots & \alpha_{1n} \\ & \ldots\ldots & \\ \alpha_{n1} & \ldots & \alpha_{nn} \end{bmatrix},$$

it will be called the *determinant* of the matrix A and denoted by $\det A$.

Determinants appear in the work of the Japanese mathematician Seki (1683) and in a letter of Leibniz (1693) to l'Hospital, but neither had any influence on later developments. The rule which expresses the solution of a system of linear equations by quotients of determinants was stated by Cramer (1750), but the study of determinants for their own sake began with Vandermonde (1771). The word 'determinant' was first used in the present sense by Cauchy (1812), who gave a systematic account of their theory. The diffusion of this theory throughout the mathematical world owes much to the clear exposition of Jacobi (1841).

For the practical solution of linear equations Cramer's rule is certainly inferior to the age-old method of elimination of variables. Even many of the theoretical uses to which determinants were once put have been replaced by simpler arguments from linear algebra, to the extent that some have advocated banning determinants from the curriculum. However, determinants have a geometrical interpretation which makes their survival desirable.

Let $M_n(\mathbb{R})$ denote the set of all $n \times n$ matrices with entries from the real field $\mathbb{R}$. If $A \in M_n(\mathbb{R})$, then the linear map $x \to Ax$ of $\mathbb{R}^n$ into itself multiplies the volume of any parallelotope by a fixed factor $\mu(A) \geq 0$. Evidently

(i)″ $\mu(AB) = \mu(A)\mu(B)$ for all $A,B \in M_n(\mathbb{R})$,

(ii)″ $\mu(D) = |\alpha|$ for any diagonal matrix $D = \text{diag } [1,...,1,\alpha] \in M_n(\mathbb{R})$.

(A matrix $A = (\alpha_{jk})$ is denoted by diag $[\alpha_{11},\alpha_{22},...,\alpha_{nn}]$ if $\alpha_{jk} = 0$ whenever $j \neq k$ and is then said to be *diagonal*.) It may be shown (e.g., by representing A as a product of elementary matrices in the manner described below) that $\mu(A) = |\det A|$. The sign of the determinant also has a geometrical interpretation: $\det A \gtreqless 0$ according as the linear map $x \to Ax$ preserves or reverses orientation.

Now let F be an arbitrary field and let $M_n = M_n(F)$ denote the set of all $n \times n$ matrices with entries from F. We intend to show that determinants, as defined above, have the properties:

(i)′ $\det (AB) = \det A \cdot \det B$ for all $A,B \in M_n$,

(ii)′ $\det D = \alpha$ for any diagonal matrix $D = \text{diag } [1,...,1,\alpha] \in M_n$,

and, moreover, that these two properties actually characterize determinants. To avoid notational complexity, we consider first the case $n = 2$.

Let $\mathcal{C}$ denote the set of all matrices $A \in M_2$ which are products of finitely many matrices of the form U_λ, V_μ, where

$$U_\lambda = \begin{bmatrix} 1 & \lambda \\ 0 & 1 \end{bmatrix}, \quad V_\mu = \begin{bmatrix} 1 & 0 \\ \mu & 1 \end{bmatrix},$$

and $\lambda,\mu \in F$. The set $\mathscr{E}$ is a group under matrix multiplication, since multiplication is associative, $I \in \mathscr{E}$, $\mathscr{E}$ is obviously closed under multiplication and U_λ, V_μ have inverses $U_{-\lambda}, V_{-\mu}$ respectively.

We are going to show that, *if $A \in M_2$ and $A \neq O$, then there exist $S,T \in \mathscr{E}$ and $\delta \in F$ such that $SAT = \mathrm{diag}\,[1,\delta]$.*

For any $\rho \neq 0$, put

$$W = \begin{bmatrix} 0 & -1 \\ 1 & 0 \end{bmatrix}, \quad R_\rho = \begin{bmatrix} \rho^{-1} & 0 \\ 0 & \rho \end{bmatrix}.$$

Then $W = U_{-1}V_1U_{-1} \in \mathscr{E}$ and also $R_\rho \in \mathscr{E}$ since, if $\sigma = 1 - \rho$, $\rho' = \rho^{-1}$ and $\tau = \rho^2 - \rho$, then

$$R_\rho = V_{-1}U_\sigma V_{\rho'}U_\tau.$$

Let

$$A = \begin{bmatrix} \alpha & \beta \\ \gamma & \delta \end{bmatrix},$$

where at least one of $\alpha,\beta,\gamma,\delta$ is nonzero. By multiplying A on the left, or on the right, or both by W we may suppose that $\alpha \neq 0$. Now, by multiplying A on the right or left by R_α, we may suppose that $\alpha = 1$. Next, by multiplying A on the right by $U_{-\beta}$, we may further suppose that $\beta = 0$. Finally, by multiplying A on the left by $V_{-\gamma}$, we may also suppose that $\gamma = 0$.

The preceding argument is valid even if F is a division ring. In what follows we will use the commutativity of multiplication in F.

We are now going to show that *if $d: \mathscr{E} \to F$ is a map such that $d(ST) = d(S)d(T)$ for all $S,T \in \mathscr{E}$, then either $d(S) = 0$ for every $S \in \mathscr{E}$ or $d(S) = 1$ for every $S \in \mathscr{E}$.*

If $d(T) = 0$ for some $T \in \mathscr{E}$, then $d(I) = d(T)d(T^{-1}) = 0$ and $d(S) = d(I)d(S) = 0$ for every $S \in \mathscr{E}$. Thus we now suppose $d(S) \neq 0$ for every $S \in \mathscr{E}$. Then, in the same way, $d(I) = 1$ and $d(S^{-1}) = d(S)^{-1}$ for every $S \in \mathscr{E}$.

It is easily verified that

$$U_\lambda U_\mu = U_{\lambda+\mu}, \quad V_\lambda V_\mu = V_{\lambda+\mu},$$

$$W^{-1} = -W, \quad W^{-1}V_\mu W = U_{-\mu}.$$

It follows that

$$d(V_\mu) = d(U_{-\mu}) = d(U_\mu)^{-1}.$$

Also, for any $\rho \neq 0$,

$$R_\rho^{-1} U_\lambda R_\rho = U_{\lambda\rho^2}.$$

Hence $d(U_{\lambda\rho^2}) = d(U_\lambda)$ and $d(U_{\lambda(\rho^2-1)}) = 1$.

If the field F contains more than three elements, then $\rho^2 - 1 \neq 0$ for some nonzero $\rho \in F$. Since $\lambda(\rho^2 - 1)$ runs through the nonzero elements of F at the same time as λ, it follows that $d(U_\lambda) = 1$ for every $\lambda \in F$. Hence also $d(V_\mu) = 1$ for every $\mu \in F$ and $d(S) = 1$ for all $S \in \mathscr{E}$.

If F contains 2 elements, then $d(S) = 1$ for every $S \in \mathscr{E}$ is the only possibility. If F contains 3 elements, then $d(S) = \pm 1$ for every $S \in \mathscr{E}$. Hence $d(S^{-1}) = d(S)$ and $d(S^2) = 1$. Since $U_2 = U_1^2$ and $U_1 = U_2^{-1}$, this implies $d(U_\lambda) = 1$ for every $\lambda \in F$, and the rest follows as before.

The preceding discussion is easily extended to higher dimensions. Put

$$U_{ij}(\lambda) \ = \ I_n + \lambda E_{ij},$$

for any $i,j \in \{1,\dots,n\}$ with $i \neq j$, where E_{ij} is the $n \times n$ matrix with all entries 0 except the (i,j)-th, which is 1, and let $SL_n(F)$ denote the set of all $A \in M_n$ which are products of finitely many matrices $U_{ij}(\lambda)$. Then $SL_n(F)$ is a group under matrix multiplication.

If $A \in M_n$ and $A \neq O$, then there exist $S,T \in SL_n(F)$ and an integer r $(1 \leq r \leq n)$ such that

$$SAT \ = \ \mathrm{diag}\,[1_{r-1},\delta,0_{n-r}]$$

for some nonzero $\delta \in F$. The matrix A is *singular* if $r < n$ and *nonsingular* if $r = n$. Hence $A = (\alpha_{jk})$ is nonsingular if and only if its transpose $A^t = (\alpha_{kj})$ is nonsingular. In the nonsingular case we need multiply A on only one side by a matrix from $SL_n(F)$ to bring it to the form

$$D_\delta \ = \ \mathrm{diag}\,[1_{n-1},\delta].$$

For if $SAT = D_\delta$, then $SA = D_\delta T^{-1}$ and this implies $SA = S'D_\delta$ for some $S' \in SL_n(F)$, since

$$\begin{aligned}
D_\delta U_{ij}(\lambda) &= U_{ij}(\lambda\delta^{-1})D_\delta && \text{if } i < j = n,\\
D_\delta U_{ij}(\lambda) &= U_{ij}(\delta\lambda)D_\delta && \text{if } j < i = n,\\
D_\delta U_{ij}(\lambda) &= U_{ij}(\lambda)D_\delta && \text{if } i,j \neq n \text{ and } i \neq j.
\end{aligned}$$

In the same way as for $n = 2$ it may be shown that, if $d: SL_n(F) \to F$ is a map such that $d(ST) = d(S)d(T)$ for all $S,T \in SL_n(F)$, then either $d(S) = 0$ for every S or $d(S) = 1$ for every S.

THEOREM 1 *There exists a unique map* $d: M_n \to F$ *such that*

(i)$'$ $d(AB) = d(A)d(B)$ *for all* $A,B \in M_n$,

(ii)$'$ *for any* $\alpha \in F$, *if* $D_\alpha = \mathrm{diag}\,[1_{n-1},\alpha]$, *then* $d(D_\alpha) = \alpha$.

Proof We consider first uniqueness. Since $d(I) = d(D_1) = 1$, we must have $d(S) = 1$ for every $S \in SL_n(F)$, by what we have just said. Also, if

$$H = \text{diag}\,[\eta_1,\ldots,\eta_{n-1},0],$$

then $d(H) = 0$, since $H = D_0 H$. In particular, $d(O) = 0$. If $A \in M_n$ and $A \neq O$, there exist $S,T \in SL_n(F)$ such that

$$SAT = \text{diag}\,[1_{r-1},\delta,0_{n-r}],$$

where $1 \leq r \leq n$ and $\delta \neq 0$. It follows that $d(A) = 0$ if $r < n$, i.e. if A is singular. On the other hand, $d(A) = \delta$ if $r = n$, i.e. if A is nonsingular. Since $d(D_\alpha) \neq d(D_\beta)$ if $\alpha \neq \beta$, this proves uniqueness.

We consider next existence. For any $A = (\alpha_{jk}) \in M_n$, define

$$\det A = \sum_{\sigma \in \mathscr{S}_n} (\text{sgn }\sigma)\, \alpha_{1\sigma 1}\alpha_{2\sigma 2} \cdots \alpha_{n\sigma n},$$

where σ is a permutation of $1,2,\ldots,n$, sgn $\sigma = 1$ or -1 according as the permutation σ is even or odd, and the summation is over the symmetric group $\mathscr{S}_n$ of all permutations. Several consequences of this definition will now be derived.

(i)   *if every entry in some row of A is* 0, *then det $A = 0$.*

Proof Every summand vanishes in the expression for det A.

(ii)   *if the matrix B is obtained from the matrix A by multiplying all entries in one row by λ, then* det $B = \lambda$ det A.

Proof This is also clear, since in the expression for det A each summand contains exactly one factor from any given row.

(iii)   *if two rows of A are the same, then det $A = 0$.*

Proof Suppose for definiteness that the first and second rows are the same, and let τ be the permutation which interchanges 1 and 2 and leaves fixed every $k > 2$. Then τ is odd and we can write

$$\det A = \sum_{\sigma \in \mathscr{A}_n} \alpha_{1\sigma 1}\alpha_{2\sigma 2} \cdots \alpha_{n\sigma n} - \sum_{\sigma \in \mathscr{A}_n} \alpha_{1\sigma\tau 1}\alpha_{2\sigma\tau 2} \cdots \alpha_{n\sigma\tau n},$$

where $\mathscr{A}_n$ is the alternating group of all even permutations. In the second sum

$$\alpha_{1\sigma\tau 1}\alpha_{2\sigma\tau 2} \cdots \alpha_{n\sigma\tau n} = \alpha_{1\sigma 2}\alpha_{2\sigma 1}\alpha_{3\sigma 3} \cdots \alpha_{n\sigma n} = \alpha_{2\sigma 2}\alpha_{1\sigma 1}\alpha_{3\sigma 3} \cdots \alpha_{n\sigma n},$$

because the first and second rows are the same. Hence the two sums cancel.

(iv)   *if the matrix B is obtained from the matrix A by adding a scalar multiple of one row to a different row, then* det $B = $ det A.

Proof Suppose for definiteness that B is obtained from A by adding λ times the second row to the first. Then

$$\det B \;=\; \sum_{\sigma \in \mathscr{S}_n} (\text{sgn } \sigma)\, \alpha_{1\sigma1}\alpha_{2\sigma2} \cdots \alpha_{n\sigma n} + \lambda \sum_{\sigma \in \mathscr{S}_n} (\text{sgn } \sigma)\, \alpha_{2\sigma1}\alpha_{2\sigma2} \cdots \alpha_{n\sigma n}.$$

The first sum is $\det A$ and the second sum is 0, by (iii), since it is the determinant of the matrix obtained from A by replacing the first row by the second.

(v) *if A is singular, then det $A = 0$.*

Proof If A is singular, then some row of A is a linear combination of the remaining rows. Thus by subtracting from this row scalar multiples of the remaining rows we can replace it by a row of 0's. For the new matrix B we have $\det B = 0$, by (i). On the other hand, $\det B = \det A$, by (iv).

(vi) *if $A = \text{diag } [\delta_1,\ldots,\delta_n]$, then det $A = \delta_1 \cdots \delta_n$. In particular,* det $D_\alpha = \alpha$.

Proof In the expression for $\det A$ the only possible nonzero summand is that for which σ is the identity permutation, and the identity permutation is even.

(vii) $\det (AB) = \det A \cdot \det B$ *for all* $A,B \in M_n$.

Proof If A is singular, then AB is also and so, by (v), $\det (AB) = 0 = \det A \cdot \det B$. Thus we now suppose that A is nonsingular. Then there exists $S \in SL_n(F)$ such that $SA = D_\delta$ for some nonzero $\delta \in F$. Since, by the definition of $SL_n(F)$, left multiplication by S corresponds to a finite number of operations of the type considered in (iv) we have

$$\det A \;=\; \det (SA) \;=\; \det D_\delta$$

and

$$\det (AB) \;=\; \det (SAB) \;=\; \det (D_\delta B).$$

But $\det D_\delta = \delta$, by (vi), and $\det (D_\delta B) = \delta \det B$, by (ii). Hence $\det (AB) = \det A \cdot \det B$.

This completes the proof of existence. $\square$

COROLLARY 2 *If $A \in M_n$ and if A^t is the transpose of A, then* $\det A^t = \det A$.

Proof The map $d: M_n \to F$ defined by $d(A) = \det A^t$ also has the properties (i)′,(ii)′. $\square$

The proof of Theorem 1 shows also that $SL_n(F)$ *is the special linear group, consisting of all $A \in M_n$ with* $\det A = 1$.

We do not propose to establish here all the properties of determinants which we may later require. However, we note that if

$$A \; = \; \begin{bmatrix} B & 0 \\ C & D \end{bmatrix}$$

is a partitioned matrix, where B and D are square matrices of smaller size, then

$$\det A = \det B \cdot \det D.$$

It follows that if $A = (\alpha_{jk})$ is *lower triangular* (i.e. $\alpha_{jk} = 0$ for all j,k with $j < k$) or *upper triangular* (i.e. $\alpha_{jk} = 0$ for all j,k with $j > k$), then

$$\det A = \alpha_{11}\alpha_{22}\cdots\alpha_{nn}.$$

2 Hadamard Matrices

We begin by obtaining an upper bound for $\det (A^tA)$, where A is an $n\times m$ real matrix. If $m = n$, then $\det (A^tA) = (\det A)^2$ and bounding $\det (A^tA)$ is the same as Hadamard's problem of bounding $|\det A|$. However, as we will see in §3, the problem is of interest also for $m < n$.

In the statement of the following result we denote by $\|v\|$ the Euclidean norm of a vector $v = (\alpha_1,...,\alpha_n) \in \mathbb{R}^n$. Thus $\|v\| \geq 0$ and $\|v\|^2 = \alpha_1{}^2 + ... + \alpha_n{}^2$. The geometrical interpretation of the result is that a parallelotope with given side lengths has maximum volume when the sides are orthogonal.

PROPOSITION 3   *Let A be an $n\times m$ real matrix with linearly independent columns $v_1,...,v_m$. Then*

$$\det (A^tA) \; \leq \; \textstyle\prod_{k=1}^{m} \|v_k\|^2,$$

with equality if and only if A^tA is a diagonal matrix.

Proof We are going to construct inductively mutually orthogonal vectors $w_1,...,w_m$ such that w_k is a linear combination of $v_1,...,v_k$ in which the coefficient of v_k is 1 ($1 \leq k \leq m$). Take $w_1 = v_1$ and supose $w_1,...,w_{k-1}$ have been determined. If we take

$$w_k \; = \; v_k - \alpha_1 w_1 - ... - \alpha_{k-1}w_{k-1},$$

where $\alpha_j = (v_k,w_j)$, then $(w_k,w_j) = 0$ $(1 \leq j < k)$. Moreover, $w_k \neq 0$, since $v_1,...,v_k$ are linearly independent. (This is the same process as in §10 of Chapter I, but without the normalization.)

If B is the matrix with columns $w_1,...,w_m$ then, by construction,

$$B^t B = \text{diag } [\delta_1,...,\delta_m]$$

is a diagonal matrix with diagonal entries $\delta_k = \|w_k\|^2$ and $AT = B$ for some upper triangular matrix T with 1's in the main diagonal. Since $\det T = 1$, we have

$$\det (A^t A) = \det (B^t B) = \Pi_{k=1}^m \|w_k\|^2.$$

But

$$\|v_k\|^2 = \|w_k\|^2 + |\alpha_1|^2 \|w_1\|^2 + ... + |\alpha_{k-1}|^2 \|w_{k-1}\|^2$$

and hence $\|w_k\|^2 \le \|v_k\|^2$, with equality only if $w_k = v_k$. The result follows. $\square$

COROLLARY 4 *Let $A = (\alpha_{jk})$ be an $n\times m$ real matrix such that $|\alpha_{jk}| \le 1$ for all j,k. Then*

$$\det (A^t A) \le n^m,$$

with equality if and only if $\alpha_{jk} = \pm 1$ for all j,k and $A^t A = nI_m$.

Proof We may assume that the columns of A are linearly independent, since otherwise $\det (A^t A) = 0$. If v_k is the k-th column of A, then $\|v_k\|^2 \le n$, with equality if and only if $|\alpha_{jk}| = 1$ for $1 \le j \le n$. The result now follows from Proposition 3. $\square$

An $n\times m$ matrix $A = (\alpha_{jk})$ will be said to be an *H-matrix* if $\alpha_{jk} = \pm 1$ for all j,k and $A^t A = nI_m$. If, in addition, $m = n$ then A will be said to be a *Hadamard matrix* of *order n*.

If A is an $n\times m$ H-matrix, then $m \le n$. Furthermore, if A is a Hadamard matrix of order n then, for any $m < n$, the submatrix formed by the first m columns of A is an *H*-matrix. (This distinction between *H*-matrices and Hadamard matrices is convenient, but not standard. It is an unproven conjecture that any *H*-matrix can be completed to a Hadamard matrix.)

The transpose A^t of a Hadamard matrix A is again a Hadamard matrix, since $A^t = nA^{-1}$ commutes with A. The 1×1 unit matrix is a Hadamard matrix, and so is the 2×2 matrix

$$\begin{bmatrix} 1 & 1 \\ 1 & -1 \end{bmatrix}.$$

There is one rather simple procedure for constructing *H*-matrices. If $A = (\alpha_{jk})$ is an $n\times m$ matrix and $B = (\beta_{i\ell})$ a $q\times p$ matrix, then the $nq\times mp$ matrix

$$\begin{bmatrix} \alpha_{11}B & \alpha_{12}B & ... & \alpha_{1m}B \\ \alpha_{21}B & \alpha_{22}B & ... & \alpha_{2m}B \\ & ... & ... & \\ \alpha_{n1}B & \alpha_{n2}B & ... & \alpha_{nm}B \end{bmatrix},$$

with entries $\alpha_{jk}\beta_{i\ell}$, is called the *Kronecker product* of A and B and is denoted by $A \otimes B$. It is easily verified that

$$(A \otimes B)(C \otimes D) \; = \; AC \otimes BD$$

and

$$(A \otimes B)^t \; = \; A^t \otimes B^t.$$

It follows directly from these rules of calculation that if A_1 is an $n_1 \times m_1$ H-matrix and A_2 an $n_2 \times m_2$ H-matrix, then $A_1 \otimes A_2$ is an $n_1 n_2 \times m_1 m_2$ H-matrix. Consequently, since there exist Hadamard matrices of orders 1 and 2, there also exist Hadamard matrices of order any power of 2. This was already known to Sylvester (1867).

PROPOSITION 5 *Let $A = (\alpha_{jk})$ be an $n \times m$ H-matrix. If $n > 1$, then n is even and any two distinct columns of A have the same entries in exactly $n/2$ rows. If $n > 2$, then n is divisible by* 4 *and any three distinct columns of A have the same entries in exactly $n/4$ rows.*

Proof   If $j \neq k$, then

$$\alpha_{1j}\alpha_{1k} + \ldots + \alpha_{nj}\alpha_{nk} \; = \; 0.$$

Since $\alpha_{ij}\alpha_{ik} = 1$ if the j-th and k-th columns have the same entry in the i-th row and $= -1$ otherwise, the number of rows in which the j-th and k-th columns have the same entry is $n/2$.

If j,k,ℓ are all different, then

$$\sum_{i=1}^{n} (\alpha_{ij} + \alpha_{ik})(\alpha_{ij} + \alpha_{i\ell}) \; = \; \sum_{i=1}^{n} \alpha_{ij}^2 \; = \; n.$$

But $(\alpha_{ij} + \alpha_{ik})(\alpha_{ij} + \alpha_{i\ell}) = 4$ if the j-th, k-th and ℓ-th columns all have the same entry in the i-th row and $= 0$ otherwise. Hence the number of rows in which the j-th, k-th and ℓ-th columns all have the same entry is exactly $n/4$. $\square$

Thus the order n of a Hadamard matrix must be divisible by 4 if $n > 2$. It is unknown if a Hadamard matrix of order n exists for every n divisible by 4. However, it is known for $n \leq 424$ and for several infinite families of n. We restrict attention here to the family of Hadamard matrices constructed by Paley (1933).

The following lemma may be immediately verified by matrix multiplication.

LEMMA 6 *Let C be an $n \times n$ matrix, with 0's on the main diagonal and all other entries 1 or -1, such that*

$$C^t C \; = \; (n-1)I_n.$$

If C is skew-symmetric, then C + I is a Hadamard matrix of order n, whereas if C is symmetric, then

$$\begin{bmatrix} C+I & C-I \\ C-I & -C-I \end{bmatrix}$$

is a Hadamard matrix of order 2n. ◻

PROPOSITION 7 *If q is a power of an odd prime, there exists a $(q + 1) \times (q + 1)$ matrix C with 0's on the main diagonal and all other entries 1 or -1, such that*

(i) $C^t C = q I_{q+1}$,

(ii)   *C is skew-symmetric if $q \equiv 3$ mod 4 and symmetric if $q \equiv 1$ mod 4.*

Proof Let F be a finite field containing q elements. Since q is odd, not all elements of F are squares. For any $a \in F$, put

$$\chi(a) = \begin{cases} 0 & \text{if } a = 0, \\ 1 & \text{if } a \neq 0 \text{ and } a = c^2 \text{ for some } c \in F, \\ -1 & \text{if } a \text{ is not a square.} \end{cases}$$

If $q = p$ is a prime, then F is the field of integers modulo p and $\chi(a) = (a/p)$ is the Legendre symbol studied in Chapter III. The following argument may be restricted to this case, if desired.

Since the multiplicative group of F is cyclic, we have

$$\chi(ab) = \chi(a)\chi(b) \text{ for all } a,b \in F.$$

Since the number of nonzero elements which are squares is equal to the number which are non-squares, we also have

$$\sum_{a \in F} \chi(a) = 0.$$

It follows that, for any $c \neq 0$,

$$\sum_{b \in F} \chi(b)\chi(b + c) = \sum_{b \neq 0} \chi(b)^2 \chi(1 + cb^{-1}) = \sum_{x \neq 1} \chi(x) = -1.$$

Let $0 = a_0, a_1, \ldots, a_{q-1}$ be an enumeration of the elements of F and define a $q \times q$ matrix $Q = (q_{jk})$ by

$$q_{jk} = \chi(a_j - a_k) \quad (0 \leq j,k < q).$$

Thus Q has 0's on the main diagonal and ± 1's elsewhere. Also, by what has been said in the previous paragraph, if J_m denotes the $m \times m$ matrix with all entries 1, then

$$QJ_q = 0, \quad Q^t Q = qI_q - J_q.$$

Furthermore, since $\chi(-1) = (-1)^{(q-1)/2}$, Q is symmetric if $q \equiv 1 \bmod 4$ and skew-symmetric if $q \equiv 3 \bmod 4$. If e_m denotes the $1 \times m$ matrix with all entries 1, it follows that the matrix

$$C = \begin{bmatrix} 0 & e_q \\ \pm e_q^{\,t} & Q \end{bmatrix},$$

where the $\pm$ sign is chosen according as $q \equiv \pm 1 \bmod 4$, satisfies all the requirements. $\square$

By combining Lemma 6 with Proposition 7 we obtain Paley's result that, for any odd prime power q, there exists a Hadamard matrix of order $q + 1$ if $q \equiv 3 \bmod 4$ and of order $2(q + 1)$ if $q \equiv 1 \bmod 4$. Together with the Kronecker product construction, this establishes the existence of Hadamard matrices for all orders $n \equiv 0 \bmod 4$ with $n \le 100$, except $n = 92$.

A Hadamard matrix of order 92 was first found by Baumert, Golomb and Hall (1962), using a computer search and the following method proposed by Williamson (1944). Let A,B,C,D be $d \times d$ matrices with entries ± 1 and let

$$H = \begin{bmatrix} A & D & B & C \\ -D & A & -C & B \\ -B & C & A & -D \\ -C & -B & D & A \end{bmatrix},$$

i.e. $H = A \otimes I + B \otimes i + C \otimes j + D \otimes k$, where the 4×4 matrices I,i,j,k are matrix representations of the unit quaternions. It may be immediately verified that H is a Hadamard matrix of order $n = 4d$ if

$$A^t A + B^t B + C^t C + D^t D = 4dI_d$$

and

$$X^t Y = Y^t X$$

for every two distinct matrices X,Y from the set $\{A,B,C,D\}$. The first infinite class of Hadamard matrices of Williamson type was found by Turyn (1972), who showed that they exist for all orders $n = 2(q + 1)$, where q is a prime power and $q \equiv 1 \bmod 4$. Lagrange's theorem that any positive integer is a sum of four squares suggests that Hadamard matrices of Williamson type may exist for all orders $n \equiv 0 \bmod 4$.

The Hadamard matrices constructed by Paley are either symmetric or of the form $I + S$, where S is skew-symmetric. It has been conjectured that in fact Hadamard matrices of both these types exist for all orders $n \equiv 0 \bmod 4$.

3 The art of weighing

It was observed by Yates (1935) that, if several quantities are to be measured, more accurate results may be obtained by measuring suitable combinations of them than by measuring each separately. Suppose, for definiteness, that we have m objects whose weights are to be determined and we perform $n \geq m$ weighings. The whole experiment may be represented by an $n \times m$ matrix $A = (\alpha_{jk})$. If the k-th object is not involved in the j-th weighing, then $\alpha_{jk} = 0$; if it is involved, then $\alpha_{jk} = +1$ or -1 according as it is placed in the left-hand or right-hand pan of the balance. The individual weights $\xi_1,...,\xi_m$ are connected with the observed results $\eta_1,...,\eta_n$ of the weighings by the system of linear equations

$$y = Ax, \tag{1}$$

where $x = (\xi_1,...,\xi_m)^t \in \mathbb{R}^m$ and $y = (\eta_1,...,\eta_n)^t \in \mathbb{R}^n$.

We will again denote by $\|y\|$ the Euclidean norm $(|\eta_1|^2 + ... + |\eta_n|^2)^{1/2}$ of the vector y. Let $\bar{x} \in \mathbb{R}^m$ have as its coordinates the correct weights and let $\bar{y} = A\bar{x}$. If, because of errors of measurement, y ranges over the ball $\|y - \bar{y}\| \leq \rho$ in $\mathbb{R}^n$, then x ranges over the ellipsoid $(x - \bar{x})^t A^t A (x - \bar{x}) \leq \rho^2$ in $\mathbb{R}^m$. Since the volume of the ellipsoid is $[\det (A^t A)]^{-1/2}$ times the volume of the ball, we may regard the best choice of the design matrix A to be that for which the ellipsoid has minimum volume. Thus we are led to the problem of maximizing $\det (A^t A)$ among all $n \times m$ matrices $A = (\alpha_{jk})$ with $\alpha_{jk} \in \{0,-1,1\}$.

A different approach to the best choice of design matrix leads (by §2) to a similar result. If $n > m$ the linear system (1) is overdetermined. However, the least squares estimate for the solution of (1) is

$$x = Cy,$$

where $C = (A^t A)^{-1} A^t$. Let $a_k \in \mathbb{R}^n$ be the k-th column of A and let $c_k \in \mathbb{R}^n$ be the k-th row of C. Since $CA = I_m$, we have $c_k a_k = 1$. If y ranges over the ball $\|y - \bar{y}\| \leq \rho$ in $\mathbb{R}^n$, then ξ_k ranges over the real interval $|\xi_k - \bar{\xi}_k| \leq \rho\|c_k\|$. Thus we may regard the optimal choice of the design matrix A for measuring ξ_k to be that for which $\|c_k\|$ is a minimum. By Schwarz's inequality (Chapter I, §4),

$$\|c_k\| \, \|a_k\| \geq 1,$$

with equality only if c_k^t is a scalar multiple of a_k. Also $\|a_k\| \leq n^{1/2}$, since all elements of A have absolute value at most 1. Hence $\|c_k\| \geq n^{-1/2}$, with equality if and only if all elements of a_k have absolute value 1 and $c_k^t = a_k/n$. It follows that the design matrix A is optimal for measuring *each* of $\xi_1,...,\xi_m$ if all elements of A have absolute value 1 and $A^tA = nI_m$. Moreover, in this case the least squares estimate for the solution of (1) is simply $x = A^ty/n$. Thus the individual weights are easily determined from the observed measurements by additions and subtractions, followed by a division by n.

Suppose, for example, that $m = 3$ and $n = 4$. If we take

$$A \;=\; \begin{bmatrix} + + + \\ + + - \\ - + + \\ + - + \end{bmatrix},$$

where $+$ and $-$ stand for 1 and -1 respectively, then $A^tA = 4I_3$. With this experimental design the individual weights may all be determined with twice the accuracy of the weighing procedure.

The next result shows, in particular, that if we wish to maximize det (A^tA) among the $n{\times}m$ matrices A with all entries 0,1 or -1, then we may restrict attention to those with all entries 1 or -1.

PROPOSITION 8 *Let α,β be real numbers with $\alpha < \beta$ and let $\mathcal{S}$ be the set of all $n{\times}m$ matrices $A = (\alpha_{jk})$ such that $\alpha \le \alpha_{jk} \le \beta$ for all j,k. Then there exists an $n{\times}m$ matrix $M = (\mu_{jk})$ such that $\mu_{jk} \in \{\alpha,\beta\}$ for all j,k and*

$$\det (M^tM) \;=\; \max_{A \in \mathcal{S}} \det (A^tA).$$

Proof For any $n{\times}m$ real matrix A, either the symmetric matrix A^tA is positive definite and det $(A^tA) > 0$, or A^tA is positive semidefinite and det $(A^tA) = 0$. Since the result is obvious if det $(A^tA) = 0$ for every $A \in \mathcal{S}$, we assume that det $(A^tA) > 0$ for some $A \in \mathcal{S}$. This implies $m \le n$. Partition such an A in the form

$$A \;=\; (v\, B),$$

where v is the first column of A and B is the remainder. Then

$$A^tA \;=\; \begin{bmatrix} v^tv & v^tB \\ B^tv & B^tB \end{bmatrix}$$

and B^tB is also a positive definite symmetric matrix. By multiplying A^tA on the left by

$$\begin{bmatrix} I & -v^tB(B^tB)^{-1} \\ O & I \end{bmatrix}$$

and taking determinants, we see that

$$\det (A^tA) = f(v) \det (B^tB),$$

where

$$f(v) = v^tv - v^tB(B^tB)^{-1}B^tv.$$

We can write $f(v) = v^tQv$, where

$$Q = I - P, \quad P = B(B^tB)^{-1}B^t.$$

From $P^t = P = P^2$ we obtain $Q^t = Q = Q^2$. Hence $Q = Q^tQ$ is a positive semidefinite symmetric matrix.

If $v = \theta v_1 + (1 - \theta)v_2$, where v_1 and v_2 are fixed vectors and $\theta \in \mathbb{R}$, then $f(v)$ is a quadratic polynomial $q(\theta)$ in θ whose leading coefficient

$$v_1{}'Qv_1 - v_2{}'Qv_1 - v_1{}'Qv_2 + v_2{}'Qv_2$$

is nonnegative, since Q is positive semidefinite. It follows that $q(\theta)$ attains its maximum value in the interval $0 \le \theta \le 1$ at an endpoint.

Put

$$\mu = \sup_{A \in \mathcal{S}} \det (A^tA).$$

Since $\det (A^tA)$ is a continuous function of the mn variables α_{jk} and $\mathcal{S}$ may be regarded as a compact set in $\mathbb{R}^{mn}$, μ is finite and there exists a matrix $A \in \mathcal{S}$ for which $\det (A^tA) = \mu$. By repeatedly applying the argument of the preceding paragraph to this A we may replace it by one for which every entry in the first column is either α or β and for which also $\det (A^tA) = \mu$. These operations do not affect the submatrix B formed by the last $m - 1$ columns of A. By interchanging the k-th column of A with the first, which does not alter the value of $\det (A^tA)$, we may apply the same argument to every other column of A. $\quad\square$

The proof of Proposition 8 actually shows that if C is a compact subset of $\mathbb{R}^n$ and if $\mathcal{S}$ is the set of all $n{\times}m$ matrices A whose columns are in C, then there exists an $n{\times}m$ matrix M whose columns are extreme points of C such that

$$\det (M^tM) = \sup_{A \in \mathcal{S}} \det (A^tA).$$

Here $e \in C$ is said to be an *extreme point* of C if there do not exist distinct $v_1, v_2 \in C$ and $\theta \in (0,1)$ such that $e = \theta v_1 + (1 - \theta)v_2$.

The preceding discussion concerns weighings by a chemical balance. If instead we use a spring balance, then we are similarly led to the problem of maximizing $\det (B^tB)$ among all $n{\times}m$ matrices $B = (\beta_{jk})$ with $\beta_{jk} = 1$ or 0 according as the k-th object is or is not involved in the j-th weighing. Moreover other types of measurement lead to the same problem. A spectrometer sorts electromagnetic radiation into bundles of rays, each bundle having a characteristic wavelength. Instead of measuring the intensity of each bundle separately, we can measure the intensity of various combinations of bundles by using masks with open or closed slots.

It will now be shown that in the case $m = n$ the chemical and spring balance problems are essentially equivalent.

LEMMA 9 *If B is an $(n-1)\times(n-1)$ matrix of 0's and 1's, and if J_n is the $n\times n$ matrix whose entries are all 1, then*

$$A = J_n - \begin{bmatrix} O & O \\ O & 2B \end{bmatrix},$$

is an $n\times n$ matrix of 1's and -1's, whose first row and column contain only 1's, such that

$$\det A = (-2)^{n-1} \det B.$$

Moreover, every $n\times n$ matrix of 1's and -1's, whose first row and column contain only 1's, is obtained in this way.

Proof Since

$$A = \begin{bmatrix} 1 & O \\ e_{n-1}^t & I \end{bmatrix} \begin{bmatrix} 1 & e_{n-1} \\ O & -2B \end{bmatrix},$$

where e_m denotes a row of m 1's, the matrix A has determinant $(-2)^{n-1} \det B$. The rest of the lemma is obvious. $\square$

Let A be an $n\times n$ matrix with entries ± 1. By multiplying rows and columns of A by -1 we can make all elements in the first row and first column equal to 1 without altering the value of $\det (A^t A)$. It follows from Lemma 9 that if α_n is the maximum of $\det (A^t A)$ among all $n\times n$ matrices $A = (\alpha_{jk})$ with $\alpha_{jk} \in \{-1,1\}$, and if β_{n-1} is the maximum of $\det (B^t B)$ among all $(n-1)\times(n-1)$ matrices $B = (\beta_{jk})$ with $\beta_{jk} \in \{0,1\}$, then

$$\alpha_n = 2^{2n-2}\beta_{n-1}.$$

4 Some matrix theory

In rectangular coordinates the equation of an ellipse with centre at the origin has the form

$$Q: = ax^2 + 2bxy + cy^2 = \text{const.} \tag{*}$$

This is not the form in which the equation of an ellipse is often written, because of the 'cross product' term $2bxy$. However, we can bring it to that form by rotating the axes, so that the major axis of the ellipse lies along one coordinate axis and the minor axis along the other. This is possible because the major and minor axes are perpendicular to one another. These assertions will now be verified analytically.

In matrix notation, $Q = z^t A z$, where

$$A = \begin{bmatrix} a & b \\ b & c \end{bmatrix}, \quad z = \begin{bmatrix} x \\ y \end{bmatrix}.$$

A rotation of coordinates has the form $z = Tw$, where

$$T = \begin{bmatrix} \cos\theta & -\sin\theta \\ \sin\theta & \cos\theta \end{bmatrix}, \quad w = \begin{bmatrix} u \\ v \end{bmatrix}.$$

Then $Q = w^t B w$, where $B = T^t A T$. Multiplying out, we obtain

$$B = \begin{bmatrix} a' & b' \\ b' & c' \end{bmatrix},$$

where

$$b' = b(\cos^2\theta - \sin^2\theta) - (a - c)\sin\theta\cos\theta.$$

To eliminate the cross product term we choose θ so that $b(\cos^2\theta - \sin^2\theta) = (a - c)\sin\theta\cos\theta$; i.e., $2b\cos 2\theta = (a - c)\sin 2\theta$, or

$$\tan 2\theta = 2b/(a - c).$$

The preceding argument applies equally well to a hyperbola, since it is also described by an equation of the form (*). We now wish to extend this result to higher dimensions. An n-dimensional conic with centre at the origin has the form

$$Q: = x^t A x = \text{const.},$$

where $x \in \mathbb{R}^n$ and A is an $n \times n$ real symmetric matrix. The analogue of a rotation is a linear transformation $x = Ty$ which preserves Euclidean lengths, i.e. $x^t x = y^t y$. This holds for all $y \in \mathbb{R}^n$ if and only if

$$T^t T = I.$$

A matrix T which satisfies this condition is said to be *orthogonal*. Then $T^t = T^{-1}$ and hence also $TT^t = I$.

The single most important fact about real symmetric matrices is the *principal axes transformation*:

THEOREM 10 *If H is an $n\times n$ real symmetric matrix, then there exists an $n\times n$ real orthogonal matrix U such that $U^t H U$ is a diagonal matrix:*

$$U^t H U \;=\; \mathrm{diag}\,[\lambda_1,\ldots,\lambda_n].$$

Proof Let $f\colon \mathbb{R}^n \to \mathbb{R}$ be the map defined by

$$f(x) \;=\; x^t H x.$$

Since f is continuous and the unit sphere $S = \{x \in \mathbb{R}^n\colon x^t x = 1\}$ is compact,

$$\lambda_1\colon \;=\; \sup_{x\,\in\, S} f(x)$$

is finite and there exists an $x_1 \in S$ such that $f(x_1) = \lambda_1$. We are going to show that, if $x \in S$ and $x^t x_1 = 0$, then also $x^t H x_1 = 0$.

For any real ε, put

$$y \;=\; (x_1 + \varepsilon x)/(1 + \varepsilon^2)^{1/2}.$$

Then also $y \in S$, since x and x_1 are orthogonal vectors of unit length. Hence $f(y) \le f(x_1)$, by the definition of x_1. But $x_1^t H x = x^t H x_1$, since H is symmetric, and hence

$$f(y) \;=\; \{f(x_1) + 2\varepsilon x^t H x_1 + \varepsilon^2 f(x)\}/(1 + \varepsilon^2).$$

For small $|\varepsilon|$ it follows that

$$f(y) \;=\; f(x_1) + 2\varepsilon x^t H x_1 + O(\varepsilon^2).$$

If $x^t H x_1$ were different from zero, we could choose ε to have the same sign as it and obtain the contradiction $f(y) > f(x_1)$.

On the intersection of the unit sphere S with the hyperplane $x^t x_1 = 0$, the function f attains its maximum value λ_2 at some point x_2. Similarly, on the intersection of the unit sphere S with the $(n-2)$-dimensional subspace of all x such that $x^t x_1 = x^t x_2 = 0$, the function f attains its maximum value λ_3 at some point x_3. Proceeding in this way we obtain n mutually orthogonal unit vectors $x_1,\ldots,x_n$. Moreover $x_j^{\,t} H x_j = \lambda_j$ and, by the argument of the previous paragraph, $x_j^{\,t} H x_k = 0$ if $j > k$. It follows that the matrix U with columns $x_1,\ldots,x_n$ satisfies all the requirements. $\square$

It should be noted that, if U is any orthogonal matrix such that $U^t H U = \mathrm{diag}\,[\lambda_1,\ldots,\lambda_n]$ then, since $U U^t = I$, the columns $x_1,\ldots,x_n$ of U satisfy

$$Hx_j = \lambda_j x_j \quad (1 \le j \le n).$$

That is, λ_j is an *eigenvalue* of H and x_j a corresponding *eigenvector* $(1 \le j \le n)$.

A real symmetric matrix A is *positive definite* if $x^t A x > 0$ for every real vector $x \ne 0$ (and *positive semi-definite* if $x^t A x \ge 0$ for every real vector x with equality for some $x \ne 0$). It follows from Theorem 10 that two real symmetric matrices can be simultaneously diagonalized, if one of them is positive definite, although the transforming matrix may not be orthogonal:

PROPOSITION 11 *If A and B are $n \times n$ real symmetric matrices, with A positive definite, then there exists an $n \times n$ nonsingular real matrix T such that $T^t A T$ and $T^t B T$ are both diagonal matrices.*

Proof By Theorem 10, there exists a real orthogonal matrix U such that $U^t A U$ is a diagonal matrix:

$$U^t A U = \text{diag } [\lambda_1, \ldots, \lambda_n].$$

Moreover, $\lambda_j > 0$ $(1 \le j \le n)$, since A is positive definite. Hence there exists $\delta_j > 0$ such that $\delta_j^2 = 1/\lambda_j$. If $D = \text{diag } [\delta_1, \ldots, \delta_n]$, then $D^t U^t A U D = I$. By Theorem 10 again, there exists a real orthogonal matrix V such that

$$V^t (D^t U^t B U D) V = \text{diag } [\mu_1, \ldots, \mu_n]$$

is a diagonal matrix. Hence we can take $T = UDV$. $\square$

Proposition 11 will now be used to obtain an inequality due to Fischer (1908):

PROPOSITION 12 *If G is a positive definite real symmetric matrix, and if*

$$G = \begin{bmatrix} G_1 & G_2 \\ G_2^{\,t} & G_3 \end{bmatrix}$$

is any partition of G, then

$$\det G \le \det G_1 \cdot \det G_3,$$

with equality if and only if $G_2 = 0$.

Proof We can write $G = Q^t H Q$, where

$$Q = \begin{bmatrix} I & 0 \\ G_3^{-1} G_2^{\,t} & I \end{bmatrix}, \; H = \begin{bmatrix} H_1 & 0 \\ 0 & G_3 \end{bmatrix},$$

and $H_1 = G_1 - G_2 G_3^{-1} G_2{}^t$. Since $\det G = \det H_1 \cdot \det G_3$, we need only show that $\det H_1 \le \det G_1$, with equality only if $G_2 = 0$.

Since G_1 and H_1 are both positive definite, they can be simultaneously diagonalized. Thus, if G_1 and H_1 are $p \times p$ matrices, there exists a nonsingular real matrix T such that

$$T^t G_1 T = \operatorname{diag}[\gamma_1,...,\gamma_p], \quad T^t H_1 T = \operatorname{diag}[\delta_1,...,\delta_p].$$

Since G_3^{-1} is positive definite, $u^t(G_1 - H_1)u \ge 0$ for any $u \in \mathbb{R}^p$. Hence $\gamma_i \ge \delta_i > 0$ for $i = 1,...,p$ and $\det G_1 \ge \det H_1$. Moreover $\det G_1 = \det H_1$ only if $\gamma_i = \delta_i$ ($i = 1,...,p$).

Hence if $\det G_1 = \det H_1$, then $G_1 = H_1$, i.e. $G_2 G_3^{-1} G_2{}^t = 0$. Thus $w^t G_3^{-1} w = 0$ for any vector $w = G_2{}^t v$. Since $w^t G_3^{-1} w = 0$ implies $w = 0$, it follows that $G_2 = 0$. $\quad\square$

From Proposition 12 we obtain by induction

PROPOSITION 13   *If $G = (\gamma_{jk})$ is an $m \times m$ positive definite real symmetric matrix, then*

$$\det G \le \gamma_{11}\gamma_{22}\cdots\gamma_{mm},$$

with equality if and only if G is a diagonal matrix. $\quad\square$

By applying Proposition 13 to the matrix $G = A^t A$, we obtain again Proposition 3. Proposition 13 may be sharpened in the following way:

PROPOSITION 14   *If $G = (\gamma_{jk})$ is an $m \times m$ positive definite real symmetric matrix, then*

$$\det G \le \gamma_{11} \prod_{j=2}^{m} (\gamma_{jj} - \gamma_{1j}{}^2/\gamma_{11}),$$

with equality if and only if $\gamma_{jk} = \gamma_{1j}\gamma_{1k}/\gamma_{11}$ for $2 \le j < k \le m$.

Proof   If

$$T = \begin{bmatrix} 1 & g \\ 0 & I_{m-1} \end{bmatrix},$$

where $g = (-\gamma_{12}/\gamma_{11}, \ldots, -\gamma_{1m}/\gamma_{11})$, then

$$T^t G T = \begin{bmatrix} \gamma_{11} & 0 \\ 0 & H \end{bmatrix},$$

where $H = (\eta_{jk})$ is an $(m-1) \times (m-1)$ positive definite real symmetric matrix with entries

$$\eta_{jk} = \gamma_{jk} - \gamma_{1j}\gamma_{1k}/\gamma_{11}. \qquad\qquad (2 \le j \le k \le m)$$

Since $\det G = \gamma_{11} \det H$, the result now follows from Proposition 13. $\square$

Some further inequalities for the determinants of positive definite matrices will now be derived, which will be applied to Hadamard's determinant problem in the next section. We again denote by J_m the $m \times m$ matrix whose entries are all 1.

LEMMA 15 *If* $C = \alpha I_m + \beta J_m$ *for some real* α, β, *then*

$$\det C = \alpha^{m-1}(\alpha + m\beta).$$

Moreover, if $\det C \ne 0$, *then* $C^{-1} = \gamma I_m + \delta J_m$, *where* $\gamma = \alpha^{-1}$ *and* $\delta = -\beta\alpha^{-1}(\alpha + m\beta)^{-1}$.

Proof Subtract the first row of C from each of the remaining rows, and then add to the first column of the resulting matrix each of the remaining columns. These operations do not alter the determinant and replace C by an upper triangular matrix with main diagonal entries $\alpha + m\beta$ (once) and α $(m - 1$ times). Hence $\det C = \alpha^{m-1}(\alpha + m\beta)$.

If $\det C \ne 0$ and if γ, δ are defined as in the statement of the lemma, then from $J_m^2 = mJ_m$ it follows directly that

$$(\alpha I_m + \beta J_m)(\gamma I_m + \delta J_m) = I_m. \quad \square$$

PROPOSITION 16 *Let* $G = (\gamma_{jk})$ *be an* $m \times m$ *positive definite real symmetric matrix such that* $|\gamma_{jk}| \ge \beta$ *for all* j,k *and* $\gamma_{jj} \le \alpha + \beta$ *for all* j, *where* $\alpha, \beta > 0$. *Then*

$$\det G \le \alpha^{m-1}(\alpha + m\beta). \qquad\qquad (2)$$

Moreover, equality holds if and only if there exists a diagonal matrix D, *with main diagonal elements* ± 1, *such that*

$$DGD = \alpha I_m + \beta J_m.$$

Proof The result is trivial if $m = 1$ and is easily verified if $m = 2$. We assume $m > 2$ and use induction on m. By replacing G by DGD, where D is a diagonal matrix whose main diagonal elements have absolute value 1, we may suppose that $\gamma_{1k} \ge 0$ for $2 \le k \le m$. Since the determinant is a linear function of its rows, we have

$$\det G = (\gamma_{11} - \beta)\delta + \eta,$$

where δ is the determinant of the matrix obtained from G by omitting the first row and column and η is the determinant of the matrix H obtained from G by replacing γ_{11} by β. By the induction hypothesis,

$$\delta \le \alpha^{m-2}(\alpha + m\beta - \beta).$$

If $\eta \le 0$, it follows that

$$\det G \le \alpha^{m-1}(\alpha + m\beta - \beta) < \alpha^{m-1}(\alpha + m\beta).$$

Thus we now suppose $\eta > 0$. Then H is positive definite, since the submatrix obtained by omitting the first row and column is positive definite. By Proposition 14,

$$\eta \le \beta \, \Pi_{j=2}^{m} \, (\gamma_{jj} - \gamma_{1j}{}^2/\beta),$$

with equality only if $\gamma_{jk} = \gamma_{1j}\gamma_{1k}/\beta$ for $2 \le j < k \le m$. Hence $\eta \le \alpha^{m-1}\beta$, with equality only if $\gamma_{jj} = \alpha + \beta$ for $2 \le j \le m$ and $\gamma_{jk} = \beta$ for $1 \le j < k \le m$. Consequently

$$\det G \le \alpha^{m-1}(\alpha + m\beta - \beta) + \alpha^{m-1}\beta = \alpha^{m-1}(\alpha + m\beta),$$

with equality only if $G = \alpha I_m + \beta J_m$. $\quad\square$

A square matrix will be called a *signed permutation matrix* if each row and column contains only one nonzero entry and this entry is 1 or -1.

PROPOSITION 17   *Let $G = (\gamma_{jk})$ be an $m \times m$ positive definite real symmetric matrix such that $\gamma_{jj} \le \alpha + \beta$ for all j and either $\gamma_{jk} = 0$ or $|\gamma_{jk}| \ge \beta$ for all j,k, where $\alpha,\beta > 0$.*
    Suppose in addition that $\gamma_{ik} = \gamma_{jk} = 0$ implies $\gamma_{ij} \ne 0$. Then

$$\det G \le \alpha^{m-2}(\alpha + m\beta/2)^2 \qquad\qquad\text{*if m is even,*}$$

$$\det G \le \alpha^{m-2}(\alpha + (m+1)\beta/2)(\alpha + (m-1)\beta/2) \quad \text{*if m is odd.*} \tag{3}$$

Moreover, equality holds if and only if there is a signed permutation matrix U such that

$$U^t G U = \begin{bmatrix} L & 0 \\ 0 & M \end{bmatrix},$$

where

$$L = M = \alpha I_{m/2} + \beta J_{m/2} \qquad\qquad \text{*if m is even,*}$$

$$L = \alpha I_{(m+1)/2} + \beta J_{(m+1)/2}, \quad M = \alpha I_{(m-1)/2} + \beta J_{(m-1)/2} \quad \text{*if m is odd.*}$$

Proof We are going to establish the inequality

$$\det G \le \alpha^{m-2}(\alpha + s\beta)(\alpha + m\beta - s\beta), \tag{4}$$

where s is the maximum number of zero elements in any row of G. Since, as a function of the real variable s, the quadratic on the right of (4) attains its maximum value for $s = m/2$, and has the same value for $s = (m + 1)/2$ as for $s = (m - 1)/2$, this will imply (3). It will also imply that if equality holds in (3), then $s = m/2$ if m is even and $s = (m + 1)/2$ or $(m - 1)/2$ if m is odd.

For $m = 2$ it is easily verified that (4) holds. We assume $m > 2$ and use induction. By performing the same signed permutation on rows and columns, we may suppose that the second row of G has the maximum number s of zero elements, and that all nonzero elements of the first row are positive and precede the zero elements. All the hypotheses of the proposition remain satisfied by the matrix G after this operation.

Let s' be the number of zero elements in the first row and put $r' = m - s'$. As in the proof of Proposition 16, we have

$$\det G \;=\; (\gamma_{11} - \beta)\delta + \eta,$$

where δ is the determinant of the matrix obtained from G by omitting the first row and column and η is the determinant of the matrix H obtained from G by replacing γ_{11} by β. We partition H in the form

$$H \;=\; \begin{bmatrix} L & N \\ N^t & M \end{bmatrix},$$

where L, M are square matrices of orders r', s' respectively. By construction all elements in the first row of L are positive and all elements in the first row of N are zero. Furthermore, by the hypotheses of the proposition, all elements of M have absolute value $\geq \beta$.

By the induction hypothesis,

$$\delta \;\leq\; \alpha^{m-3}(\alpha + s\beta)(\alpha + m\beta - \beta - s\beta).$$

If $\eta \leq 0$, it follows immediately that (4) holds with strict inequality. Thus we now suppose $\eta > 0$. Then H is positive definite and hence, by Fischer's inequality (Proposition 12), $\eta \leq \det L \cdot \det M$, with equality only if $N = 0$. But, by Proposition 14,

$$\det L \;\leq\; \beta \prod_{j=2}^{r'} (\gamma_{jj} - \gamma_{1j}^2/\beta) \;\leq\; \alpha^{r'-1}\beta$$

and, by Proposition 16,

$$\det M \;\leq\; \alpha^{s'-1}(\alpha + s'\beta).$$

Hence

$$\det G \;\leq\; \alpha^{m-2}(\alpha + s\beta)(\alpha + m\beta - \beta - s\beta) + \alpha^{m-2}\beta(\alpha + s'\beta),$$

Since $s' \leq s$, it follows that (4) holds and actually with strict inequality if $s' \neq s$.

If equality holds in (4) then, by Proposition 14, we must have $L = \alpha I_{r'} + \beta J_{r'}$, and by Proposition 16 after normalization we must also have $M = \alpha I_{s'} + \beta J_{s'}$. $\square$

5 Application to Hadamard's determinant problem

We have seen that, if A is an $n \times m$ real matrix with all entries ± 1, then $\det (A^t A) \leq n^m$, with strict inequality if $n > 2$ and n is not divisible by 4. The question arises, what is the maximum value of $\det (A^t A)$ in such a case? In the present section we use the results of the previous section to obtain some answers to this question. We consider first the case where n is odd.

PROPOSITION 18 *Let $A = (\alpha_{jk})$ be an $n \times m$ matrix with $\alpha_{jk} = \pm 1$ for all j,k. If n is odd, then*

$$\det (A^t A) \leq (n - 1)^{m-1}(n - 1 + m).$$

Moreover, equality holds if and only if $n \equiv 1 \bmod 4$ and, after changing the signs of some columns of A,

$$A^t A = (n - 1)I_m + J_m.$$

Proof We may assume $\det (A^t A) \neq 0$ and thus $m \leq n$. Then $A^t A = G = (\gamma_{jk})$ is a positive definite real symmetric matrix. For all j,k,

$$\gamma_{jk} = \alpha_{1j}\alpha_{1k} + \ldots + \alpha_{nj}\alpha_{nk}$$

is an integer and $\gamma_{jj} = n$. Moreover γ_{jk} is odd for all j,k, being the sum of an odd number of ± 1's. Hence the matrix G satisfies the hypotheses of Proposition 16 with $\alpha = n - 1$ and $\beta = 1$. Everything now follows from Proposition 16, except for the remark that if equality holds we must have $n \equiv 1 \bmod 4$.

But if $G = (n - 1)I_m + J_m$, then $\gamma_{jk} = 1$ for $j \neq k$. It now follows, by the argument used in the proof of Proposition 5, that any two distinct columns of A have the same entries in exactly $(n + 1)/2$ rows, and any three distinct columns of A have the same entries in exactly $(n + 3)/4$ rows. Thus $n \equiv 1 \bmod 4$. $\square$

Even if $n \equiv 1 \bmod 4$ there is no guarantee that that the upper bound in Proposition 18 is attained. However the question may be reduced to the existence of H-matrices if $m \neq n$. For suppose $m \leq n - 1$ and there exists an $(n - 1) \times m$ H-matrix B. If we put

$$A = \begin{bmatrix} B \\ e_m \end{bmatrix},$$

where e_m again denotes a row of m 1's, then $A^t A = (n - 1)I_m + J_m$.

On the other hand if $m = n$, then equality in Proposition 18 can hold only under very restrictive conditions. For in this case

$$(\det A)^2 = \det A^t A = (n - 1)^{n-1}(2n - 1)$$

and, since n is odd, it follows that $2n - 1$ is the square of an integer. It is an open question whether the upper bound in Proposition 18 is always attained when $m = n$ and $2n - 1$ is a square. However the nature of an extremal matrix, if one exists, can be specified rather precisely:

PROPOSITION 19 *If $A = (\alpha_{jk})$ is an $n \times n$ matrix with $n > 1$ odd and $\alpha_{jk} = \pm 1$ for all j,k, then*

$$\det (A^t A) \leq (n - 1)^{n-1}(2n - 1).$$

Moreover if equality holds, then $n \equiv 1 \bmod 4$, $2n - 1 = s^2$ for some integer s and, after changing the signs of some rows and columns of A, the matrix A must satisfy

$$A^t A = (n - 1)I_n + J_n, \quad AJ_n = sJ_n.$$

Proof By Proposition 18 and the preceding remarks, it only remains to show that if there exists an A such that $A^t A = (n - 1)I_n + J_n$ then, by changing the signs of some rows, we can ensure that also $AJ_n = sJ_n$.

Since $\det (AA^t) = \det (A^t A)$, it follows from Proposition 18 that there exists a diagonal matrix D with $D^2 = I_n$ such that

$$DAA^tD = (n - 1)I_n + J_n = A^t A.$$

Replacing A by DA, we obtain $AA^t = A^t A$. Then A commutes with $A^t A$ and hence also with J_n. Thus the rows and columns of A all have the same sum s and $AJ_n = sJ_n = A^t J_n$. Moreover $s^2 = 2n - 1$, since

$$s^2 J_n = sA^t J_n = A^t A J_n = (2n - 1)J_n. \quad \square$$

The maximum value of $\det (A^t A)$ when $n \equiv 3 \bmod 4$ is still something of a mystery. We now consider the remaining case when n is even, but not divisible by 4.

PROPOSITION 20 *Let $A = (\alpha_{jk})$ be an $n \times m$ matrix with $2 \leq m \leq n$ and $\alpha_{jk} = \pm 1$ for all j,k. If $n \equiv 2 \bmod 4$ and $n > 2$, then*

$$\det (A^t A) \leq (n - 2)^{m-2}(n - 2 + m)^2 \qquad \textit{if m is even,}$$
$$\det (A^t A) \leq (n - 2)^{m-2}(n - 1 + m)(n - 3 + m) \quad \textit{if m is odd.}$$

Moreover, equality holds if and only if there is a signed permutation matrix U such that

$$U^t A^t A U = \begin{bmatrix} L & 0 \\ 0 & M \end{bmatrix},$$

where

$$L = M = (n-2)I_{m/2} + 2J_{m/2} \qquad \textit{if m is even,}$$
$$L = (n-2)I_{(m+1)/2} + 2J_{(m+1)/2}, \quad M = (n-2)I_{(m-1)/2} + 2J_{(m-1)/2} \textit{ if m is odd.}$$

Proof We need only show that $G = A^t A$ satisfies the hypotheses of Proposition 17 with $\alpha = n - 2$ and $\beta = 2$. We certainly have $\gamma_{jj} = n$. Moreover all γ_{jk} are even, since n is even and

$$\gamma_{jk} = \alpha_{1j}\alpha_{1k} + \ldots + \alpha_{nj}\alpha_{nk}.$$

Hence $|\gamma_{jk}| \geq 2$ if $\gamma_{jk} \neq 0$. Finally, if j,k,ℓ are all different and $\gamma_{j\ell} = \gamma_{k\ell} = 0$, then

$$\sum_{i=1}^{n} (\alpha_{ij} + \alpha_{ik})(\alpha_{ij} + \alpha_{i\ell}) = n + \gamma_{jk}.$$

Since $n \equiv 2 \bmod 4$, it follows that also $\gamma_{jk} \equiv 2 \bmod 4$ and thus $\gamma_{jk} \neq 0$. $\quad\square$

Again there is no guarantee that the upper bound in Proposition 20 is attained. However the question may be reduced to the existence of H-matrices if $m \neq n, n - 1$. For suppose $m \leq n - 2$ and there exists an $(n-2)\times m$ H-matrix B. If we put

$$A = \begin{bmatrix} B \\ C \end{bmatrix},$$

where

$$C = \begin{bmatrix} e_r & e_s \\ e_r & -e_s \end{bmatrix},$$

and $r + s = m$, then

$$A^t A = \begin{bmatrix} (n-2)I_r + 2J_r & 0 \\ 0 & (n-2)I_s + 2J_s \end{bmatrix}.$$

Thus the upper bound in Proposition 20 is attained by taking $r = s = m/2$ when m is even and $r = (m + 1)/2, s = (m - 1)/2$ when m is odd.

Suppose now that $m = n$ and

$$A^t A = \begin{bmatrix} L & 0 \\ 0 & L \end{bmatrix},$$

where $L = (n-2)I_{n/2} + 2J_{n/2}$. If B is the $n\times(n-1)$ submatrix of A obtained by omitting the last column, then

$$B^tB = \begin{bmatrix} L & 0 \\ 0 & M \end{bmatrix},$$

where $M = (n-2)I_{n/2-1} + 2J_{n/2-1}$. Thus if the upper bound in Proposition 20 is attained for $m = n$, then it is also attained for $m = n - 1$. Furthermore, since

$$\det(AA^t) = \det(A^tA),$$

it follows from Proposition 20 that there exists a signed permutation matrix U such that

$$UAA^tU^t = A^tA.$$

Replacing A by UA, we obtain $AA^t = A^tA$. Then A commutes with A^tA. If

$$A = \begin{bmatrix} X & Y \\ Z & W \end{bmatrix},$$

is the partition of A into square submatrices of order $n/2$, it follows that X,Y,Z,W all commute with L and hence with $J_{n/2}$. This means that the entries in any row or any column of X have the same sum, which we will denote by x. Similarly the entries in any row or any column of Y,Z,W have the same sum, which will be denoted by y,z,w respectively. We may assume $x,y,w \geq 0$ by replacing A by

$$\begin{bmatrix} I_{n/2} & 0 \\ 0 & \pm I_{n/2} \end{bmatrix} A \begin{bmatrix} \pm I_{n/2} & 0 \\ 0 & \pm I_{n/2} \end{bmatrix},$$

We have

$$X^tX + Z^tZ = Y^tY + W^tW = L, \quad X^tY + Z^tW = 0,$$

and

$$XX^t + YY^t = ZZ^t + WW^t = L, \quad XZ^t + YW^t = 0.$$

Postmultiplying by J, we obtain

$$x^2 + z^2 = y^2 + w^2 = 2n - 2, \quad xy + zw = 0,$$

and

$$x^2 + y^2 = z^2 + w^2 = 2n - 2, \quad xz + yw = 0.$$

Adding, we obtain $x^2 = w^2$ and hence $x = w$. Thus $z^2 = y^2$ and actually $z = -y$, since $xy + zw = 0$.

This shows, in particular, that if the upper bound in Proposition 20 is attained for $m = n \equiv 2 \bmod 4$, then $2n - 2 = x^2 + y^2$, where x and y are integers. By Proposition II.39, such a representation is possible if and only if, for every prime $p \equiv 3 \bmod 4$, the highest power of p which divides $n - 1$ is even. Hence the upper bound in Proposition 20 is never attained if $m = n = 22$. On the other hand if $m = n = 6$, then $2n - 2 = 10 = 9 + 1$ and an extremal matrix A is obtained by taking $W = X = J_3$ and $Z = -Y = 2I_3 - J_3$.

It is an open question whether the upper bound in Proposition 20 is always attained when $m = n$ and $2n - 2$ is a sum of two squares. It is also unknown if, when an extremal matrix exists, one can always take $W = X$ and $Z = -Y$.

6 Designs

A *design* (in the most general sense) is a pair $(P,\mathscr{B})$, where P is a finite set of elements, called *points*, and $\mathscr{B}$ is a collection of subsets of P, called *blocks*. If $p_1,...,p_v$ are the points of the design and $B_1,...,B_b$ the blocks, then the *incidence matrix* of the design is the $v \times b$ matrix $A = (\alpha_{ij})$ of 0's and 1's defined by

$$\alpha_{ij} = \begin{cases} 1 & \text{if } p_i \in B_j, \\ 0 & \text{if } p_i \notin B_j. \end{cases}$$

Conversely, any $v \times b$ matrix $A = (\alpha_{ij})$ of 0's and 1's defines in this way a design. However, two such matrices define the same design if one can be obtained from the other by permutations of the rows and columns.

We will be interested in designs with rather more structure. A *2-design* or, especially in older literature, a 'balanced incomplete block design' (*BIBD*) is a design, with more than one point and more than one block, in which each block contains the same number k of points, each point belongs to the same number r of blocks, and every pair of distinct points occurs in the same number λ of blocks.

Thus each column of the incidence matrix contains k 1's and each row contains r 1's. Counting the total number of 1's in two ways, by columns and by rows, we obtain

$$bk = vr.$$

Similarly, by counting in two ways the 1's which lie below the 1's in the first row, we obtain

$$r(k - 1) = \lambda(v - 1).$$

Thus if v,k,λ are given, then r and b are determined and we may speak of a $2\text{-}(v,k,\lambda)$ design. Since $v > 1$ and $b > 1$, we have

$$1 < k < v, \quad 1 \le \lambda < r.$$

A $v \times b$ matrix $A = (\alpha_{ij})$ of 0's and 1's is the incidence matrix of a 2-design if and only if, for some positive integers k,r,λ,

$$\textstyle\sum_{i=1}^{v} \alpha_{ij} = k, \quad \sum_{k=1}^{b} \alpha_{ik}^2 = r, \quad \sum_{k=1}^{b} \alpha_{ik}\alpha_{jk} = \lambda \text{ if } i \ne j \quad (1 \le i,j \le v),$$

or in other words,

$$e_v A = ke_b, \quad AA^t = (r - \lambda)I_v + \lambda J_v, \tag{5}$$

where e_n is the $1 \times n$ matrix with all entries 1, I_n is the $n \times n$ unit matrix and J_n is the $n \times n$ matrix with all entries 1.

Designs have been used extensively in the design of agricultural and other experiments. To compare the yield of v varieties of a crop on b blocks of land, it would be expensive to test each variety separately on each block. Instead we can divide each block into k plots and use a 2-(v,k,λ) design, where $\lambda = bk(k - 1)/v(v - 1)$. Then each variety is used exactly $r = bk/v$ times, no variety is used more than once in any block, and any two varieties are used together in exactly λ blocks. As an example, take $v = 4$, $b = 6$, $k = 2$ and hence $\lambda = 1$, $r = 3$.

Some examples of 2-designs are the finite projective planes. In fact a *projective plane* of *order n* may be defined as a 2-(v,k,λ) design with

$$v = n^2 + n + 1, \quad k = n + 1, \quad \lambda = 1.$$

It follows that $b = v$ and $r = k$. The blocks in this case are called 'lines'. The projective plane of order 2, or *Fano plane*, is illustrated in Figure 1. There are seven points and seven blocks, the blocks being the six triples of collinear points and the triple of points on the circle.

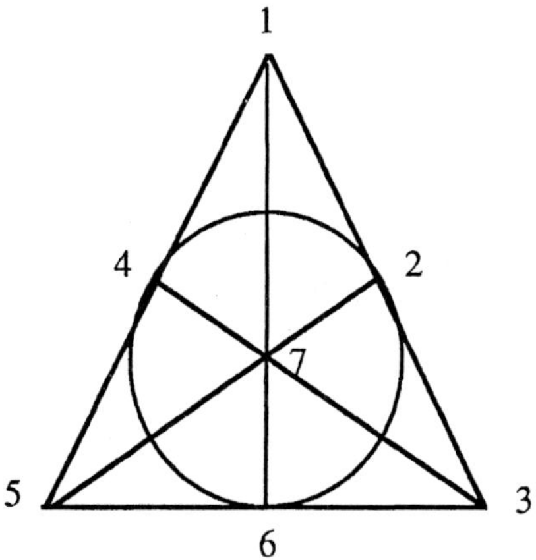

Figure 1: The Fano plane

Consider now an arbitrary 2-(v,k,λ) design. By (5) and Lemma 15,

$$\det(AA^t) = (r - \lambda)^{v-1}(r - \lambda + \lambda v) > 0,$$

since $r > \lambda$. This implies the inequality $b \geq v$, due to Fisher (1940), since AA^t would be singular if $b < v$.

A 2-design is said to be *square* or (more commonly, but misleadingly) 'symmetric' if $b = v$, i.e. if the number of blocks is the same as the number of points. Thus any projective plane is a square 2-design.

For a square 2-(v,k,λ) design, $k = r$ and the incidence matrix A is itself nonsingular. The first relation (5) is now equivalent to $J_v A = k J_v$. Since $k = r$, the sum of the entries in any row of A is also k and thus

$$J_v A^t = k J_v.$$

By multiplying the second relation (5) on the left by A^{-1} and on the right by A, we further obtain

$$A^t A = (r - \lambda)I_v + \lambda J_v.$$

Thus A^t is also the incidence matrix of a square 2-(v,k,λ) design, the *dual* of the given design.

The preceding partly combinatorial argument may be replaced by a purely matrix one:

LEMMA 21   *A nonsingular real $v \times v$ matrix A satisfies*

$$AA^t = (k - \lambda)I + \lambda J, \quad JA = kJ, \tag{6}$$

if and only if $k(k - 1) = \lambda(v - 1)$ and

$$A^t A = (k - \lambda)I + \lambda J, \quad JA^t = kJ. \tag{7}$$

Proof   It is enough to prove that (6) implies (7), since A may then be replaced by A^t. Moreover we may suppose $v > 1$, since the result is easily verified for $v = 1$. By Lemma 15,

$$\det(AA^t) = (k - \lambda)^{v-1}(k - \lambda + \lambda v).$$

Since A is nonsingular, it follows that $k \neq \lambda$ and $k - \lambda + \lambda v \neq 0$. Furthermore $k \neq 0$, since $J = kJA^{-1}$. From (6) we obtain

$$JA^t = (k - \lambda)JA^{-1} + \lambda JA^{-1}J.$$

Since $JA^{-1} = k^{-1}J$ and $J^2 = vJ$, it follows that

$$JA^t = k^{-1}(k - \lambda + \lambda v)J$$

and

$$JA^tJ = k^{-1}v(k - \lambda + \lambda v)J.$$

Thus JA^tJ is symmetric. Since $(JA^tJ)^t = JAJ = kvJ$, this implies

$$k^{-1}(k - \lambda + \lambda v) = k$$

and hence $JA^t = kJ$. Thus $AJ = (JA^t)^t = kJ$ and $A^{-1}J = k^{-1}J$. From (6) we now obtain

$$A^tA = (k - \lambda)I + \lambda A^{-1}JA = (k - \lambda)I + \lambda J. \quad \square$$

In Chapter VII we will derive necessary and sufficient conditions for the existence of a nonsingular *rational* $v \times v$ matrix A such that $AA^t = (k - \lambda)I + \lambda J$, and thus obtain some basic restrictions on the parameters v, k, λ for the existence of a square 2-(v,k,λ) design. These were first obtained by Bruck, Ryser and Chowla (1949/50).

We now consider the relationship between designs and Hadamard's determinant problem. By passing from A to $B = (J_n - A^t)/2$, it may be seen immediately that equality holds in Proposition 19 if and only if there exists a 2-(n,k,λ) design, where $k = (n - s)/2$, $\lambda = (n + 1 - 2s)/4$ and $s^2 = 2n - 1$.

We now show that with any Hadamard matrix $A = (\alpha_{jk})$ of order $n = 4d$ there is associated a 2-$(4d - 1, 2d - 1, d - 1)$ design. Assume without loss of generality that all elements in the first row and column of A are 1. We take $P = \{2,\dots,n\}$ as the set of points and $\mathcal{B} = \{B_2,\dots,B_n\}$ as the set of blocks, where $B_k = \{j \in P: \alpha_{jk} = 1\}$. Then B_k has cardinality $|B_k| = n/2 - 1$ for $k = 2,\dots,n$. Moreover, if T is any subset of P with $|T| = 2$, then the number of blocks containing T is $n/4 - 1$. The argument may also be reversed to show that any 2-$(4d - 1, 2d - 1, d - 1)$ design is associated in this way with a Hadamard matrix of order $4d$.

In particular, for $d = 2$, the 2-$(7,3,1)$ design associated with the Hadamard matrix $H_2 \otimes H_2 \otimes H_2$, where

$$H_2 = \begin{bmatrix} 1 & 1 \\ 1 & -1 \end{bmatrix},$$

is the projective plane of order 2 (Fano plane) illustrated in Figure 1.

The connection between Hadamard matrices and designs may also be derived by a matrix argument. If

$$A = \begin{bmatrix} 1 & e_{n-1} \\ e_{n-1}^t & \tilde{A} \end{bmatrix},$$

is a Hadamard matrix of order $n = 4d$, normalized so that its first row and column contain only 1's, then $B = (J_{n-1} + \tilde{A})/2$ is a matrix of 0's and 1's such that

$$J_{4d-1}B = (2d-1)J_{4d-1}, \quad BB^t = dI_{4d-1} + (d-1)J_{4d-1}.$$

The optimal spring balance design of order $4d-1$, obtained by taking $C = (J_{n-1} - \tilde{A})/2$, is a 2-$(4d-1,2d,d)$ design, since

$$J_{4d-1}C = 2dJ_{4d-1}, \quad CC^t = dI_{4d-1} + dJ_{4d-1}.$$

The notion of 2-design will now be generalized. Let t,v,k,λ be positive integers with $v \geq k \geq t$. A t-(v,k,λ) *design*, or simply a *t-design*, is a pair $(P,\mathcal{B})$, where P is a set of cardinality v and $\mathcal{B}$ is a collection of subsets of P, each of cardinality k, such that any subset of P of cardinality t is contained in exactly λ elements of $\mathcal{B}$. The elements of P will be called *points* and the elements of $\mathcal{B}$ will be called *blocks*. A t-(v,k,λ) design with $\lambda = 1$ is known as a *Steiner system*. The *automorphism group* of a t-design is the group of all permutations of the points which map blocks onto blocks.

If $t = 1$, then each point is contained in exactly λ blocks and so the number of blocks is $\lambda v/k$. Suppose now that $t > 1$. Let S be a fixed subset of P of cardinality $t-1$ and let λ' be the number of blocks which contain S. Consider the number of pairs (T,B), where $B \in \mathcal{B}$, $S \subseteq T \subseteq B$ and $|T| = t$. By first fixing B and varying T we see that this number is $\lambda'(k-t+1)$. On the other hand, by first fixing T and varying B we see that this number is $\lambda(v-t+1)$. Hence

$$\lambda' = \lambda(v-t+1)/(k-t+1)$$

does not depend on the choice of S and a t-(v,k,λ) design $(P,\mathcal{B})$ is also a $(t-1)$-(v,k,λ') design. By repeating this argument, we see that each point is contained in exactly r blocks, where

$$r = \lambda(v-t+1)...(v-1)/(k-t+1)...(k-1),$$

and the total number of blocks is $b = rv/k$. In particular, any t-design with $t > 2$ is also a 2-design.

With any Hadamard matrix $A = (\alpha_{jk})$ of order $n = 4d$ there is, in addition, associated a 3-$(4d,2d,d-1)$ design. For assume without loss of generality that all elements in the first column of A are 1. We take $P = \{1,2,...,n\}$ as the set of points and $\{B_2,...,B_n,B_2',...,B_n'\}$ as the set of blocks, where $B_k = \{j \in P: \alpha_{jk} = 1\}$ and $B_k' = \{j \in P: \alpha_{jk} = -1\}$. Then, by Proposition 5, $|B_k| = |B_k'| = n/2$ for $k = 2,...,n$. If T is any subset of P with $|T| = 3$, say $T = \{i,j,\ell\}$, then the number of blocks containing T is the number of $k > 1$ such that $\alpha_{ik} = \alpha_{jk} = \alpha_{\ell k}$. But, by Proposition 5 again, the number of columns of A which have the same entries in rows i,j,ℓ is $n/4$ and this includes the first column. Hence T is contained in exactly

$n/4 - 1$ blocks. Again the argument may be reversed to show that any $3\text{-}(4d,2d,d-1)$ design is associated in this way with a Hadamard matrix of order $4d$.

7 Groups and codes

A group is said to be *simple* if it contains more than one element and has no *normal* subgroups besides itself and the subgroup containing only the identity element. The finite simple groups are in some sense the building blocks from which all finite groups are constructed. There are several infinite families of them: the cyclic groups C_p of prime order p, the alternating groups $\mathcal{A}_n$ of all even permutations of n objects ($n \geq 5$), the groups $PSL(n,q)$ derived from the general linear groups of all invertible linear transformations of an n-dimensional vector space over a finite field of $q = p^m$ elements ($n \geq 2$ and $q > 3$ if $n = 2$), and some other families similar to the last which are analogues for a finite field of the simple Lie groups.

In addition to these infinite families there are 26 *sporadic* finite simple groups. (The *classification theorem* states that there are no other finite simple groups besides those already mentioned. The proof of the classification theorem at present occupies thousands of pages, scattered over a variety of journals, and some parts are actually still unpublished.) All except five of the sporadic groups were found in the years 1965-1981. However, the first five were found by Mathieu (1861,1873): M_{12} is a 5-fold transitive group of permutations of 12 objects of order $12 \cdot 11 \cdot 10 \cdot 9 \cdot 8$ and M_{11} the subgroup of all permutations in M_{12} which fix one of the objects; M_{24} is a 5-fold transitive group of permutations of 24 objects of order $24 \cdot 23 \cdot 22 \cdot 21 \cdot 20 \cdot 48$, M_{23} the subgroup of all permutations in M_{24} which fix one of the objects and M_{22} the subgroup of all permutations which fix two of the objects. The Mathieu groups may be defined in several ways, but the definitions by means of Hadamard matrices that we are going to give are certainly competitive with the others.

Two $n \times n$ Hadamard matrices H_1, H_2 are said to be *equivalent* if one may be obtained from the other by interchanging two rows or two columns, or by changing the sign of a row or a column, or by any finite number of such operations. Otherwise expressed, $H_2 = PH_1Q$, where P and Q are signed permutation matrices. An *automorphism* of a Hadamard matrix H is an equivalence of H with itself: $H = PHQ$. Since $P = HQ^{-1}H^{-1}$, the automorphism is uniquely determined by Q. Under matrix multiplication all admissible Q form a group $\mathcal{G}$, the *automorphism group* of the Hadamard matrix H. Evidently $-I \in \mathcal{G}$ and $-I$ commutes with all elements of $\mathcal{G}$. The factor group $\mathcal{G}/\{\pm I\}$, obtained by identifying Q and $-Q$, may be called the *reduced automorphism group* of H.

To illustrate these concepts we will show that all Hadamard matrices of order 12 are equivalent. In fact rather more is true:

PROPOSITION 22 *Any Hadamard matrix of order* 12 *may be brought to the form*

$$
\begin{array}{cccc}
{+}{+}{+} & {+}{+}{+} & {+}{+}{+} & {+}{+}{+} \\
{+}{+}{+} & {+}{+}{+} & {-}{-}{-} & {-}{-}{-} \\
{+}{+}{+} & {-}{-}{-} & {+}{+}{+} & {-}{-}{-} \\
{+}{-}{+} & {-}{+}{-} & {-}{+}{-} & {+}{-}{+} \\
{+}{+}{-} & {-}{-}{+} & {-}{-}{+} & {+}{-}{+} \\
{-}{+}{+} & {+}{-}{-} & {+}{-}{-} & {+}{-}{+} \\
{+}{-}{+} & {-}{-}{+} & {+}{-}{-} & {+}{+}{-} \\
{+}{-}{+} & {+}{-}{-} & {-}{-}{+} & {-}{+}{+} \\
{+}{+}{-} & {-}{+}{-} & {+}{-}{-} & {-}{+}{+} \\
{-}{+}{+} & {-}{+}{-} & {-}{-}{+} & {+}{+}{-} \\
{+}{+}{-} & {+}{-}{-} & {-}{+}{-} & {+}{+}{-} \\
{-}{+}{+} & {-}{-}{+} & {-}{+}{-} & {-}{+}{+}
\end{array}
\qquad (*)
$$

(where + stands for 1 and − for −1) by changing the signs of some rows and columns, by permuting the columns, and by permuting the first three rows and the last seven rows.

Proof Let $A = (\alpha_{jk})$ be a Hadamard matrix of order 12. By changing the signs of some columns we may assume that all elements of the first row are $+1$. Then, by the orthogonality relations, half the elements of any other row are $+1$. By permuting the columns we may assume that all elements in the first half of the second row are $+1$. It now follows from the orthogonality relations that in any row after the second the sum of all elements in each half is zero. Hence, by permuting the columns within each half we may assume that the third row is the same as the third row of the array $(*)$ displayed above. In the r-th row, where $r > 3$, let ρ_k be the sum of the entries in the k-th block of three columns ($k = 1,2,3,4$). The orthogonality relations now imply that

$$\rho_1 = \rho_4 = -\rho_2 = -\rho_3.$$

In the s-th row, where $s > 3$ and $s \neq r$, let σ_k be the sum of the entries in the k-th block of three columns. Then also

$$\sigma_1 = \sigma_4 = -\sigma_2 = -\sigma_3.$$

If $\rho_1 = \pm 3$, then all elements of the same triple of columns in the r-th row have the same sign and orthogonality to the s-th row implies $\sigma_1 = 0$, which is impossible because σ_1 is odd. Hence $\rho_1 = \pm 1$. By changing the signs of some rows we may assume that $\rho_1 = 1$ for every $r > 3$. By permuting columns within each block of three we may also normalize the 4-th row, so that the first four rows are now the same as the first four rows of the array (*).

In any row after the third, within a given block of three columns two elements have the same sign and the third element the opposite sign. Moreover, these signs depend only on the block and not on the row, since $\rho_1 = 1$. The scalar product of the triples from two different rows belonging to the same block of columns is 3 if the exceptional elements have the same position in the triple and is -1 otherwise. Since the two rows are orthogonal, the exceptional elements must have the same position in exactly one of the four blocks of columns. Thus if two rows after the 4-th have the same triple of elements in the k-th block as the 4-th row, then they have no other triple in common with the 4-th row or with one another. But this implies that if one of the two rows is given, then the other is uniquely determined. Hence no other row besides these two has the same triple of elements in the k-th block as the 4-th row. Since there are eight rows after the 4-th, and since each has exactly one triple in common with the 4-th row, it follows that, for each $k \in \{1,2,3,4\}$, exactly two of them have the same triple in the k-th block as the 4-th row.

The first four rows are unaltered by the following operations:

(i) interchange of the first and last columns of any triple of columns,

(ii) interchange of the second and third triple of columns, and then interchange of the second and third rows,

(iii) interchange of the first and fourth triple of columns, then interchange of the second and third rows and change of sign of these two rows,

(iv) interchange of the second and fourth triple of columns and change of their signs, then interchange of the first and third rows.

If we denote the elements of the r-th row $(r > 4)$ by $\xi_1,...,\xi_{12}$, then we have

$$\xi_1 + \xi_2 + \xi_3 = 1 = \xi_{10} + \xi_{11} + \xi_{12},$$
$$\xi_4 + \xi_5 + \xi_6 = -1 = \xi_7 + \xi_8 + \xi_9,$$
$$\xi_2 - \xi_5 - \xi_8 + \xi_{11} = 2.$$

In particular in the 5-th row we have $\alpha_{52} - \alpha_{55} - \alpha_{58} + \alpha_{5,11} = 2$. Thus α_{52} and $\alpha_{5,11}$ cannot both be -1 and by an operation (iii) we may assume that $\alpha_{52} = 1$. Similarly α_{55} and α_{58} cannot both be 1 and by an operation (ii) we may assume that $\alpha_{58} = -1$. Then $\alpha_{55} = \alpha_{5,11}$ and by an operation (iv) we may assume that $\alpha_{55} = \alpha_{5,11} = -1$. By operations (i) we may finally assume that the 5-th row is the same as the 5-th row of the array (*).

As we have already shown, exactly one row after the 5-th row has the same triple $+{-}+$ in the last block of columns as the 4-th and 5-th rows and this row must be the same as the 6-th row of the array (*). By permuting the last seven rows we may assume that this row is also the 6-th row of the given matrix, that the 7-th and 8-th rows have the same first triple of elements as the 4-th row, that the 9-th and 10-th rows have the same second triple of elements as the 4-th row, and that the 11-th and 12-th rows have the same third triple of elements as the 4-th row.

In any row after the 6-th we have, in addition to the relations displayed above, $\xi_{11} = 1$, $\xi_{10} + \xi_{12} = 0$ and

$$\xi_1 - \xi_4 - \xi_7 = \xi_2 - \xi_5 - \xi_8 = \xi_3 - \xi_6 - \xi_9 = 1.$$

In the 7-th and 8-th rows we have $\xi_1 = \xi_3 = 1$, $\xi_2 = -1$, and hence $\xi_5 = \xi_8 = -1$, $\xi_4 = -\xi_6 = -\xi_7 = \xi_9$. Since the first six rows are still unaltered by an operation (ii), and also by interchanging the first and third columns of the last block, we may assume that $\alpha_{74} = -1$, $\alpha_{7,10} = 1$. The 7-th and 8-th rows are now uniquely determined and are the same as the 7-th and 8-th rows of the array (*).

In any row after the 8-th we have

$$\xi_2 - \xi_6 - \xi_7 + \xi_{12} = 2 = \xi_2 - \xi_4 - \xi_9 + \xi_{10}.$$

In the 9-th and 10-th rows we have $\xi_5 = \xi_{11} = 1$ and $\xi_4 = \xi_6 = -1$. Hence $\xi_2 = -\xi_8 = 1$, $\xi_1 = \xi_7 = -\xi_3 = -\xi_9$, and finally $\xi_9 = \xi_{10} = -\xi_{12}$. Thus the 9-th and 10-th rows are together uniquely determined and may be ordered so as to coincide with the corresponding rows of the array (*). Similarly the 11-th and 12-th rows are together uniquely determined and may be ordered so as to coincide with the corresponding rows of the displayed array. $\square$

It follows from Proposition 22 that, for any five distinct rows of a Hadamard matrix of order 12, there exists exactly one pair of columns which either agree in all these rows or disagree in all these rows. Indeed, by permuting the rows we may arrange that the five given rows are the first five rows. Now, by Proposition 22, we may assume that the matrix has the form (*). But it is evident that in this case there is exactly one pair of columns which either agree or disagree in all the first five rows, namely the 10-th and 12-th columns.

Hence a 5-(12,6,1) design is obtained by taking the points to be elements of the set $P = \{1,\ldots,12\}$ and the blocks to be the $12 \cdot 11$ subsets B_{jk}, B_{jk}' with $j,k \in P$ and $j \neq k$, where

$$B_{jk} = \{i \in P: \alpha_{ij} = \alpha_{ik}\}, \quad B_{jk}' = \{i \in P: \alpha_{ij} \neq \alpha_{ik}\}.$$

The Mathieu group M_{12} may be defined as the automorphism group of this design or as the reduced automorphism group of any Hadamard matrix of order 12.

It is certainly not true in general that all Hadamard matrices of the same order n are equivalent. For example, there are 60 equivalence classes of Hadamard matrices of order 24. The Mathieu group M_{24} is connected with the Hadamard matrix of order 24 which is constructed by Paley's method, described in §2. The connection is not as immediate as for M_{12}, but the ideas involved are of general significance, as we now explain.

A sequence $x = (\xi_1,...,\xi_n)$ of n 0's and 1's may be regarded as a vector in the n-dimensional vector space $V = \mathbb{F}_2{}^n$ over the field of two elements. If we define the *weight* $|x|$ of the vector x to be the number of nonzero coordinates ξ_k, then

(i) $|x| \geq 0$ with equality if and only if $x = 0$,

(ii) $|x + y| \leq |x| + |y|$.

The vector space V acquires the structure of a metric space if we define the *(Hamming) distance* between the vectors x and y to be $d(x,y) = |x - y|$.

A *binary linear code* is a subspace U of the vector space V. If U has dimension k, then a *generator matrix* for the code is a $k \times n$ matrix G whose rows form a basis for U. The *automorphism group* of the code is the group of all permutations of the n coordinates which map U onto itself. An $[n,k,d]$-*binary code* is one for which V has dimension n, U has dimension k and d is the least weight of any nonzero vector in U.

There are useful connections between codes and designs. Corresponding to any design with incidence matrix A there is the binary linear code generated over $\mathbb{F}_2$ by the rows of A. Given a binary linear code U, on the other hand, a theorem of Assmus and Mattson (1969) provides conditions under which the nonzero vectors in U with minimum weight form the rows of the incidence matrix of a t-design.

Suppose now that H is a Hadamard matrix of order n, normalized so that all elements in the first row are 1. Then $A = (H + J_n)/2$ is a matrix of 0's and 1's with all elements in the first row 1. The code $C(H)$ *defined* by the Hadamard matrix H is the subspace generated by the rows of A, considered as vectors in the n-dimensional vector space $V = \mathbb{F}_2{}^n$.

In particular, take $H = H_{24}$ to be the Hadamard matrix of order 24 formed by Paley's construction:

$$H_{24} = I_{24} + \begin{bmatrix} 0 & e_{23} \\ -e_{23}{}^t & Q \end{bmatrix},$$

where $Q = (q_{jk})$ with $q_{jk} = 0$ if $j = k$ and otherwise $= 1$ or -1 according as $j - k$ is or is not a square mod 23 $(0 \leq j,k \leq 22)$. It may be shown that the *extended binary Golay code*

$G_{24} = C(H_{24})$ is a 12-dimensional subspace of $\mathbb{F}_2^{24}$, that the minimum weight of any nonzero vector in G_{24} is 8, and that the sets of nonzero coordinates of the vectors $x \in G_{24}$ with $|x| = 8$ form the blocks of a 5-(24,8,1) design. The Mathieu group M_{24} may be defined as the automorphism group of this design or as the automorphism group of the code G_{24}.

Again, suppose that $H^{(m)}$ is the Hadamard matrix of order $n = 2^m$ defined by

$$H^{(m)} = H_2 \otimes \ ... \ \otimes H_2 \quad (m \text{ factors}),$$

where

$$H_2 = \begin{bmatrix} 1 & 1 \\ 1 & -1 \end{bmatrix}.$$

The *first-order Reed–Muller code* $R(1,m) = C(H^{(m)})$ is an $(m + 1)$-dimensional subspace of $\mathbb{F}_2^n$ and the minimum weight of any nonzero vector in $R(1,m)$ is 2^{m-1}. It may be mentioned that the 3-$(2^m, 2^{m-1}, 2^{m-2} - 1)$ design associated with the Hadamard matrix $H^{(m)}$ has a simple geometrical interpretation. Its points are the points of m-dimensional affine space over the field of two elements, and its blocks are the hyperplanes of this space (not necessarily containing the origin).

In electronic communication a message is sent as a sequence of 'bits' (an abbreviation for *binary digits*), which may be realised physically by *off* or *on* and which may be denoted mathematically by 0 or 1. On account of noise the message received may differ slightly from that transmitted, and in some situations it is extremely important to detect and correct the errors. One way of doing so would be to send the same message many times, but it is an inefficient way. Instead suppose the message is composed of *codewords* of length n, taken from a subspace U of the vector space $V = \mathbb{F}_2^n$. There are 2^k different codewords, where k is the dimension of U. If the minimum weight of any nonzero vector in U is d, then any two distinct codewords differ in at least d places. Hence if a codeword $u \in U$ is transmitted and the received vector $v \in V$ contains less than $d/2$ errors, then v will be closer to u than to any other codeword. Thus if we are confident that any transmitted codeword will contain less than $d/2$ errors, we can correct them all by replacing each received vector by the codeword nearest to it.

The Golay code and the first-order Reed–Muller codes are of considerable practical importance in this connection. For the first-order Reed–Muller codes there is a fast algorithm for finding the nearest codeword to any received vector. Photographs of Mars taken by the Mariner 9 spacecraft were transmitted to Earth, using the code $R(1,5)$.

Other *error-correcting codes* are used with compact discs to ensure high quality sound reproduction by eliminating imperfections due, for example, to dust particles.

8 Further remarks

Kowalewski [22] gives a useful traditional account of determinants. Muir [28] is a storehouse of information on special types of determinants; the early Japanese work is described in Mikami [27].

Another approach to determinants, based on the work of Grassmann (1844), should be mentioned here, as it provides easy access to their formal properties and is used in the theory of differential forms. If V is an n-dimensional vector space over a field F, then there exists an associative algebra E, of dimension 2^n as a vector space over F, such that

(a) $V \subseteq E$,

(b) $v^2 = 0$ for every $v \in V$,

(c) V generates E, i.e. each element of E can be expressed as a sum of a scalar multiple of the unit element 1 and of a finite number of products of elements of V.

The associative algebra E, which is uniquely determined by these properties, is called the *Grassmann algebra* or *exterior algebra* of the vector space V. It is easily seen that any two products of n elements of V differ only by a scalar factor. Hence, for any linear transformation $A: V \to V$, there exists $d(A) \in F$ such that

$$(Av_1)\cdots(Av_n) = d(A)\, v_1\cdots v_n \quad \text{for all } v_1,\dots,v_n \in V.$$

Evidently $d(AB) = d(A)d(B)$ and in fact $d(A) = \det A$, if we identify A with its matrix with respect to some fixed basis of V. This approach to determinants is developed in Bourbaki [6]; see also Barnabei *et al.* [4].

Dieudonné (1943) has extended the notion of determinant to matrices with entries from a division ring; see Artin [1] and Cohn [9]. For a very different method, see Gelfand and Retakh [13].

The original paper of Hadamard (1893) is reproduced in [16]. Surveys on Hadamard matrices have been given by Hedayat and Wallis [19], Seberry and Yamada [34], and Craigen and Wallis [11]. Weighing designs are treated in Raghavarao [31]. For applications of Hadamard matrices to spectrometry, see Harwit and Sloane [18]. The proof of Proposition 8 is due to Shahriari [35].

Our proof of Theorem 10 is a pure existence proof. A more constructive approach was proposed by Jacobi (1846). If one applies to $n\times n$ matrices the method which we used for 2×2 matrices, one can annihilate a symmetric pair of off-diagonal entries. By choosing at each step an off-diagonal pair with maximum absolute value, one obtains a sequence of orthogonal transforms of the given symmetric matrix which converges to a diagonal matrix.

Calculating the eigenvalues of a real symmetric matrix has important practical applications, e.g. to problems of small oscillations in dynamical systems. Householder [21] and Golub and van Loan [14] give accounts of the various computational methods available.

Gantmacher [12] and Horn and Johnson [20] give general treatments of matrix theory, including the inequalities of Hadamard and Fischer. Our discussion of the Hadamard determinant problem for matrices of order not divisible by 4 is mainly based on Wojtas [37]. Further references are given in Neubauer and Ratcliffe [29].

Results of Brouwer (1983) are used in [29] to show that the upper bound in Proposition 19 is attained for infinitely many values of n. It follows that the upper bound in Proposition 20, with $m = n$, is also attained for infinitely many values of n. For if the $n{\times}n$ matrix A satisfies

$$A^t A = (n - 1)I_n + J_n,$$

then the $2n{\times}2n$ matrix

$$\tilde{A} = \begin{bmatrix} A & A \\ A & -A \end{bmatrix}$$

satisfies

$$\tilde{A}^t \tilde{A} = \begin{bmatrix} L & O \\ O & L \end{bmatrix},$$

where $L = 2\,A^t A = (2n - 2)I_n + 2J_n$.

There are introductions to design theory in Ryser [33], Hall [17], and van Lint and Wilson [25]. For more detailed information, see Brouwer [7], Lander [23] and Beth *et al.* [5]. Applications of design theory are treated in Chapter XIII of [5].

We mention two interesting results which are proved in Chapter 16 of Hall [17]. Given positive integers v,k,λ with $\lambda < k < v$:

(i) If $k(k - 1) = \lambda(v - 1)$ and if there exists a $v{\times}v$ matrix A of rational numbers such that

$$AA^t = (k - \lambda)I + \lambda J,$$

then A may be chosen so that in addition $JA = kJ$.

(ii) If there exists a $v{\times}v$ matrix A of integers such that

$$AA^t = (k - \lambda)I + \lambda J, \quad JA = kJ,$$

then every entry of A is either 0 or 1, and thus A is the incidence matrix of a square 2-design.

For introductions to the classification theorem for finite simple groups, see Aschbacher [2] and Gorenstein [15]. Detailed information about the finite simple groups is given in Conway *et al.* [10]. There is a remarkable connection between the largest sporadic simple group, nicknamed the 'Monster', and modular forms; see Ray [32].

Good introductions to coding theory are given by van Lint [24] and Pless [30]. MacWilliams and Sloane [26] is more comprehensive, but less up-to-date. Assmus and Mattson [3] is a useful survey article. Connections between codes, designs and graphs are treated in Cameron and van Lint [8]. The historical account in Thompson [36] recaptures the excitement of scientific discovery.

9 Selected references

[1] E. Artin, *Geometric algebra*, reprinted, Wiley, New York, 1988. [Original edition, 1957]

[2] M. Aschbacher, The classification of the finite simple groups, *Math. Intelligencer* **3** (1980/81), 59-65.

[3] E.F. Assmus Jr. and H.F. Mattson Jr., Coding and combinatorics, *SIAM Rev.* **16** (1974), 349-388.

[4] M. Barnabei, A. Brini and G.-C. Rota, On the exterior calculus of invariant theory, *J. Algebra* **96** (1985), 120-160.

[5] T. Beth, D. Jungnickel and H. Lenz, *Design theory*, 2nd ed., 2 vols., Cambridge University Press, 1999.

[6] N. Bourbaki, *Algebra I, Chapters 1-3*, Hermann, Paris, 1974. [French original, 1948]

[7] A.E. Brouwer, Block designs, *Handbook of combinatorics* (ed. R.L. Graham, M. Grötschel and L. Lóvasz), Vol. I, pp. 693-745, Elsevier, Amsterdam, 1995.

[8] P.J. Cameron and J.H. van Lint, *Designs, graphs, codes and their links*, Cambridge University Press, 1991.

[9] P.M. Cohn, *Algebra*, 2nd ed., Vol. 3, Wiley, Chichester, 1991.

[10] J.H. Conway *et al.*, *Atlas of finite groups*, Clarendon Press, Oxford, 1985.

[11] R. Craigen and W.D. Wallis, Hadamard matrices: 1893-1993, *Congr. Numer.* **97** (1993), 99-129.

[12] F.R. Gantmacher, *The theory of matrices*, English transl. by K. Hirsch, 2 vols., Chelsea, New York, 1960.

[13] I.M. Gelfand and V.S. Retakh, A theory of noncommutative determinants and characteristic functions of graphs, *Functional Anal. Appl.* **26** (1992), 231-246.

[14] G.H. Golub and C.F. van Loan, *Matrix computations*, 3rd ed., Johns Hopkins University Press, Baltimore, MD, 1996.

[15] D. Gorenstein, Classifying the finite simple groups, *Bull. Amer. Math. Soc. (N.S.)* **14** (1986), 1-98.

[16] J. Hadamard, Résolution d'une question relative aux determinants, *Selecta*, pp. 136-142, Gauthier-Villars, Paris, 1935 and *Oeuvres*, Tome I, pp. 239-245, CNRS, Paris, 1968.

[17] M. Hall, *Combinatorial theory*, 2nd ed., Wiley, New York, 1986.

[18] M. Harwit and N.J.A. Sloane, *Hadamard transform optics*, Academic Press, New York, 1979.

[19] A. Hedayat and W.D. Wallis, Hadamard matrices and their applications, *Ann. Statist.* **6** (1978), 1184-1238.

[20] R.A. Horn and C.A. Johnson, *Matrix analysis*, Cambridge University Press, 1985.

[21] A.S. Householder, *The theory of matrices in numerical analysis*, Blaisdell, New York, 1964.

[22] G. Kowalewski, *Einführung in die Determinantentheorie*, 4th ed., de Gruyter, Berlin, 1954.

[23] E.S. Lander, *Symmetric designs: an algebraic approach*, London Mathematical Society Lecture Note Series **74**, Cambridge University Press, 1983.

[24] J.H. van Lint, *Introduction to coding theory*, 3rd ed., Springer, Berlin, 2000.

[25] J.H. van Lint and R.M. Wilson, *A course in combinatorics*, Cambridge University Press, 1992.

[26] F.J. MacWilliams and N.J.A. Sloane, *The theory of error-correcting codes*, 2 vols., North-Holland, Amsterdam, 1977.

[27] Y. Mikami, *The development of mathematics in China and Japan*, 2nd ed., Chelsea, New York, 1974.

[28] T. Muir, *The theory of determinants in the historical order of development*, reprinted in 2 vols, Dover, New York, 1960.

[29] M.G. Neubauer and A.J. Ratcliffe, The maximum determinant of ± 1 matrices, *Linear Algebra Appl.* **257** (1997), 289-306.

[30] V. Pless, *Introduction to the theory of error-correcting codes*, 3rd ed., Wiley, New York, 1998.

[31] D. Raghavarao, *Constructions and combinatorial problems in design of experiments*, Wiley, New York, 1971.

[32] U. Ray, Generalized Kac–Moody algebras and some related topics, *Bull. Amer. Math. Soc. (N.S.)* **38** (2001), 1-42.

[33] H.J. Ryser, *Combinatorial mathematics*, Mathematical Association of America, 1963.

[34] J. Seberry and M. Yamada, Hadamard matrices, sequences and block designs, *Contemporary design theory* (ed. J.H. Dinitz and D.R. Stinson), pp. 431-560, Wiley, New York, 1992.

[35] S. Shahriari, On maximizing det $X^t X$, *Linear and multilinear algebra* **36** (1994), 275-278.

[36] T.M. Thompson, *From error-correcting codes through sphere packings to simple groups*, Mathematical Association of America, 1983.

[37] M. Wojtas, On Hadamard's inequality for the determinants of order non-divisible by 4, *Colloq. Math.* **12** (1964), 73-83.

VI
Hensel's p-adic numbers

The ring $\mathbb{Z}$ of all integers has a very similar algebraic structure to the ring $\mathbb{C}[z]$ of all polynomials in one variable with complex coefficients. This similarity extends to their fields of fractions, the field $\mathbb{Q}$ of rational numbers and the field $\mathbb{C}(z)$ of rational functions in one variable with complex coefficients. Hensel (1899) had the bold idea of pushing this analogy even further. For any $\zeta \in \mathbb{C}$, the ring $\mathbb{C}[z]$ may be embedded in the ring $\mathbb{C}_\zeta[[z]]$ of all functions $f(z) = \sum_{n \geq 0} \alpha_n (z - \zeta)^n$ with complex coefficients α_n which are holomorphic at ζ, and the field $\mathbb{C}(z)$ may be embedded in the field $\mathbb{C}_\zeta((z))$ of all functions $f(z) = \sum_{n \in \mathbb{Z}} \alpha_n (z - \zeta)^n$ with complex coefficients α_n which are meromorphic at ζ, i.e. $\alpha_n \neq 0$ for at most finitely many $n < 0$. Hensel constructed, for each prime p, a ring $\mathbb{Z}_p$ of all 'p-adic integers' $\sum_{n \geq 0} \alpha_n p^n$, where $\alpha_n \in \{0,1,\dots,p-1\}$, and a field $\mathbb{Q}_p$ of all 'p-adic numbers' $\sum_{n \in \mathbb{Z}} \alpha_n p^n$, where $\alpha_n \in \{0,1,\dots,p-1\}$ and $\alpha_n \neq 0$ for at most finitely many $n < 0$. This led him to arithmetic analogues of various analytic results and even to analytic methods of proving them. Hensel's idea of concentrating attention on one prime at a time has proved very fruitful for algebraic number theory. Furthermore, his methods enable the theory of algebraic numbers and the theory of algebraic functions of one variable to be developed completely in parallel.

Hensel simply defined p-adic integers by their power series expansions. We will adopt a more general approach, due to Kürschák (1913), which is based on absolute values.

1 Valued fields

Let F be an arbitrary field. An *absolute value* on F is a map $| \ | : F \to \mathbb{R}$ with the following properties:

(V1) $|0| = 0$, $|a| > 0$ *for all* $a \in F$ *with* $a \neq 0$;
(V2) $|ab| = |a|\,|b|$ *for all* $a,b \in F$;
(V3) $|a + b| \leq |a| + |b|$ *for all* $a,b \in F$.

A field with an absolute value will be called simply a *valued field*.

A *non-archimedean absolute value* on F is a map $| \ | : F \to \mathbb{R}$ with the properties **(V1)**, **(V2)** and

(V3)′ $|a + b| \leq \max(|a|,|b|)$ *for all* $a,b \in F$.

A non-archimedean absolute value is indeed an absolute value, since **(V1)** implies that **(V3)′** is a strengthening of **(V3)**. An absolute value is said to be *archimedean* if it is not non-archimedean.

The inequality **(V3)** is usually referred to as the *triangle inequality* and **(V3)′** as the 'strong triangle', or *ultrametric*, inequality.

If F is a field with an absolute value $| \ |$, then the set of real numbers $|a|$ for all nonzero $a \in F$ is clearly a subgroup of the multiplicative group of positive real numbers. This subgroup will be called the *value group* of the valued field.

Here are some examples to illustrate these definitions:

(i) An arbitrary field F has a *trivial* non-archimedean absolute value defined by

$$|0| = 0, \ |a| = 1 \text{ if } a \neq 0.$$

(ii) The ordinary absolute value

$$|a| = a \text{ if } a \geq 0, \ |a| = -a \text{ if } a < 0,$$

defines an archimedean absolute value on the field $\mathbb{Q}$ of rational numbers. We will denote this absolute value by $| \ |_{\infty}$ to avoid confusion with other absolute values on $\mathbb{Q}$ which will now be defined.

If p is a fixed prime, any rational number $a \neq 0$ can be uniquely expressed in the form $a = ep^v m/n$, where $e = \pm 1$, $v = v_p(a)$ is an integer and m,n are relatively prime positive integers which are not divisible by p. It is easily verified that a non-archimedean absolute value is defined on $\mathbb{Q}$ by putting

$$|0|_p = 0, \ |a|_p = p^{-v_p(a)} \text{ if } a \neq 0.$$

We call this the *p-adic absolute value*.

(iii) Let $F = K(t)$ be the field of all rational functions in one indeterminate with coefficients from some field K. Any rational function $f \neq 0$ can be uniquely expressed in the form $f = g/h$, where

g and h are relatively prime polynomials with coefficients from K and h is *monic* (i.e., has leading coefficient 1). If we denote the degrees of g and h by $\partial(g)$ and $\partial(h)$, then a non-archimedean absolute value is defined on F by putting, for a fixed $q > 1$,

$$|0|_\infty = 0, \quad |f|_\infty = q^{\partial(g) - \partial(h)} \quad \text{if } f \neq 0.$$

Other absolute values on F can be defined in the following way. If $p \in K[t]$ is a fixed irreducible polynomial, then any rational function $f \neq 0$ can be uniquely expressed in the form $f = p^v g/h$, where $v = v_p(f)$ is an integer, g and h are relatively prime polynomials with coefficients from K which are not divisible by p, and h is monic. It is easily verified that a non-archimedean absolute value is defined on F by putting, for a fixed $q > 1$,

$$|0|_p = 0, \quad |f|_p = q^{-\partial(p) v_p(f)} \quad \text{if } f \neq 0.$$

(iv)   Let $F = K((t))$ be the field of all formal Laurent series $f(t) = \sum_{n \in \mathbb{Z}} \alpha_n t^n$ with coefficients $\alpha_n \in K$ such that $\alpha_n \neq 0$ for at most finitely many $n < 0$. A non-archimedean absolute value is defined on F by putting, for a fixed $q > 1$,

$$|0| = 0, \quad |f| = q^{-v(f)} \quad \text{if } f \neq 0,$$

where $v(f)$ is the least integer n such that $\alpha_n \neq 0$.

(v)   Let $F = \mathbb{C}_\zeta((z))$ be the field of all complex-valued functions $f(z) = \sum_{n \in \mathbb{Z}} \alpha_n (z - \zeta)^n$ which are meromorphic at $\zeta \in \mathbb{C}$. Any $f \in F$ which is not identically zero can be uniquely expressed in the form $f(z) = (z - \zeta)^v g(z)$, where $v = v_\zeta(f)$ is an integer, g is holomorphic at ζ and $g(\zeta) \neq 0$. A non-archimedean absolute value is defined on F by putting, for a fixed $q > 1$,

$$|0|_\zeta = 0, \quad |f|_\zeta = q^{-v_\zeta(f)} \quad \text{if } f \neq 0.$$

It should be noted that in examples (iii) and (iv) the restriction of the absolute value to the ground field K is the trivial absolute value, and the same holds in example (v) for the restriction of the absolute value to $\mathbb{C}$. For all the absolute values considered in examples (iii)–(v) the value group is an infinite cyclic group.

We now derive some simple properties common to all absolute values. The notation in the statement of the following lemma is a bit sloppy, since we use the same symbol to denote the unit elements of both F and $\mathbb{R}$ (as we have already done for the zero elements).

LEMMA 1 *In any field F with an absolute value $|\ |$ the following properties hold:*

(i) $|1| = 1, |-1| = 1$ *and, more generally, $|a| = 1$ for every $a \in F$ which is a root of unity;*

(ii) $|-a| = |a|$ *for every $a \in F$;*

(iii) $\big||a| - |b|\big|_\infty \le |a - b|$ *for all $a,b \in F$, where $|\ |_\infty$ is the ordinary absolute value on $\mathbb{R}$;*

(iv) $|a^{-1}| = |a|^{-1}$ *for every $a \in F$ with $a \ne 0$.*

Proof By taking $a = b = 1$ in **(V2)** and using **(V1)**, we obtain $|1| = 1$. If $a^n = 1$ for some positive integer n, it now follows from **(V2)** that $\alpha = |a|$ satisfies $\alpha^n = 1$. Since $\alpha > 0$, this implies $\alpha = 1$. In particular, $|-1| = 1$. Taking $b = -1$ in **(V2)**, we now obtain (ii).

Replacing a by $a - b$ in **(V3)**, we obtain

$$|a| - |b| \le |a - b|.$$

Since a and b may be interchanged, by (ii), this implies (iii). Finally, if we take $b = a^{-1}$ in **(V2)** and use (i), we obtain (iv). $\square$

It follows from Lemma 1(i) that a finite field admits only the trivial absolute value.

We show next how non-archimedean and archimedean absolute values may be distinguished from one another. The notation in the statement of the following proposition is very sloppy, since we use the same symbol to denote both the positive integer n and the sum $1 + 1 + \ldots + 1$ (n summands), although the latter may be 0 if the field has prime characteristic.

PROPOSITION 2 *Let F be a field with an absolute value $|\ |$. Then the following properties are equivalent:*

(i) $|2| \le 1$;

(ii) $|n| \le 1$ *for every positive integer n;*

(iii)   *the absolute value $|\ |$ is non-archimedean.*

Proof It is trivial that (iii) $\Rightarrow$ (i). Suppose now that (i) holds. Then $|2^k| = |2|^k \le 1$ for any positive integer k. An arbitrary positive integer n can be written to the base 2 in the form

$$n = a_0 + a_1 2 + \ldots + a_g 2^g,$$

where $a_i \in \{0,1\}$ for all $i < g$ and $a_g = 1$. Then

$$|n| \le |a_0| + |a_1| + \ldots + |a_g| \le g + 1.$$

Now consider the powers n^k. Since $n < 2^{g+1}$, we have $n^k < 2^{k(g+1)}$ and hence

$$n^k = b_0 + b_1 2 + \dots + b_h 2^h,$$

where $b_j \in \{0,1\}$ for all $j < h$, $b_h = 1$ and $h < k(g + 1)$. Thus

$$|n|^k = |n^k| \le h + 1 \le k(g + 1).$$

Taking k-th roots and letting $k \to \infty$, we obtain $|n| \le 1$, since $k^{1/k} = e^{(\log k)/k} \to 1$ and likewise $(g + 1)^{1/k} = e^{(\log (g+1))/k} \to 1$. Thus (i) $\Rightarrow$ (ii).

Suppose next that (ii) holds. Then, since the binomial coefficients are positive integers,

$$|x + y|^n = |(x + y)^n| = \left| \sum_{k=0}^{n} \binom{n}{k} x^k y^{n-k} \right|$$

$$\le \sum_{k=0}^{n} |x|^k |y|^{n-k}$$

$$\le (n + 1)\rho^n,$$

where $\rho = \max(|x|,|y|)$. Taking n-th roots and letting $n \to \infty$, we obtain $|x + y| \le \rho$. Thus (ii) $\Rightarrow$ (iii). $\square$

It follows from Proposition 2 that for an archimedean absolute value the sequence $(|n|)$ is unbounded, since $|2^k| \to \infty$ as $k \to \infty$. Consequently, for any $a,b \in F$ with $a \ne 0$, there is a positive integer n such that $|na| > |b|$. The name 'archimedean' derives from the analogy between this property and the archimedean axiom of geometry. It follows also from Proposition 2 that any absolute value on a field of prime characteristic is non-archimedean, since there are only finitely many distinct values of $|n|$.

2 Equivalence

If λ,μ,α are positive real numbers with $\alpha < 1$, then

$$\left(\frac{\lambda}{\lambda + \mu} \right)^{\alpha} + \left(\frac{\mu}{\lambda + \mu} \right)^{\alpha} > \frac{\lambda}{\lambda + \mu} + \frac{\mu}{\lambda + \mu} = 1$$

and hence

$$\lambda^{\alpha} + \mu^{\alpha} > (\lambda + \mu)^{\alpha}.$$

It follows that if $|\,|$ is an absolute value on a field F and if $0 < \alpha < 1$, then $|\,|^{\alpha}$ is also an absolute value, since

$$|a + b|^{\alpha} \le (|a| + |b|)^{\alpha} \le |a|^{\alpha} + |b|^{\alpha}.$$

Actually, if $|\ |$ is a *non-archimedean* absolute value on a field F, then it follows directly from the definition that, for any $\alpha > 0$, $|\ |^\alpha$ is also a non-archimedean absolute value on F. However, if $|\ |$ is an *archimedean* absolute value on F then, for all large $\alpha > 0$, $|\ |^\alpha$ is not an absolute value on F. For $|2| > 1$ and hence, if $\alpha > \log 2/\log |2|$,

$$|1 + 1|^\alpha > 2 = |1|^\alpha + |1|^\alpha.$$

PROPOSITION 3 *Let* $|\ |_1$ *and* $|\ |_2$ *be absolute values on a field* F *such that* $|a|_2 < 1$ *for any* $a \in F$ *with* $|a|_1 < 1$. *If* $|\ |_1$ *is nontrivial, then there exists a real number* $\rho > 0$ *such that*

$$|a|_2 = |a|_1{}^\rho \text{ for every } a \in F.$$

Proof By taking inverses we see that also $|a|_2 > 1$ for any $a \in F$ with $|a|_1 > 1$. Choose $b \in F$ with $|b|_1 > 1$. For any nonzero $a \in F$ we have $|a|_1 = |b|_1{}^\gamma$, where

$$\gamma = \log |a|_1/\log |b|_1.$$

Let m,n be integers with $n > 0$ such that $m/n > \gamma$. Then $|a|_1{}^n < |b|_1{}^m$ and hence $|a^n/b^m|_1 < 1$. Therefore also $|a^n/b^m|_2 < 1$ and by reversing the argument we obtain

$$m/n > \log |a|_2/\log |b|_2.$$

Similarly if m',n' are integers with $n' > 0$ such that $m'/n' < \gamma$, then

$$m'/n' < \log |a|_2/\log |b|_2.$$

It follows that

$$\log |a|_2/\log |b|_2 = \gamma = \log |a|_1/\log |b|_1.$$

Thus if we put $\rho = \log |b|_2/\log |b|_1$, then $\rho > 0$ and $|a|_2 = |a|_1{}^\rho$. This holds trivially also for $a = 0$. $\square$

Two absolute values, $|\ |_1$ and $|\ |_2$, on a field F are said to be *equivalent* when, for any $a \in F$,

$$|a|_1 < 1 \text{ if and only if } |a|_2 < 1.$$

This implies that $|a|_1 > 1$ if and only if $|a|_2 > 1$ and hence also that $|a|_1 = 1$ if and only if $|a|_2 = 1$. Thus if one absolute value is trivial, so also is the other. It now follows from Proposition 3 that *two absolute values,* $|\ |_1$ *and* $|\ |_2$, *on a field* F *are equivalent if and only if there exists a real number* $\rho > 0$ *such that* $|a|_2 = |a|_1{}^\rho$ *for every* $a \in F$.

We have seen that the field $\mathbb{Q}$ of rational numbers admits the p-adic absolute values $|\ |_p$ in addition to the ordinary absolute value $|\ |_\infty$. These absolute values are all inequivalent since, if p and q are distinct primes,

$$|p|_p < 1, \quad |p|_q = 1, \quad |p|_\infty = p > 1.$$

It was first shown by Ostrowski (1918) that these are essentially the only absolute values on $\mathbb{Q}$:

PROPOSITION 4 *Every nontrivial absolute value $|\ |$ of the rational field $\mathbb{Q}$ is equivalent either to the ordinary absolute value $|\ |_\infty$ or to a p-adic absolute value $|\ |_p$ for some prime p.*

Proof Let b,c be integers > 1. By writing c to the base b, we obtain

$$c = c_m b^m + c_{m-1} b^{m-1} + \ldots + c_0,$$

where $0 \le c_j < b$ $(j = 0,\ldots,m)$ and $c_m \ne 0$. Then $m \le \log c/\log b$, since $c_m \ge 1$. If we put $\mu = \max_{1 \le d < b} |d|$, it follows from the triangle inequality that

$$|c| \le \mu(1 + \log c/\log b)\{\max(1,|b|)\}^{\log c/\log b}.$$

Taking $c = a^n$ we obtain, for any $a > 1$,

$$|a| \le \mu^{1/n}(1 + n \log a/\log b)^{1/n}\{\max(1,|b|)\}^{\log a/\log b}$$

and hence, letting $n \to \infty$,

$$|a| \le \{\max(1,|b|)\}^{\log a/\log b}.$$

Suppose first that $|a| > 1$ for some $a > 1$. It follows that $|b| > 1$ for every $b > 1$ and

$$|b|^{1/\log b} \ge |a|^{1/\log a}.$$

In fact, since a and b may now be interchanged,

$$|b|^{1/\log b} = |a|^{1/\log a}.$$

Thus $\rho = \log |a|/\log a$ is a positive real number independent of $a > 1$ and $|a| = a^\rho$. It follows that $|a| = |a|_\infty^\rho$ for every rational number a. Thus the absolute value is equivalent to the ordinary absolute value.

Suppose next that $|a| \le 1$ for every $a > 1$ and so for every $a \in \mathbb{Z}$. Since the absolute value on $\mathbb{Q}$ is nontrivial, we must have $|a| < 1$ for some integer $a \ne 0$. The set M of all $a \in \mathbb{Z}$ such that $|a| < 1$ is a proper ideal in $\mathbb{Z}$ and hence is generated by an integer $p > 1$. We will show that p must be a prime. Suppose $p = bc$, where b and c are positive integers. Since $|b||c| = |p| < 1$, we may assume without loss of generality that $|b| < 1$. Then $b \in M$ and thus $b = pd$ for some $d \in \mathbb{Z}$. Hence $cd = 1$ and so $c = 1$. Thus p has no nontrivial factorization.

Every rational number $a \neq 0$ can be expressed in the form $a = p^v b/c$, where v is an integer and b,c are integers not divisible by p. Hence $|b| = |c| = 1$ and $|a| = |p|^v$. We can write $|p| = p^{-\rho}$, for some real number $\rho > 0$. Then $|a| = p^{-v\rho} = |a|_p{}^\rho$, and thus the absolute value is equivalent to the p-adic absolute value. $\square$

Similarly, the absolute values on the field $F = K(t)$ considered in example (iii) of §1 are all inequivalent and it may be shown that any nontrivial absolute value on F whose restriction to K is trivial is equivalent to one of these absolute values.

In example (ii) of §1 we have made a specific choice in each class of equivalent absolute values. The choice which has been made ensures the validity of the *product formula*: for any nonzero $a \in \mathbb{Q}$,

$$|a|_\infty \, \Pi_p |a|_p \; = \; 1,$$

where $|a|_p \neq 1$ for at most finitely many p.

Similarly, in example (iii) of §1 the absolute values have been chosen so that, for any nonzero $f \in K(t)$, $|f|_\infty \, \Pi_p |f|_p = 1$, where $|f|_p \neq 1$ for at most finitely many p.

The following *approximation theorem*, due to Artin and Whaples (1945), treats several absolute values simultaneously. For p-adic absolute values of the rational field $\mathbb{Q}$ the result also follows from the Chinese remainder theorem (Corollary II.38).

PROPOSITION 5 *Let* $|\;|_1,...,|\;|_m$ *be nontrivial pairwise inequivalent absolute values of an arbitrary field F and let $x_1,...,x_m$ be any elements of F. Then for each real $\varepsilon > 0$ there exists an $x \in F$ such that*

$$|x - x_k|_k < \varepsilon \quad for \; 1 \leq k \leq m.$$

Proof During the proof we will more than once use the fact that if $f_n(x) = x^n(1 + x^n)^{-1}$, then $|f_n(a)| \to 0$ or 1 as $n \to \infty$ according as $|a| < 1$ or $|a| > 1$.

We show first that *there exists an $a \in F$ such that*

$$|a|_1 > 1, \quad |a|_k < 1 \; for \; 2 \leq k \leq m.$$

Since $|\;|_1$ and $|\;|_2$ are nontrivial and inequivalent, there exist $b,c \in F$ such that

$$|b|_1 < 1, \quad |b|_2 \geq 1,$$
$$|c|_1 \geq 1, \quad |c|_2 < 1.$$

If we put $a = b^{-1}c$, then $|a|_1 > 1$, $|a|_2 < 1$. This proves the assertion for $m = 2$. We now assume $m > 2$ and use induction. Then there exist $b,c \in F$ such that

$$|b|_1 > 1, \ |b|_k < 1 \text{ for } 1 < k < m,$$
$$|c|_1 > 1, \ |c|_m < 1.$$

If $|b|_m < 1$ we can take $a = b$. If $|b|_m = 1$ we can take $a = b^n c$ for sufficiently large n. If $|b|_m > 1$ we can take $a = f_n(b)c$ for sufficiently large n.

Thus for each $i \in \{1,...,m\}$ we can choose $a_i \in F$ so that

$$|a_i|_i > 1, \ |a_i|_k < 1 \text{ for all } k \neq i.$$

Then

$$x = x_1 f_n(a_1) + ... + x_m f_n(a_m)$$

satisfies the requirements of the proposition for sufficiently large n. $\square$

It follows from Proposition 5, that if $|\ |_1,...,|\ |_m$ are nontrivial pairwise inequivalent absolute values of a field F, then there exists an $a \in F$ such that $|a|_k > 1$ $(k = 1,...,m)$. Consequently the absolute values are *multiplicatively independent*, i.e. if $\rho_1,...,\rho_m$ are nonnegative real numbers, not all zero, then for some nonzero $a \in F$,

$$|a|_1^{\rho_1} \cdots |a|_m^{\rho_m} \neq 1.$$

3   Completions

Any field F with an absolute value $|\ |$ has the structure of a metric space, with the metric

$$d(a,b) = |a - b|,$$

and thus has an associated topology. Since $|a| < 1$ if and only if $a^n \to 0$ as $n \to \infty$, it follows that two absolute values are equivalent if and only if the induced topologies are the same.

When we use topological concepts in connection with valued fields we will always refer to the topology induced by the metric space structure. In this sense addition and multiplication are continuous operations, since

$$|(a + b) - (a_0 + b_0)| \leq |a - a_0| + |b - b_0|,$$
$$|ab - a_0 b_0| \leq |a - a_0||b| + |a_0||b - b_0|.$$

Inversion is also continuous at any point $a_0 \neq 0$, since if $|a - a_0| < |a_0|/2$ then $|a_0| < 2|a|$ and

$$|a^{-1} - a_0^{-1}| \ = \ |a - a_0||a|^{-1}|a_0|^{-1} \ < \ 2|a_0|^{-2}|a - a_0|.$$

Thus a valued field is a *topological field*.

It will now be shown that the procedure by which Cantor extended the field of rational numbers to the field of real numbers can be generalized to any valued field.

Let F be a field with an absolute value $|\ |$. A sequence (a_n) of elements of F is said to *converge* to an element a of F, and a is said to be the *limit* of the sequence (a_n), if for each real $\varepsilon > 0$ there is a corresponding positive integer $N = N(\varepsilon)$ such that

$$|a_n - a| < \varepsilon \ \text{ for all } n \geq N.$$

It is easily seen that the limit of a convergent sequence is uniquely determined.

A sequence (a_n) of elements of F is said to be a *fundamental sequence* if for each $\varepsilon > 0$ there is a corresponding positive integer $N = N(\varepsilon)$ such that

$$|a_m - a_n| < \varepsilon \ \text{ for all } m,n \geq N.$$

Any convergent sequence is a fundamental sequence, since

$$|a_m - a_n| \ \leq \ |a_m - a| + |a_n - a|,$$

but the converse need not hold. However, any fundamental sequence is bounded since, if $m = N(1)$, then for $n \geq m$ we have

$$|a_n| \ \leq \ |a_m - a_n| + |a_m| \ < \ 1 + |a_m|.$$

Thus $|a_n| \leq \mu$ for all n, where $\mu = \max\{|a_1|,...,|a_{m-1}|,1 + |a_m|\}$.

The preceding definitions are specializations of the definitions for an arbitrary metric space (cf. Chapter I, §4). We now take advantage of the algebraic structure of F. Let $A = (a_n)$ and $B = (b_n)$ be two fundamental sequences. We write $A = B$ if $a_n = b_n$ for all n, and we define the sum and product of A and B to be the sequences

$$A + B \ = \ (a_n + b_n), \quad AB \ = \ (a_n b_n).$$

These are again fundamental sequences. For we can choose $\mu \geq 1$ so that $|a_n| \leq \mu$, $|b_n| \leq \mu$ for all n and then choose a positive integer N so that

$$|a_m - a_n| < \varepsilon/2\mu, \quad |b_m - b_n| < \varepsilon/2\mu \ \text{ for all } m,n \geq N.$$

It follows that, for all $m,n \geq N$,

$$|(a_m + b_m) - (a_n + b_n)| \leq |a_m - a_n| + |b_m - b_n| < \varepsilon/2\mu + \varepsilon/2\mu \leq \varepsilon,$$

and similarly

$$|a_m b_m - a_n b_n| \leq |a_m - a_n||b_m| + |a_n||b_m - b_n| < (\varepsilon/2\mu)\mu + (\varepsilon/2\mu)\mu = \varepsilon.$$

It is easily seen that the set $\mathscr{F}$ of all fundamental sequences is a commutative ring with respect to these operations. The subset of all constant sequences (a), i.e. $a_n = a$ for all n, forms a field isomorphic to F. Thus we may regard F as embedded in $\mathscr{F}$.

Let $\mathscr{N}$ denote the subset of $\mathscr{F}$ consisting of all sequences (a_n) which converge to 0. Evidently $\mathscr{N}$ is a subring of $\mathscr{F}$ and actually an ideal, since any fundamental sequence is bounded. We will show that $\mathscr{N}$ is even a maximal ideal.

Let (a_n) be a fundamental sequence which is not in $\mathscr{N}$. Then there exists $\mu > 0$ such that $|a_\nu| \geq \mu$ for infinitely many ν. Since $|a_m - a_n| < \mu/2$ for all $m,n \geq N$, it follows that $|a_n| > \mu/2$ for all $n \geq N$. Put $b_n = a_n^{-1}$ if $a_n \neq 0$, $b_n = 0$ if $a_n = 0$. Then (b_n) is a fundamental sequence since, for $m,n \geq N$,

$$|b_m - b_n| = |(a_n - a_m)/a_m a_n| \leq 4\mu^{-2}|a_n - a_m|.$$

Since $(1) - (b_n a_n) \in \mathscr{N}$, the ideal generated by (a_n) and $\mathscr{N}$ contains the constant sequence (1) and hence every sequence in $\mathscr{F}$. Since this holds for each sequence $(a_n) \in \mathscr{F} \setminus \mathscr{N}$, the ideal $\mathscr{N}$ is maximal.

Consequently (see Chapter I, §8) the quotient $\overline{F} = \mathscr{F}/\mathscr{N}$ is a field. Since (0) is the only constant sequence in $\mathscr{N}$, by mapping each constant sequence into the coset of $\mathscr{N}$ which contains it we obtain a field in $\overline{F}$ isomorphic to F. Thus we may regard F as embedded in $\overline{F}$.

It follows from Lemma 1(iii), and from the completeness of the field of real numbers, that $|A| = \lim_{n \to \infty} |a_n|$ exists for any fundamental sequence $A = (a_n)$. Moreover,

$$|A| \geq 0, \quad |AB| = |A||B|, \quad |A + B| \leq |A| + |B|.$$

Furthermore $|A| = 0$ if and only if $A \in \mathscr{N}$. It follows that $|B| = |C|$ if $B - C \in \mathscr{N}$, since

$$|B| \leq |B - C| + |C| = |C| \leq |C - B| + |B| = |B|.$$

Thus we may consider $|\ |$ as defined on $\overline{F} = \mathscr{F}/\mathscr{N}$, and it is then an absolute value on the field $\overline{F}$ which coincides with the original absolute value when restricted to the field F.

If $A = (a_n)$ is a fundamental sequence, and if A_m is the constant sequence (a_m), then $|A - A_m|$ can be made arbitrarily small by taking m sufficiently large. It follows that F is *dense* in $\overline{F}$, i.e. for any $\alpha \in \overline{F}$ and any $\varepsilon > 0$ there exists $a \in F$ such that $|\alpha - a| < \varepsilon$.

We show finally that $\overline{F}$ is *complete* as a metric space, i.e. every fundamental sequence of elements of $\overline{F}$ converges to an element of $\overline{F}$. For let (α_n) be a fundamental sequence in $\overline{F}$. Since F is dense in $\overline{F}$, for each n we can choose $a_n \in F$ so that $|\alpha_n - a_n| < 1/n$. Since

$$|a_m - a_n| \leq |a_m - \alpha_m| + |\alpha_m - \alpha_n| + |\alpha_n - a_n|,$$

it follows that (a_n) is also a fundamental sequence. Thus there exists $\alpha \in \overline{F}$ such that $\lim_{n \to \infty} |a_n - \alpha| = 0$. Since

$$|\alpha_n - \alpha| \leq |\alpha_n - a_n| + |a_n - \alpha|,$$

we have also $\lim_{n \to \infty} |\alpha_n - \alpha| = 0$. Thus the sequence (α_n) converges to α.

Summing up, we have proved

PROPOSITION 6 *If F is a field with an absolute value $|\ |$, then there exists a field $\overline{F}$ containing F, with an absolute value $|\ |$ extending that of F, such that $\overline{F}$ is complete and F is dense in $\overline{F}$.* $\square$

It is easily seen that $\overline{F}$ is uniquely determined, up to an isomorphism which preserves the absolute value. The field $\overline{F}$ is called the *completion* of the valued field F. The density of F in $\overline{F}$ implies that the absolute value on the completion $\overline{F}$ is non-archimedean or archimedean according as the absolute value on F is non-archimedean or archimedean.

It is easy to see that in example (iv) of §1 the valued field $F = K((t))$ of all formal Laurent series is complete, i.e. it is its own completion. For let $\{f^{(k)}\}$ be a fundamental sequence in F. Given any positive integer N, there is a positive integer $M = M(N)$ such that $|f^{(k)} - f^{(j)}| < q^{-N}$ for $j,k \geq M$. Thus we can write

$$f^{(k)}(t) = \sum_{n \leq N} \alpha_n t^n + \sum_{n > N} \alpha_n^{(k)} t^n \quad \text{for all } k \geq M.$$

If $f(t) = \sum_{n \in \mathbb{Z}} \alpha_n t^n$, then $\lim_{k \to \infty} |f^{(k)} - f| = 0$.

On the other hand, given any $f(t) = \sum_{n \in \mathbb{Z}} \alpha_n t^n \in K((t))$, we have $\lim_{k \to \infty} |f^{(k)} - f| = 0$, where $f^{(k)}(t) = \sum_{n \leq k} \alpha_n t^n \in K(t)$. It follows that $K((t))$ is the completion of the field $K(t)$ of rational functions considered in example (iii) of §1, with the absolute value $|\ |_t$ corresponding to the irreducible polynomial $p(t) = t$ (for which $\partial(p) = 1$).

The completion of the rational field $\mathbb{Q}$ with respect to the p-adic absolute value $|\ |_p$ will be denoted by $\mathbb{Q}_p$, and the elements of $\mathbb{Q}_p$ will be called *p-adic numbers*.

The completion of the rational field $\mathbb{Q}$ with respect to the ordinary absolute value $|\ |_\infty$ is of course the real field $\mathbb{R}$. In §6 we will show that the only fields with a complete archimedean absolute value are the real field $\mathbb{R}$ and the complex field $\mathbb{C}$, and the absolute value has the form

$| \ |_\infty^\rho$ for some $\rho > 0$. (In fact $\rho \le 1$, since $2^\rho \le 1^\rho + 1^\rho = 2$.) Thus an arbitrary archimedean valued field is equivalent to a subfield of $\mathbb{C}$ with the usual absolute value. (Hence, for a field with an archimedean absolute value $| \ |$, $|n| > 1$ for every $n > 1$ and $|n| \to \infty$ as $n \to \infty$.) Since this case may be considered well-known, we will in the following devote our attention primarily to the peculiarities of non-archimedean valued fields.

We will later be concerned with extending an absolute value on a field F to a field E which is a finite extension of F. Since all that matters for some purposes is that E is a vector space over F, it is useful to introduce the following definition.

Let F be a field with an absolute value $| \ |$ and let E be a vector space over F. A *norm* on E is a map $\| \ \|: E \to \mathbb{R}$ with the following properties:

(i) $\|a\| > 0$ for every $a \in E$ with $a \ne 0$;
(ii) $\|\alpha a\| = |\alpha| \, \|a\|$ for all $\alpha \in F$ and $a \in E$;
(iii) $\|a + b\| \le \|a\| + \|b\|$ for all $a,b \in E$.

It follows from (ii) that $\|O\| = 0$. We will require only one result about normed vector spaces:

LEMMA 7 *Let F be a complete valued field and let E be a finite-dimensional vector space over F. If $\| \ \|_1$ and $\| \ \|_2$ are both norms on E, then there exist positive constants σ, μ such that*

$$\sigma \|a\|_1 \ \le \ \|a\|_2 \ \le \ \mu \|a\|_1 \quad \text{for every } a \in E.$$

Proof   Let $e_1,\ldots,e_n$ be a basis for the vector space E. Then any $a \in E$ can be uniquely represented in the form

$$a \ = \ \alpha_1 e_1 + \ldots + \alpha_n e_n,$$

where $\alpha_1,\ldots,\alpha_n \in F$. It is easily seen that

$$\|a\|_0 = \max_{1 \le i \le n} |\alpha_i|$$

is a norm on E, and it is sufficient to prove the proposition for $\| \ \|_2 = \| \ \|_0$. Since

$$\|a\|_1 \ \le \ \|a\|_0 \, (\|e_1\|_1 + \ldots + \|e_n\|_1),$$

we can take $\sigma = (\|e_1\|_1 + \ldots + \|e_n\|_1)^{-1}$. To establish the existence of μ we assume $n > 1$ and use induction, since the result is obviously true for $n = 1$.

Assume, contrary to the assertion, that there exists a sequence $a^{(k)} \in E$ such that

$$\|a^{(k)}\|_1 \ < \ \varepsilon_k \|a^{(k)}\|_0,$$

where $\varepsilon_k > 0$ and $\varepsilon_k \to 0$ as $k \to \infty$. We may suppose, without loss of generality, that

$$|\alpha_n^{(k)}| = \|a^{(k)}\|_0$$

and also, by replacing $a^{(k)}$ by $(\alpha_n^{(k)})^{-1}a^{(k)}$, that $\alpha_n^{(k)} = 1$. Thus $a^{(k)} = b^{(k)} + e_n$, where

$$b^{(k)} = \alpha_1^{(k)}e_1 + ... + \alpha_{n-1}^{(k)}e_{n-1},$$

and $\|a^{(k)}\|_1 \to 0$ as $k \to \infty$. The sequences $\alpha_i^{(k)}$ ($i = 1,...,n-1$) are fundamental sequences in F, since

$$\|b^{(j)} - b^{(k)}\|_1 \leq \|b^{(j)} + e_n\|_1 + \|b^{(k)} + e_n\|_1 = \|a^{(j)}\|_1 + \|a^{(k)}\|_1$$

and, by the induction hypothesis,

$$|\alpha_i^{(j)} - \alpha_i^{(k)}| \leq \mu_{n-1} \|b^{(j)} - b^{(k)}\|_1 \quad (i = 1,...,n-1).$$

Hence, since F is complete, there exist $\alpha_i \in F$ such that $|\alpha_i^{(k)} - \alpha_i| \to 0$ ($i = 1,...,n-1$). Put

$$b = \alpha_1 e_1 + ... + \alpha_{n-1}e_{n-1}.$$

Since $\|b^{(k)} - b\|_1 \leq \sigma_{n-1}^{-1} \|b^{(k)} - b\|_0$, it follows that $\|b^{(k)} - b\|_1 \to 0$. But if $a = b + e_n$, then

$$\|a\|_1 \leq \|a - a^{(k)}\|_1 + \|a^{(k)}\|_1 = \|b - b^{(k)}\|_1 + \|a^{(k)}\|_1.$$

Letting $k \to \infty$, we obtain $a = 0$, which contradicts the definition of a. $\square$

4 Non-archimedean valued fields

Throughout this section we denote by F a field with a non-archimedean absolute value $|\ |$. A basic property of such fields is the following simple lemma. It may be interpreted as saying that in ultrametric geometry every triangle is isosceles.

LEMMA 8 *If $a,b \in F$ and $|a| < |b|$, then $|a + b| = |b|$.*

Proof We certainly have

$$|a + b| \leq \max\{|a|,|b|\} = |b|.$$

On the other hand, since $b = (a + b) - a$, we have

$$|b| \leq \max\{|a + b|,|-a|\}$$

and, since $|-a| = |a| < |b|$, this implies $|b| \leq |a + b|$. $\square$

It may be noted that if $a \neq 0$ and $b = -a$, then $|a| = |b|$ and $|a + b| < |b|$. From Lemma 8 it follows by induction that if $a_1,\ldots,a_n \in F$ and $|a_k| < |a_1|$ for $1 < k \leq n$, then

$$|a_1 + \ldots + a_n| = |a_1|.$$

As an application we show that *if a field E is a finite extension of a field F, then the trivial absolute value on E is the only extension to E of the trivial absolute value on F.* By Proposition 2, any extension to E of the trivial absolute value on F must be non-archimedean. Suppose $\alpha \in E$ and $|\alpha| > 1$. Then α satisfies a polynomial equation

$$\alpha^n + c_{n-1}\alpha^{n-1} + \ldots + c_0 = 0$$

with coefficients $c_k \in F$. Since $|c_k| = 0$ or 1 and since $|\alpha^k| < |\alpha^n|$ if $k < n$, we obtain the contradiction $|\alpha^n| = |\alpha^n + c_{n-1}\alpha^{n-1} + \ldots + c_0| = 0$.

As another application we prove

PROPOSITION 9 *If a field F has a non-archimedean absolute value $|\ |$, then the valuation on F can be extended to the polynomial ring $F[t]$ by defining the absolute value of* $f(t) = a_0 + a_1 t + \ldots + a_n t^n$ *to be* $|f| = \max\{|a_0|,\ldots,|a_n|\}$.

Proof We need only show that $|fg| = |f||g|$, since it is evident that $|f| = 0$ if and only if $f = 0$ and that $|f + g| \leq |f| + |g|$. Let $g(t) = b_0 + b_1 t + \ldots + b_m t^m$. Then $f(t)g(t) = c_0 + c_1 t + \ldots + c_l t^l$, where

$$c_i = a_0 b_i + a_1 b_{i-1} + \ldots + a_i b_0.$$

If r is the least integer such that $|a_r| = |f|$ and s the least integer such that $|b_s| = |g|$, then $a_r b_s$ has strictly greatest absolute value among all products $a_j b_k$ with $j + k = r + s$. Hence $|c_{r+s}| = |a_r||b_s|$ and $|fg| \geq |f||g|$. On the other hand,

$$|fg| = \max_i |c_i| \leq \max_{j,k} |a_j||b_k| = |f||g|.$$

Consequently $|fg| = |f||g|$. Clearly also $|f| = |a|$ if $f = a \in F$. (The absolute value on F can be further extended to the field $F(t)$ of rational functions by defining $|f(t)/g(t)|$ to be $|f|/|g|$.) $\square$

It also follows at once from Lemma 8 that if a sequence (a_n) of elements of F converges to a limit $a \neq 0$, then $|a_n| = |a|$ for all large n. Hence the value group of the field F is the same as the value group of its completion $\overline{F}$. The next lemma has an especially appealing corollary.

LEMMA 10 *Let F be a field with a non-archimedean absolute value $|\ |$. Then a sequence (a_n) of elements of F is a fundamental sequence if and only if* $\lim_{n \to \infty} |a_{n+1} - a_n| = 0$.

Proof If $|a_{n+1} - a_n| \to 0$, then for each $\varepsilon > 0$ there is a corresponding positive integer $N = N(\varepsilon)$ such that

$$|a_{n+1} - a_n| < \varepsilon \ \text{ for } n \geq N.$$

For any integer $k > 1$,

$$a_{n+k} - a_n \ = \ (a_{n+1} - a_n) + (a_{n+2} - a_{n+1}) + \ldots + (a_{n+k} - a_{n+k-1})$$

and hence

$$|a_{n+k} - a_n| \ \leq \ \max\{|a_{n+1} - a_n|, |a_{n+2} - a_{n+1}|, \ldots, |a_{n+k} - a_{n+k-1}|\} \ < \ \varepsilon \ \text{ for } n \geq N.$$

Thus (a_n) is a fundamental sequence. The converse follows at once from the definition of a fundamental sequence. $\square$

COROLLARY 11 *In a field F with a complete non-archimedean absolute value $|\ |$, an infinite series $\sum_{n=1}^{\infty} a_n$ of elements of F is convergent if and only if $|a_n| \to 0$.* $\square$

Let F be a field with a nontrivial non-archimedean absolute value $|\ |$ and put

$$R \ = \ \{a \in F : |a| \leq 1\},$$
$$M \ = \ \{a \in F : |a| < 1\},$$
$$U \ = \ \{a \in F : |a| = 1\}.$$

Then R is the union of the disjoint nonempty subsets M and U. It follows from the definition of a non-archimedean absolute value that R is a (commutative) ring containing the unit element of F and that, for any nonzero $a \in F$, either $a \in R$ or $a^{-1} \in R$ (or both). Moreover M is an ideal of R and U is a multiplicative group, consisting of all $a \in R$ such that also $a^{-1} \in R$. Thus a proper ideal of R cannot contain an element of U and hence M is the unique maximal ideal of R. Consequently (see again Chapter I, §8) the quotient R/M is a field.

We call R the *valuation ring*, M the *valuation ideal*, and R/M the *residue field* of the valued field F.

We draw attention to the fact that the 'closed unit ball' R is both open and closed in the topology induced by the absolute value. For if $a \in R$ and $|b - a| < 1$, then also $b \in R$. Furthermore, if $a_n \in R$ and $a_n \to a$ then $a \in R$, since $|a_n| = |a|$ for all large n. Similarly, the 'open unit ball' M is also both open and closed.

In particular, let $F = \mathbb{Q}$ be the field of rational numbers and $|\ | = |\ |_p$ the p-adic absolute value. In this case the valuation ring $R = R_p$ is the set of all rational numbers m/n, where m and

n are relatively prime integers, $n > 0$ and p does not divide n. The valuation ideal is $M = pR_p$ and the residue field $\mathbb{F}_p = R_p/pR_p$ is the finite field with p elements.

As another example, let $F = K(t)$ be the field of rational functions with coefficients from an arbitrary field K and let $|\,| = |\,|_t$ be the absolute value considered in example (iii) of §1 for the irreducible polynomial $p(t) = t$. In this case the valuation ring R is the set of all rational functions $f = g/h$, where g and h are relatively prime polynomials and h has nonzero constant term. The valuation ideal is $M = tR$ and the residue field R/M is isomorphic to K, since $f(t) \equiv f(0) \bmod M$.

Let $\overline{F}$ be the completion of F. If $\overline{R}$ and $\overline{M}$ are the valuation ring and valuation ideal of $\overline{F}$, then evidently

$$R = \overline{R} \cap F, \quad M = \overline{M} \cap F.$$

Moreover R is *dense* in $\overline{R}$ since, if $0 < \varepsilon \leq 1$, for any $\alpha \in \overline{R}$ there exists $a \in F$ such that $|\alpha - a| < \varepsilon$ and then $a \in R$ (and $\alpha - a \in \overline{M}$). Furthermore the residue fields R/M and $\overline{R}/\overline{M}$ are isomorphic. For the map $a + M \to a + \overline{M}$ $(a \in R)$ is an isomorphism of R/M onto a subfield of $\overline{R}/\overline{M}$ and this subfield is not proper (by the preceding bracketed remark).

The valuation ring of the field $\mathbb{Q}_p$ of p-adic numbers will be denoted by $\mathbb{Z}_p$ and its elements will be called *p-adic integers*. The ring $\mathbb{Z}$ of ordinary integers is dense in $\mathbb{Z}_p$, and the residue field of $\mathbb{Q}_p$ is the finite field $\mathbb{F}_p$ with p elements, since this is the residue field of $\mathbb{Q}$.

Similarly, the valuation ring of the field $K((t))$ of all formal Laurent series is the ring $K[[t]]$ of all formal power series $\sum_{n \geq 0} \alpha_n t^n$. The polynomial ring $K[t]$ is dense in $K[[t]]$, and the residue field of $K((t))$ is K, since this is the residue field of $K(t)$ with the absolute value $|\,|_t$.

A non-archimedean absolute value $|\,|$ on a field F will be said to be *discrete* if there exists some $\delta \in (0,1)$ such that $a \in F$ and $|a| \neq 1$ implies either $|a| < 1 - \delta$ or $|a| > 1 + \delta$. (This situation cannot arise for archimedean absolute values.)

A non-archimedean absolute value need not be discrete, but the examples of non-archimedean absolute values which we have given are all discrete.

LEMMA 12 *Let F be a field with a nontrivial non-archimedean absolute value $|\,|$, and let R and M be the corresponding valuation ring and valuation ideal. Then the absolute value is discrete if and only if M is a principal ideal. In this case the only nontrivial proper ideals of R are the powers M^k $(k = 1,2,\ldots)$.*

Proof Suppose first that the absolute value $|\,|$ is discrete and put $\mu = \sup_{a \in M} |a|$. Then $0 < \mu < 1$ and the supremum is attained, since $|a_n| \to \mu$ implies $|a_{n+1} a_n^{-1}| \to 1$. Thus $\mu = |\pi|$

for some $\pi \in M$. For any $a \in M$ we have $|a\pi^{-1}| \le 1$ and hence $a = \pi a'$, where $a' \in R$. Thus M is a principal ideal with generating element π.

Suppose next that M is a principal ideal with generating element π. If $|a| < 1$, then $a \in M$. Thus $a = \pi a'$, where $a' \in R$, and hence $|a| \le |\pi|$. Similarly if $|a| > 1$, then $a^{-1} \in M$. Thus $|a^{-1}| \le |\pi|$ and hence $|a| \ge |\pi|^{-1}$. This proves that the absolute value is discrete.

We now show that, for any nonzero $a \in M$, there is a positive integer k such that $|a| = |\pi|^k$. In fact we can choose k so that

$$|\pi|^{k+1} < |a| \le |\pi|^k.$$

Then $|\pi| < |a\pi^{-k}| \le 1$, which implies $|a\pi^{-k}| = 1$ and hence $|a| = |\pi|^k$. Thus the value group of the valued field F is the infinite cyclic group generated by $|\pi|$. The final statement of the lemma follows immediately. $\square$

It is clear that if an absolute value $|\,|$ on a field F is discrete, then its extension to the completion $\overline{F}$ of F is also discrete. Moreover, if π is a generating element for the valuation ideal of F, then it is also a generating element for the valuation ideal of $\overline{F}$.

Suppose now that not only is $M = (\pi)$ a principal ideal, but the residue field $k = R/M$ is finite. Then there exists a finite set $S \subseteq R$, with the same cardinality as k, such that for each $a \in R$ there is a unique $\alpha \in S$ for which $|\alpha - a| < 1$. Since the elements of k are the cosets $\alpha + M$, where $\alpha \in S$, we call S a *set of representatives* in R of the residue field. It is convenient to choose $\alpha = 0$ as the representative for M itself.

Under these hypotheses a rather explicit representation for the elements of the valued field can be derived:

PROPOSITION 13 *Let F be a field with a non-archimedean absolute value $|\,|$, and let R and M be the corresponding valuation ring and valuation ideal. Suppose the absolute value is discrete, i.e. $M = (\pi)$ is a principal ideal. Suppose also that the residue field $k = R/M$ is finite, and let $S \subseteq R$ be a set of representatives of k with $0 \in S$.*

Then for each $a \in F$ there exists a unique bi-infinite sequence $(\alpha_n)_{n \in \mathbb{Z}}$, where $\alpha_n \in S$ for all $n \in \mathbb{Z}$ and $\alpha_n \ne 0$ for at most finitely many $n < 0$, such that

$$a = \sum_{n \in \mathbb{Z}} \alpha_n \pi^n.$$

If N is the least integer n such that $\alpha_n \ne 0$, then $|a| = |\pi|^N$. In particular, $a \in R$ if and only if $\alpha_n = 0$ for all $n < 0$.

If F is complete then, for any such bi-infinite sequence (α_n), the series $\sum_{n \in \mathbb{Z}} \alpha_n \pi^n$ is convergent with sum $a \in F$.

Proof Suppose $a \in F$ and $a \neq 0$. Then $|a| = |\pi|^N$ for some $N \in \mathbb{Z}$ and hence $|a\pi^{-N}| = 1$. There is a unique $\alpha_N \in S$ such that $|a\pi^{-N} - \alpha_N| < 1$. Then $|\alpha_N| = 1$, $|a\pi^{-N} - \alpha_N| \leq |\pi|$ and

$$a\pi^{-N} = \alpha_N + a_1\pi,$$

where $a_1 \in R$. Similarly there is a unique $\alpha_{N+1} \in S$ such that

$$a_1 = \alpha_{N+1} + a_2\pi,$$

where $a_2 \in R$. Continuing in this way we obtain, for any positive integer n,

$$a = \alpha_N\pi^N + \alpha_{N+1}\pi^{N+1} + \dots + \alpha_{N+n}\pi^{N+n} + a_{n+1}\pi^{N+n+1},$$

where $\alpha_N,\alpha_{N+1},\dots,\alpha_{N+n} \in S$ and $a_{n+1} \in R$. Since $|a_{n+1}\pi^{N+n+1}| \to 0$ as $n \to \infty$, the series $\sum_{n \geq N} \alpha_n\pi^n$ converges with sum a.

On the other hand, it is clear that if $a = \sum_{n \geq N} \alpha_n\pi^n$, where $\alpha_n \in S$ and $\alpha_N \neq 0$, then the coefficients α_n must be determined in the above way.

If F is complete then, by Corollary 11, any series $\sum_{n \geq N} \alpha_n\pi^n$ is convergent, since $|\alpha_n\pi^n| \to 0$ as $n \to \infty$. $\square$

COROLLARY 14 *Every* $a \in \mathbb{Q}_p$ *can be uniquely expressed in the form*

$$a = \sum_{n \in \mathbb{Z}} \alpha_n p^n,$$

where $\alpha_n \in \{0,1,\dots,p-1\}$ *and* $\alpha_n \neq 0$ *for at most finitely many* $n < 0$. *Conversely, any such series is convergent with sum* $a \in \mathbb{Q}_p$. *Furthermore* $a \in \mathbb{Z}_p$ *if and only if* $\alpha_n = 0$ *for all* $n < 0$. $\square$

Thus we have now arrived at Hensel's starting-point. It is not difficult to show that if $a = \sum_{n \in \mathbb{Z}} \alpha_n p^n \in \mathbb{Q}_p$, then actually $a \in \mathbb{Q}$ if and only if the sequence of coefficients (α_n) is *eventually periodic*, i.e. there exist integers $h > 0$ and m such that $\alpha_{n+h} = \alpha_n$ for all $n \geq m$.

From Corollary 14 we can deduce again that the ring $\mathbb{Z}$ of ordinary integers is dense in the ring $\mathbb{Z}_p$ of p-adic integers. For, if

$$a = \sum_{n \geq 0} \alpha_n p^n \in \mathbb{Z}_p,$$

where $\alpha_n \in \{0,1,\dots,p-1\}$, then

$$a_k = \sum_{n=0}^{k} \alpha_n p^n \in \mathbb{Z}$$

and $|a - a_k| < p^{-k}$.

5   Hensel's lemma

The analogy between p-adic absolute values and ordinary absolute values suggests that methods well-known in analysis may be applied also to arithmetic problems. We will illustrate this by showing how Newton's method for finding the real or complex roots of an equation can also be used to find p-adic roots. In fact the ultrametric inequality makes it possible to establish a stronger convergence criterion than in the classical case. The following proposition is modestly known as 'Hensel's lemma'.

PROPOSITION 15 *Let F be a field with a complete non-archimedean absolute value $|\,\,|$ and let R be its valuation ring. Let*

$$f(x) \; = \; c_n x^n + c_{n-1} x^{n-1} + \dots + c_0$$

be a polynomial with coefficients $c_0,\dots,c_n \in R$ and let

$$f_1(x) \; = \; nc_n x^{n-1} + (n-1)c_{n-1} x^{n-2} + \dots + c_1$$

be its formal derivative. If $|f(a_0)| < |f_1(a_0)|^2$ for some $a_0 \in R$, then the equation $f(a) = 0$ has a unique solution $a \in R$ such that $|a - a_0| < |f_1(a_0)|$.

Proof We consider first the existence of a and postpone discussion of its uniqueness. Put

$$\sigma: = |f_1(a_0)| > 0, \quad \theta_0: = \sigma^{-2}|f(a_0)| < 1,$$

and let D_θ denote the set

$$\{a \in R: |f_1(a)| = \sigma, \; |f(a)| \le \theta\sigma^2\}.$$

Thus $a_0 \in D_{\theta_0}$ and $D_{\theta'} \subseteq D_\theta$ if $\theta' \le \theta$. We are going to show that, if $\theta \in (0,1)$, then the 'Newton' map

$$Ta \; = \; a^*: = a - f(a)/f_1(a)$$

maps D_θ into D_{θ^2}.

    We can write

$$f(x + y) \; = \; f(x) + f_1(x)y + \dots + f_n(x)y^n,$$

where $f_1(x)$ has already been defined and $f_2(x),\dots,f_n(x)$ are also polynomials with coefficients from R. We substitute

$$x = a, \quad y = b: = -f(a)/f_1(a),$$

where $a \in D_\theta$. Then $|f_j(a)| \le 1$, since $a \in R$ and $f_j(x) \in R[x]$ ($j = 1,...,n$). Furthermore

$$|b| = \sigma^{-1}|f(a)| \le \theta\sigma < \sigma.$$

Thus $b \in R$. Since $f(a) + f_1(a)b = 0$, it follows that $a^* = a + b$ satisfies

$$|f(a^*)| \le \max_{2 \le j \le n} |f_j(a)b^j| \le |b|^2 = \sigma^{-2}|f(a)|^2 \le \theta^2\sigma^2.$$

Similarly, since $f_1(a + b) - f_1(a)$ can be written as a polynomial in b with coefficients from R and with no constant term,

$$|f_1(a + b) - f_1(a)| \le |b| < \sigma = |f_1(a)|$$

and hence $|f_1(a^*)| = \sigma$. This completes the proof that $TD_\theta \subseteq D_{\theta^2}$.

Now put $a_k = T^k a_0$, so that

$$a_{k+1} - a_k = -f(a_k)/f_1(a_k).$$

It follows by induction from what we have proved that

$$|f(a_k)| \le \theta_0^{2^k}\sigma^2.$$

Since $\theta_0 < 1$ and $|a_{k+1} - a_k| = \sigma^{-1}|f(a_k)|$, this shows that $\{a_k\}$ is a fundamental sequence. Hence, since F is complete, $a_k \to a$ for some $a \in R$. Evidently $f(a) = 0$ and $|f_1(a)| = \sigma$. Since, for every $k \ge 1$,

$$|a_k - a_0| \le \max_{1 \le j \le k} |a_j - a_{j-1}| \le \theta_0\sigma,$$

we also have $|a - a_0| \le \theta_0\sigma < \sigma$.

To prove uniqueness, assume $f(\tilde{a}) = 0$ for some $\tilde{a} \ne a$ such that $|\tilde{a} - a_0| < \sigma$. If we put $b = \tilde{a} - a$, then

$$0 = f(\tilde{a}) - f(a) = f_1(a)b + ... + f_n(a)b^n.$$

From $b = \tilde{a} - a_0 - (a - a_0)$ we obtain $|b| < \sigma$. Since $b \ne 0$ and $|f_j(a)| \le 1$, it follows that, for $j \ge 2$,

$$|f_j(a)b^j| \le |b|^2 < \sigma|b| = |f_1(a)b|.$$

But this implies

$$|f(\tilde{a}) - f(a)| = |f_1(a)b| > 0,$$

which is a contradiction. $\square$

As an application of Proposition 15 we will determine which elements of the field $\mathbb{Q}_p$ of p-adic numbers are squares. Since $b = a^2$ implies $b = p^{2\nu}b'$, where $\nu \in \mathbb{Z}$ and $|b'|_p = 1$, we may restrict attention to the case $|b|_p = 1$.

PROPOSITION 16 *Suppose $b \in \mathbb{Q}_p$ and $|b|_p = 1$.*

If $p \neq 2$, then $b = a^2$ for some $a \in \mathbb{Q}_p$ if and only if $|b - a_0^2|_p < 1$ for some $a_0 \in \mathbb{Z}$.

If $p = 2$, then $b = a^2$ for some $a \in \mathbb{Q}_2$ if and only if $|b - 1|_2 \leq 2^{-3}$.

Proof Suppose first that $p \neq 2$. If $b = a^2$ for some $a \in \mathbb{Q}_p$, then $|a|_p = 1$ and $|a - a_0|_p < 1$ for some $a_0 \in \mathbb{Z}$, since $\mathbb{Z}$ is dense in $\mathbb{Z}_p$. Hence $|a_0|_p = 1$ and

$$|b - a_0^2|_p \ = \ |a - a_0|_p \, |a + a_0|_p \ \leq \ |a - a_0|_p \ < \ 1.$$

Conversely, suppose $|b - a_0^2|_p < 1$ for some $a_0 \in \mathbb{Z}$. Then $|a_0^2|_p = 1$ and so $|a_0|_p = 1$. In Proposition 15 take $F = \mathbb{Q}_p$ and $f(x) = x^2 - b$. The hypotheses of the proposition are satisfied, since $|f(a_0)|_p < 1$ and $|f_1(a_0)|_p = |2a_0|_p = 1$, and hence $b = a^2$ for some $a \in \mathbb{Q}_p$.

Suppose next that $p = 2$. If $b = a^2$ for some $a \in \mathbb{Q}_2$, then $|a|_2 = 1$ and $|a - a_0|_2 \leq 2^{-3}$ for some $a_0 \in \mathbb{Z}$, since $\mathbb{Z}$ is dense in $\mathbb{Z}_2$. Hence $|a_0|_2 = 1$ and

$$|b - a_0^2|_2 \ = \ |a - a_0|_2 \, |a + a_0|_2 \ \leq \ |a - a_0|_2 \ \leq 2^{-3}.$$

Since a_0 is odd, we have $a_0 \equiv \pm 1 \bmod 4$ and $a_0^2 \equiv 1 \bmod 8$. Hence

$$|b - 1|_2 \ \leq \ \max\{|b - a_0^2|_2, |a_0^2 - 1|_2\} \ \leq \ 2^{-3}.$$

Conversely, suppose $|b - 1|_2 \leq 2^{-3}$. In Proposition 15 take $F = \mathbb{Q}_2$ and $f(x) = x^2 - b$. The hypotheses of the proposition are satisfied, since $|f(1)|_2 \leq 2^{-3}$ and $|f_1(1)|_2 = 2^{-1}$, and hence $b = a^2$ for some $a \in \mathbb{Q}_2$. $\square$

COROLLARY 17 *Let b be an integer not divisible by the prime p.*

If $p \neq 2$, then $b = a^2$ for some $a \in \mathbb{Q}_p$ if and only if b is a quadratic residue mod p.

If $p = 2$, then $b = a^2$ for some $a \in \mathbb{Q}_2$ if and only if $b \equiv 1 \bmod 8$. $\square$

It follows from Corollary 17 that $\mathbb{Q}_p$ cannot be given the structure of an *ordered* field. For, if p is odd, then $1 - p = a^2$ for some $a \in \mathbb{Q}_p$ and hence

$$a^2 + 1 + \dots + 1 \ = \ 0,$$

where there are $p - 1$ 1's. Similarly, if $p = 2$, then $1 - 2^3 = a^2$ for some $a \in \mathbb{Q}_2$ and the same relation holds with 7 1's.

Suppose again that F is a field with a complete non-archimedean absolute value $|\ |$. Let R and M be the corresponding valuation ring and valuation ideal, and let $k = R/M$ be the residue field. For any $a \in R$ we will denote by $\bar{a}$ the corresponding element $a + M$ of k, and for any polynomial

$$f(x) \; = \; c_n x^n + c_{n-1} x^{n-1} + \ldots + c_0$$

with coefficients $c_0, \ldots, c_n \in R$, we will denote by

$$\bar{f}(x) \; = \; \bar{c}_n x^n + \bar{c}_{n-1} x^{n-1} + \ldots + \bar{c}_0$$

the polynomial whose coefficients are the corresponding elements of k.

The hypotheses of Proposition 15 are certainly satisfied if $|f(a_0)| < 1 = |f_1(a_0)|$. In this case Proposition 15 says that if

$$\bar{f}(x) \; = \; (x - \bar{a}_0)\bar{h}_0(x),$$

where $a_0 \in R$, $h_0(x) \in R[x]$ and $h_0(a_0) \notin M$, then

$$f(x) \; = \; (x - a)h(x),$$

where $a - a_0 \in M$, and $h(x) \in R[x]$. In other words, the factorization of $\bar{f}(x)$ in $k[x]$ can be 'lifted' to a factorization of $f(x)$ in $R[x]$. This form of Hensel's lemma can be generalized to factorizations where neither factor is linear, and the result is again known as Hensel's lemma!

PROPOSITION 18 *Let F be a field with a complete non-archimedean absolute value $|\;|$. Let R and M be the valuation ring and valuation ideal of F, and $k = R/M$ the residue field.*

Let $f \in R[x]$ be a polynomial with coefficients in R and suppose there exist relatively prime polynomials $\phi, \psi \in k[x]$, with ϕ monic and $\partial(\phi) > 0$, such that $\bar{f} = \phi\psi$.

Then there exist polynomials $g, h \in R[x]$, with g monic and $\partial(g) = \partial(\phi)$, such that $\bar{g} = \phi$, $\bar{h} = \psi$ and $f = gh$.

Proof Put $n = \partial(f)$ and $m = \partial(\phi)$. Then $\partial(\psi) = \partial(\bar{f}) - \partial(\phi) \leq n - m$. There exist polynomials $g_1, h_1 \in R[x]$, with g_1 monic, $\partial(g_1) = m$ and $\partial(h_1) \leq n - m$, such that $\bar{g}_1 = \phi$, $\bar{h}_1 = \psi$. Since ϕ, ψ are relatively prime, there exist polynomials $\chi, \omega \in k[x]$ such that

$$\chi\phi + \omega\psi \; = \; 1,$$

and there exist polynomials $u, v \in R[x]$ such that $\bar{u} = \chi$, $\bar{v} = \omega$. Thus

$$f - g_1 h_1 \in M[x], \quad ug_1 + vh_1 - 1 \in M[x].$$

If $f = g_1 h_1$, there is nothing more to do. Otherwise, let π be the coefficient of $f - g_1 h_1$ or of $ug_1 + vh_1 - 1$ which has maximum absolute value. Then

$$f - g_1 h_1 \in \pi R[x], \quad u g_1 + v h_1 - 1 \in \pi R[x].$$

We are going to construct inductively polynomials $g_j, h_j \in R[x]$ such that

(i) $\overline{g}_j = \phi, \;\; \overline{h}_j = \psi$;

(ii) g_j is monic and $\partial(g_j) = m, \; \partial(h_j) \le n - m$;

(iii) $g_j - g_{j-1} \in \pi^{j-1} R[x], \; h_j - h_{j-1} \in \pi^{j-1} R[x]$;

(iv) $f - g_j h_j \in \pi^j R[x]$.

This holds already for $j = 1$ with $g_0 = h_0 = 0$. Assume that, for some $k \ge 2$, it holds for all $j < k$ and put $f - g_j h_j = \pi^j \ell_j$, where $\ell_j \in R[x]$. Since g_1 is monic, the Euclidean algorithm provides polynomials $q_k, r_k \in R[x]$ such that

$$\ell_{k-1} v \;=\; q_k g_1 + r_k, \;\; \partial(r_k) < \partial(g_1) = m.$$

Let $w_k \in R[x]$ be a polynomial of minimal degree such that all coefficients of $\ell_{k-1} u + q_k h_1 - w_k$ have absolute value at most $|\pi|$. Then

$$w_k g_1 + r_k h_1 - \ell_{k-1} \;=\; (u g_1 + v h_1 - 1)\ell_{k-1} - (\ell_{k-1} u + q_k h_1 - w_k) g_1 \in \pi R[x].$$

We will show that $\partial(w_k) \le n - m$. Indeed otherwise

$$\partial(w_k g_1) \;>\; n \;\ge\; \partial(r_k h_1 - \ell_{k-1})$$

and hence, since g_1 is monic, $w_k g_1 + r_k h_1 - \ell_{k-1}$ has the same leading coefficient as w_k. Consequently the leading coefficient of w_k is in πR. Thus the polynomial obtained from w_k by omitting the term of highest degree satisfies the same requirements as w_k, which is a contradiction.

If we put

$$g_k = g_{k-1} + \pi^{k-1} r_k, \;\; h_k = h_{k-1} + \pi^{k-1} w_k,$$

then (i)–(iii) are evidently satisfied for $j = k$. Moreover

$$f - g_k h_k \;=\; -\pi^{k-1}(w_k g_{k-1} + r_k h_{k-1} - \ell_{k-1}) - \pi^{2k-2} r_k w_k$$

and

$$w_k g_{k-1} + r_k h_{k-1} - \ell_{k-1} \;=\; w_k g_1 + r_k h_1 - \ell_{k-1} + w_k(g_{k-1} - g_1) + r_k(h_{k-1} - h_1) \in \pi R[x].$$

Hence also (iv) is satisfied for $j = k$.

Put

$$g_j(x) = x^m + \sum_{i=0}^{m-1} \alpha_i^{(j)} x^i, \;\; h_j(x) = \sum_{i=0}^{n-m} \beta_i^{(j)} x^i.$$

By (iii), the sequences $(\alpha_i^{(j)})$ and $(\beta_i^{(j)})$ are fundamental sequences for each i and hence, since F is complete, there exist $\alpha_i, \beta_i \in R$ such that

$$\alpha_i^{(j)} \to \alpha_i, \ \beta_i^{(j)} \to \beta_i \ \text{ as } j \to \infty.$$

If

$$g(x) = x^m + \sum_{i=0}^{m-1} \alpha_i x^i, \quad h_j(x) = \sum_{i=0}^{n-m} \beta_i x^i,$$

then, for each $j \geq 1$,

$$g - g_j \in \pi^j R[x], \quad h - h_j \in \pi^j R[x].$$

Since

$$f - gh = f - g_j h_j - (g - g_j)h - g_j(h - h_j),$$

it follows that $f - gh \in \pi^j R[x]$ for each $j \geq 1$. Hence $f = gh$. It is obvious that g and h have the other required properties. $\square$

As an application of this form of Hensel's lemma we prove

PROPOSITION 19 *Let F be a field with a complete non-archimedean absolute value $| \ |$ and let*

$$f(t) = c_n t^n + c_{n-1} t^{n-1} + \dots + c_0 \in F[t].$$

If $c_0 c_n \neq 0$ and, for some m such that $0 < m < n$,

$$|c_0| \leq |c_m|, \quad |c_n| \leq |c_m|,$$

with at least one of the two inequalities strict, then f is reducible over F.

Proof Suppose first that $|c_0| < |c_m|$ and $|c_n| \leq |c_m|$. Evidently we may choose m so that $|c_m| = \max_{0 < i < n} |c_i|$ and $|c_i| < |c_m|$ for $0 \leq i < m$. By multiplying f by c_m^{-1} we may further assume that, if R is the valuation ring of F, then $f(t) \in R[t]$, $c_m = 1$ and $|c_i| < 1$ for $0 \leq i < m$. Hence

$$\bar{f}(t) = t^m(\bar{c}_n t^{n-m} + \bar{c}_{n-1} t^{n-m-1} + \dots + 1).$$

Since the two factors are relatively prime, it follows from Proposition 18 that f is reducible.

If $|c_n| < |c_m|$ and $|c_0| \leq |c_m|$, the same argument applies to the polynomial $t^n f(t^{-1})$. $\square$

Proposition 19 shows that if a quadratic $at^2 + bt + c$ is irreducible, then $|b| \leq \max\{|a|, |c|\}$, with strict inequality if $|a| \neq |c|$. Proposition 19 will now be used to extend an absolute value on a given field to a finite extension of that field.

PROPOSITION 20 *Let F be a field with a complete non-archimedean absolute value $|\ |$. If the field E is a finite extension of F, then the absolute value on F can be extended to an absolute value on E.*

Proof We will not only show that an extension of the absolute value exists, but we will provide an explicit expression for it.

Regard E as a vector space over F of finite dimension n, and with any $a \in E$ associate the linear transformation $L_a: E \to E$ defined by $L_a(x) = ax$. Then $\det L_a \in F$ and we claim that *an extended absolute value is given by the formula*

$$|a| = |\det L_a|^{1/n}.$$

Evidently $|a| \geq 0$, and equality holds only if $a = 0$, since $ax = 0$ for some $x \neq 0$ implies $a = 0$. Furthermore $|ab| = |a||b|$, since $L_{ab} = L_a L_b$ and hence $\det L_{ab} = (\det L_a)(\det L_b)$. If $a \in F$, then $L_a = aI_n$ and hence the proposed absolute value coincides with the original absolute value on F. It only remains to show that

$$|a - b| \leq \max(|a|,|b|) \quad \text{for all } a,b \in F.$$

In fact we may suppose $|a| \leq |b|$ and then, by dividing by b, we see that it is sufficient to show that $0 < |a| \leq 1$ implies $|1 - a| \leq 1$.

To simplify notation, write $A = L_a$ and let

$$f(t) = \det(tI - A) = t^n + c_{n-1}t^{n-1} + \dots + c_0$$

be the *characteristic polynomial* of A. Then $c_i \in F$ for all i and $c_0 = (-1)^n \det A$. Let $g(t)$ be the monic polynomial in $F[t]$ of least positive degree such that $g(a) = 0$. Then $g(t)$ is irreducible, since the field E has no zero divisors. Evidently $g(t)$ is also the *minimal polynomial* of A. But, for an arbitrary linear transformation of an n-dimensional vector space, the characteristic polynomial divides the n-th power of the minimal polynomial. It follows in the present case that $f(t) = g(t)^r$ for some positive integer r.

Suppose

$$g(t) = t^m + b_{m-1}t^{m-1} + \dots + b_0$$

and let $a \in E$ satisfy $|a| \leq 1$ with respect to the proposed absolute value. Then $|c_0| = |\det A| \leq 1$ and hence, since $b_0^r = c_0$, $|b_0| \leq 1$. Since g is irreducible, it follows from Proposition 19 that $|b_j| \leq 1$ for all j. Since

$$g(1) = 1 + b_{m-1} + \dots + b_0,$$

this implies $|g(1)| \le 1$ and hence $|f(1)| \le 1$. Since $f(1) = \det(I - A)$, this proves that $|1 - a| \le 1$.
$\square$

Finally we show that there is no other extension to E of the given absolute value on F besides the one constructed in the proof of Proposition 20.

PROPOSITION 21 *Let F be a complete field with respect to the absolute value $|\,|$ and let the field E be a finite extension of F. Then there is at most one extension of the absolute value on F to an absolute value on E, and E is necessarily complete with respect to the extended absolute value.*

Proof Let $e_1,...,e_n$ be a basis for E, regarded as a vector space over F. Then any $a \in E$ can be uniquely expressed in the form

$$a = \alpha_1 e_1 + ... + \alpha_n e_n,$$

where $\alpha_1,...,\alpha_n \in F$. By Lemma 7, for any extended absolute value there exist positive real numbers σ,μ such that

$$\sigma\,|a| \le \max_i |\alpha_i| \le \mu\,|a| \quad \text{for every } a \in E.$$

It follows at once that E is complete. For if $a^{(k)}$ is a fundamental sequence, then $\alpha_i^{(k)}$ is a fundamental sequence in F for $i = 1,...,n$. Since F is complete, there exist $\alpha_i \in F$ such that $\alpha_i^{(k)} \to \alpha_i$ $(i = 1,...,n)$ and then $a^{(k)} \to a$, where $a = \alpha_1 e_1 + ... + \alpha_n e_n$.

It will now be shown that there is at most one extension to E of the absolute value on F. Since we saw in §4 that the trivial absolute value on E is the only extension of the trivial absolute value on F, we may assume that the given absolute value on F is nontrivial. For a fixed $a \in E$, consider the powers $a,a^2,...$. For each k we can write

$$a^k = \alpha_1^{(k)} e_1 + ... + \alpha_n^{(k)} e_n.$$

Since $|a| < 1$ if and only if $|a^k| \to 0$, it follows from the remarks at the beginning of the proof that $|a| < 1$ if and only if $|\alpha_i^{(k)}| \to 0$ $(i = 1,...,n)$. This condition is independent of the absolute value on E. Thus if there exist two absolute values, $|\,|_1$ and $|\,|_2$, which extend the absolute value on F, then $|a|_1 < 1$ if and only if $|a|_2 < 1$. Hence, by Proposition 3, there exists a positive real number ρ such that

$$|a|_2 = |a|_1^\rho \quad \text{for every } a \in E.$$

In fact $\rho = 1$, since for some $a \in F$ we have $|a|_2 = |a|_1 > 1$. $\square$

6   Locally compact valued fields

We prove first a theorem of Ostrowski (1918):

THEOREM 22 *A complete archimedean valued field F is (isomorphic to) either the real field $\mathbb{R}$ or the complex field $\mathbb{C}$, and its absolute value is equivalent to the usual absolute value.*

Proof   Since the valuation on it is archimedean, the field F has characteristic 0 and thus contains $\mathbb{Q}$. Since an archimedean absolute value on $\mathbb{Q}$ is equivalent to the usual absolute value, by replacing the given absolute value on F by an equivalent one we may assume that it reduces to the usual absolute value on $\mathbb{Q}$. Since the valuation on F is complete, it now follows that F contains (a copy of) $\mathbb{R}$ and that the absolute value on F reduces to the usual absolute value on $\mathbb{R}$. If F contains an element i such that $i^2 = -1$, then F contains (a copy of) $\mathbb{C}$ and, by Proposition 21, the absolute value on F reduces to the usual absolute value on $\mathbb{C}$.

We now show that if $a \in F$ and $|a| < 1$, then $1 - a$ is a square in F. Let B be the set of all $x \in F$ such that $|x| \leq |a|$ and, for any $x \in B$, put

$$Tx = (x^2 + a)/2.$$

Then also $Tx \in B$, since

$$|Tx| \leq (|x|^2 + |a|)/2 \leq (|a|^2 + |a|)/2 \leq |a|.$$

Moreover, the map T is a contraction since, for all $x,y \in B$,

$$|Tx - Ty| = |x^2 - y^2|/2 = |x - y||x + y|/2 \leq |a||x - y|.$$

Since F is complete and B is a closed subset of F, it follows from the contraction principle (Proposition I.26) that the map T has a fixed point $\bar{x} \in B$. Then $\bar{x} = (\bar{x}^2 + a)/2$ and

$$1 - a = 1 - 2\bar{x} + \bar{x}^2 = (1 - \bar{x})^2.$$

We show next that, if the polynomial $t^2 + 1$ does not have a root in F, then the valuation on F can be extended to the field $E = F(i)$, where $i^2 = -1$. Each $\gamma \in E$ has a unique representation $\gamma = a + ib$, where $a,b \in F$. We claim that $|\gamma| = \sqrt{|a^2 + b^2|}$ is an extension to E of the given valuation on F.

The only part of this claim which is not easily established is the triangle inequality. To prove it, we need only show that

$$|1 + \gamma| \leq 1 + |\gamma| \quad \text{for every } \gamma \in E.$$

That is, we need only show that

$$|(1 + a)^2 + b^2| \;\leq\; 1 + 2\sqrt{|a^2 + b^2|} + |a^2 + b^2| \quad \text{for all } a,b \in F.$$

Since, by the triangle inequality in F,

$$|(1 + a)^2 + b^2| \;\leq\; 1 + 2|a| + |a^2 + b^2|,$$

it is enough to show that

$$|a| \;\leq\; \sqrt{|a^2 + b^2|} \quad \text{for all } a,b \in F$$

or, since we may suppose $a \neq 0$,

$$1 \leq |1 + c^2| \quad \text{for every } c \in F.$$

Assume, on the contrary, that $|1 + c^2| < 1$ for some $c \in F$. Then, by the previous part of the proof,

$$-c^2 \;=\; 1 - (1 + c^2) \;=\; x^2 \quad \text{for some } x \in F.$$

Since $c \neq 0$, this implies that $-1 = i^2$ for some $i \in F$, which is a contradiction.

Now $E = F(i)$ contains $\mathbb{C}$ and the absolute value on E reduces to the usual absolute value on $\mathbb{C}$. To prove the theorem it is enough to show that $E = \mathbb{C}$. For then $\mathbb{R} \subseteq F \subseteq \mathbb{C}$ and F has dimension 1 or 2 as a vector space over $\mathbb{R}$ according as $i \notin F$ or $i \in F$.

Assume on the contrary that there exists $\zeta \in E \setminus \mathbb{C}$. Consider the function $\varphi \colon \mathbb{C} \to \mathbb{R}$ defined by

$$\varphi(z) \;=\; |z - \zeta|$$

and put $r = \inf_{z \in \mathbb{C}} \varphi(z)$. Since $\varphi(0) = |\zeta|$ and $\varphi(z) > |\zeta|$ for $|z| > 2|\zeta|$, and since φ is continuous, the compact set $\{z \in \mathbb{C} \colon |z| \leq 2|\zeta|\}$ contains a point w such that $\varphi(w) = r$.

Thus if we put $\omega = \zeta - w$, then $\omega \neq 0$ and

$$0 < r = |\omega| \;\leq\; |\omega - z| \quad \text{for every } z \in \mathbb{C}.$$

We will show that $|\omega - z| = r$ for every $z \in \mathbb{C}$ such that $|z| < r$.

If $\varepsilon = e^{2\pi i/n}$, then

$$\omega^n - z^n \;=\; (\omega - z)(\omega - \varepsilon z)\cdots(\omega - \varepsilon^{n-1}z)$$

and hence

$$|\omega^n - z^n| \;\geq\; r^{n-1}\,|\omega - z|.$$

Thus $|\omega - z| \leq r\,|1 - z^n/\omega^n|$. Since $|z| < |\omega|$, by letting $n \to \infty$ we obtain $|\omega - z| \leq r$. But this is possible only if $|\omega - z| = r$.

Thus if $0 < |z| < r$, then ω may be replaced by $\omega - z$. It follows that $|\omega - nz| = r$ for every positive integer n. Hence $r \geq n|z| - r$, which yields a contradiction for sufficiently large n. $\square$

If a field F is locally compact with respect to an archimedean absolute value, then it is certainly complete and so, by Theorem 22, it is equivalent either to $\mathbb{R}$ or to $\mathbb{C}$ with the usual absolute value. It will now be shown that a field F is locally compact with respect to a non-archimedean absolute value if and only if it is a complete field of the type discussed in Proposition 13. It should be observed that a non-archimedean valued field F is locally compact if and only if its valuation ring R is compact, since then any closed ball in F is compact.

PROPOSITION 23 *Let F be a field with a non-archimedean absolute value $|\ |$. Then F is locally compact with respect to the topology induced by the absolute value if and only if the following three conditions are satisfied:*

(i)   *F is complete,*
(ii)   *the absolute value $|\ |$ is discrete,*
(iii) *the residue field is finite.*

Proof   As we have just observed, F is locally compact if and only if its valuation ring R is compact. Moreover, since R is a subset of the metric space F, it is compact if and only if any sequence of elements of R has a convergent subsequence.

The field F is certainly complete if it is locally compact, since any fundamental sequence is bounded. If the residue field is infinite, then there exists an infinite sequence (a_k) of elements of R such that $|a_k - a_j| = 1$ for $j \neq k$. Since the sequence (a_k) has no convergent subsequence, R is not compact. If the absolute value $|\ |$ is not discrete, then there exists an infinite sequence (a_k) of elements of R with

$$|a_1| < |a_2| < \ldots$$

and $|a_k| \to 1$ as $k \to \infty$. If $k > j$, then $|a_k - a_j| = |a_k|$ and again the sequence (a_k) has no convergent subsequence. Thus the conditions (i)-(iii) are all necessary for F to be locally compact.

Suppose now that the conditions (i)-(iii) are satisfied and let $\sigma = (a_k)$ be a sequence of elements of R. In the notation of Proposition 13, let

$$a_k = \sum_{n \geq 0} \alpha_n^{(k)} \pi^n,$$

where $\alpha_n^{(k)} \in S$. Since S is finite, there exists $\alpha_0 \in S$ such that $\alpha_0^{(k)} = \alpha_0$ for infinitely many $a_k \in \sigma$. If σ_0 is the subsequence of σ containing those a_k for which $\alpha_0^{(k)} = \alpha_0$, then there exists $\alpha_1 \in S$ such that $\alpha_1^{(k)} = \alpha_1$ for infinitely many $a_k \in \sigma_0$. Similarly, if σ_1 is the subsequence of σ_0 containing those a_k for which $\alpha_1^{(k)} = \alpha_1$, then there exists $\alpha_2 \in S$ such that $\alpha_2^{(k)} = \alpha_2$ for infinitely many $a_k \in \sigma_1$. And so on. If $a^{(j)} \in \sigma_j$, then

$$a^{(j)} = \alpha_0 + \alpha_1 \pi + \dots + \alpha_j \pi^j + \sum_{n \geq 0} \alpha_n(j)\pi^{j+1+n}.$$

But $a = \sum_{n \geq 0} \alpha_n \pi^n \in F$, since F is complete, and $|a^{(j)} - a| \leq |\pi|^{j+1}$. Thus the subsequence $(a^{(j)})$ of σ converges to a. $\square$

COROLLARY 24 *The field* $\mathbb{Q}_p$ *of p-adic numbers is locally compact, and the ring* $\mathbb{Z}_p$ *of p-adic integers is compact.* $\square$

COROLLARY 25 *If K is a finite field, then the field $K((t))$ of all formal Laurent series is locally compact, and the ring $K[[t]]$ of all formal power series is compact.* $\square$

It will now be shown that all locally compact valued fields F with a non-archimedean absolute value can in fact be explicitly determined. It is convenient to treat the cases where F has prime characteristic and zero characteristic separately, since the arguments in the two cases are quite different.

LEMMA 26 *Let F be a locally compact valued field with a nontrivial valuation. A normed vector space E over F is locally compact if and only if it is finite-dimensional.*

Proof Suppose first that E is finite-dimensional over F. If $e_1,\dots,e_n$ is a basis for the vector space E, then any $a \in E$ can be uniquely represented in the form

$$a = \alpha_1 e_1 + \dots + \alpha_n e_n,$$

where $\alpha_1,\dots,\alpha_n \in F$, and

$$\|a\|_0 = \max_{1 \leq i \leq n} |\alpha_i|$$

is a norm on E. Since the field F is locally compact, it is also complete. Hence, by Lemma 7, there exist positive real constants σ, μ such that

$$\sigma \|a\|_0 \leq \|a\| \leq \mu \|a\|_0 \quad \text{for every } a \in E.$$

Consequently, if $\{a_k\}$ is a bounded sequence of elements of E then, for each $j \in \{1,\dots,n\}$, the corresponding coefficients $\{\alpha_{kj}\}$ form a bounded sequence of elements of F. Hence, since F is locally compact, there exists a subsequence $\{a_{k_v}\}$ such that each of the sequences $\{\alpha_{k_v j}\}$ converges in F, with limit β_j say $(j = 1,\dots,n)$. It follows that the subsequence $\{a_{k_v}\}$ converges in E with limit $b = \beta_1 e_1 + \dots + \beta_n e_n$. Thus E is locally compact.

Suppose next that E is infinite-dimensional over F. Since the valuation on F is nontrivial, there exists $\alpha \in F$ such that $r = |\alpha|$ satisfies $0 < r < 1$. Let V be any finite-dimensional subspace of E, let $u' \in E \setminus V$ and let

$$d = \inf_{v \in V} \|u' - v\|.$$

Since V is locally compact, $d > 0$ and $d = \|u' - v'\|$ for some $v' \in V$. Choose $k \in \mathbb{Z}$ so that $r^{k+1} < d \le r^k$ and put $w' = \alpha^{-k}(u' - v')$. For any $v \in V$,

$$\|\alpha^k v + v' - u'\| \ge d$$

and hence

$$\|w' - v\| \ge d r^{-k} > r.$$

On the other hand,

$$\|w'\| = d r^{-k} \le 1.$$

We now define a sequence $\{w_m\}$ of elements of E in the following way. Taking $V = \{O\}$ we obtain a vector w_1 with $r < \|w_1\| \le 1$. Suppose we have defined $w_1,\dots,w_m \in E$ so that, for $1 \le j \le m$, $\|w_j\| \le 1$ and $\|w_j - v_j\| > r$ for all v_j in the vector subspace V_{j-1} of E spanned by $w_1,\dots,w_{j-1}$. Then, taking $V = V_m$, we obtain a vector w_{m+1} such that $\|w_{m+1}\| \le 1$ and $\|w_{m+1} - v_{m+1}\| > r$ for all $v_{m+1} \in V_m$. Thus the process can be continued indefinitely. Since $\|w_m\| \le 1$ for all m and $\|w_m - w_j\| > r$ for $1 \le j < m$, the bounded sequence $\{w_m\}$ has no convergent subsequence. Thus E is not locally compact. $\square$

PROPOSITION 27 *A non-archimedean valued field E with zero characteristic is locally compact if and only if, for some prime p, E is isomorphic to a finite extension of the field $\mathbb{Q}_p$ of p-adic numbers.*

Proof If E is a finite extension of the p-adic field $\mathbb{Q}_p$, then, since $\mathbb{Q}_p$ is locally compact, so also is E, by Lemma 26.

Suppose on the other hand that E is a locally compact valued field with zero characteristic. Then $\mathbb{Q} \subseteq E$. By Proposition 23, the residue field $k = R/M$ is finite and thus has prime characteristic p. It follows from Proposition 4 that the restriction to $\mathbb{Q}$ of the absolute value on E is (equivalent to) the p-adic absolute value. Hence, since E is necessarily complete, $\mathbb{Q}_p \subseteq E$. If E were infinite-dimensional as a vector space over $\mathbb{Q}_p$ then, by Lemma 26, it would not be locally compact. Hence E is a finite extension of $\mathbb{Q}_p$. $\square$

We consider next locally compact valued fields of prime characteristic.

PROPOSITION 28 *A valued field F with prime characteristic p is locally compact if and only if F is isomorphic to the field $K((t))$ of formal Laurent series over a finite field K of characteristic p, with the absolute value defined in example* (iv) *of* §1. *The finite field K is the residue field of F.*

Proof We need only prove the necessity of the condition, since its sufficiency has already been established (Corollary 25). Since F has prime characteristic, the absolute value on F is non-archimedean. Hence, by Proposition 23 and Lemma 12, the absolute value on F is discrete and the valuation ideal M is a principal ideal. Let π be a generating element for M. By Proposition 23 also, the residue field $k = R/M$ is finite. Evidently the characteristic of k must also be p. Let $q = p^f$ be the number of elements in k. Since F has characteristic p, for any $a,b \in F$,

$$(b-a)^p = b^p - a^p$$

and hence, by induction,

$$(b-a)^{p^n} = b^{p^n} - a^{p^n} \quad \text{for all } n \geq 1.$$

The multiplicative group of k is a cyclic group of order $q - 1$. Choose $a \in R$ so that $a + M$ generates this cyclic group. Then $|a^q - a| < 1$. By what we have just proved,

$$a^{q^{n+1}} - a^{q^n} = (a^q - a)^{q^n},$$

and hence (a^{q^n}) is a fundamental sequence, by Lemma 10. Since F is complete, by Proposition 23, it follows that $a^{q^n} \to \alpha \in R$. Moreover $\alpha^q = \alpha$, since

$$\lim_{n \to \infty} (a^{q^n})^q = \lim_{n \to \infty} a^{q^{n+1}},$$

and $\alpha - a \in M$, since $a^{q^{n+1}} - a^{q^n} \in M$ for every $n \geq 0$. Hence $\alpha \neq 0$ and $\alpha^{q-1} = 1$. Moreover $\alpha^j \neq 1$ for $1 \leq j < q - 1$, since $\alpha^j \equiv a^j \bmod M$. It follows that the set S consisting of 0 and the powers $1, \alpha, \ldots, \alpha^{q-1}$ is a set of representatives in R of the residue field k.

Since F has prime characteristic, α generates a finite subfield K of F. Evidently $\beta^q = \beta$ for every $\beta \in K$. But the polynomial $x^q - x$ of degree q has at most q roots in K. Since $S \subseteq K$, we conclude that $K = S$. Thus K has q elements and is isomorphic to the residue field k.

Every element a of F has a unique representation

$$a = \sum_{n \in \mathbb{Z}} \alpha_n \pi^n,$$

where π is a generating element for the principal ideal M, $\alpha_n \in S$ and $\alpha_n \neq 0$ for at most finitely many $n < 0$. The map

$$a' = \sum_{n \in \mathbb{Z}} \alpha_n t^n \to a = \sum_{n \in \mathbb{Z}} \alpha_n \pi^n$$

is a bijection of the field $K((t))$ onto F. Since S is closed under addition this map preserves sums, and since S is also closed under multiplication it also preserves products. Finally, if N is the least integer such that $\alpha_N \neq 0$, then $|a| = |\pi|^N$ and $|a'| = \rho^{-N}$ for some fixed $\rho > 1$. Hence the map is an isomorphism of the valued field $K((t))$ onto F. $\square$

7 Further remarks

Valued fields are discussed in more detail in the books of Cassels [1], Endler [3] and Ribenboim [5].

For still more forms of Hensel's lemma, see Ribenboim [6]. There are also generalizations to polynomials in several variables and to power series. The algorithmic implementation of Hensel's lemma is studied in von zur Gathen [4]. Newton's method for finding real or complex zeros is discussed, for example, in Stoer and Bulirsch [7].

Proposition 20 continues to hold if the word 'complete' is omitted from its statement. However, the formula given in the proof of Proposition 20 defines an absolute value on E if and only if there is a *unique* extension of the absolute value on F to an absolute value on E; see Viswanathan [8].

Ostrowski's Theorem 22 has been generalized by weakening the requirement $|ab| = |a||b|$ to $|ab| \leq |a||b|$. Mazur (1938) proved that the only normed associative division algebras over $\mathbb{R}$ are $\mathbb{R}$, $\mathbb{C}$ and $\mathbb{H}$, and that the only normed associative division algebra over $\mathbb{C}$ is $\mathbb{C}$ itself. An elegant functional-analytic proof of the latter result was given by Gelfand (1941). See Chapter 8 (by Koecher and Remmert) of Ebbinghaus *et al.* [2].

8 Selected references

[1] J.W.S. Cassels, *Local fields*, Cambridge University Press, 1986.

[2] H.-D. Ebbinghaus *et al.*, *Numbers*, English transl. of 2nd German ed. by H.L.S. Orde, Springer-Verlag, New York, 1990.

[3] O. Endler, *Valuation theory*, Springer-Verlag, Berlin, 1972.

[4] J. von zur Gathen, Hensel and Newton methods in valuation rings, *Math. Comp.* **42** (1984), 637-661.

[5] P. Ribenboim, *The theory of classical valuations*, Springer-Verlag, New York, 1999.

[6] P. Ribenboim, Equivalent forms of Hensel's lemma, *Exposition. Math.* **3** (1985), 3-24.

[7] J. Stoer and R. Bulirsch, *Introduction to numerical analysis*, 3rd ed. (English transl.), Springer-Verlag, New York, 2002.

[8] T.M. Viswanathan, A characterisation of Henselian valuations via the norm, *Bol. Soc. Brasil. Mat.* **4** (1973), 51-53.

Notations

The *Landau order symbols* are defined in the following way: if $I = [t_0,\infty)$ is a half-line and if $f,g: I \to \mathbb{R}$ are real-valued functions with $g(t) > 0$ for all $t \in I$, we write

$f = O(g)$ if there exists a constant $C > 0$ such that $|f(t)|/g(t) \leq C$ for all $t \in I$;

$f = o(g)$ if $f(t)/g(t) \to 0$ as $t \to \infty$;

$f \sim g$ if $f(t)/g(t) \to 1$ as $t \to \infty$.

The *end of a proof* is denoted by $\square$.

Axioms

Index